Springer Complexity

Springer Complexity is an interdisciplinary program publishing the best research and academic-level teaching on both fundamental and applied aspects of complex systems—cutting across all traditional disciplines of the natural and life sciences, engineering, economics, medicine, neuroscience, social and computer science.

Complex Systems are systems that comprise many interacting parts with the ability to generate a new quality of macroscopic collective behavior the manifestations of which are the spontaneous formation of distinctive temporal, spatial or functional structures. Models of such systems can be successfully mapped onto quite diverse "real-life" situations like the climate, the coherent emission of light from lasers, chemical reaction-diffusion systems, biological cellular networks, the dynamics of stock markets and of the Internet, earthquake statistics and prediction, freeway traffic, the human brain, or the formation of opinions in social systems, to name just some of the popular applications.

Although their scope and methodologies overlap somewhat, one can distinguish the following main concepts and tools: self-organization, nonlinear dynamics, synergetics, turbulence, dynamical systems, catastrophes, instabilities, stochastic processes, chaos, graphs and networks, cellular automata, adaptive systems, genetic algorithms and computational intelligence.

The three major book publication platforms of the Springer Complexity program are the monograph series "Understanding Complex Systems" focusing on the various applications of complexity, the "Springer Series in Synergetics", which is devoted to the quantitative theoretical and methodological foundations, and the "Springer Briefs in Complexity" which are concise and topical working reports, case studies, surveys, essays and lecture notes of relevance to the field. In addition to the books in these two core series, the program also incorporates individual titles ranging from textbooks to major reference works.

Series Editors

Understanding Complex Systems

Founding Editor: S. Kelso

Future scientific and technological developments in many fields will necessarily depend upon coming to grips with complex systems. Such systems are complex in both their composition – typically many different kinds of components interacting simultaneously and nonlinearly with each other and their environments on multiple levels – and in the rich diversity of behavior of which they are capable.

The Springer Series in Understanding Complex Systems series (UCS) promotes new strategies and paradigms for understanding and realizing applications of complex systems research in a wide variety of fields and endeavors. UCS is explicitly transdisciplinary. It has three main goals: First, to elaborate the concepts, methods and tools of complex systems at all levels of description and in all scientific fields, especially newly emerging areas within the life, social, behavioral, economic, neuro- and cognitive sciences (and derivatives thereof); second, to encourage novel applications of these ideas in various fields of engineering and computation such as robotics, nano-technology and informatics; third, to provide a single forum within which commonalities and differences in the workings of complex systems may be discerned, hence leading to deeper insight and understanding.

UCS will publish monographs, lecture notes and selected edited contributions aimed at communicating new findings to a large multidisciplinary audience.

More information about this series at http://www.springer.com/series/5394

Anna Zakharova

Chimera Patterns in Networks

Interplay between Dynamics, Structure, Noise, and Delay

Springer

Anna Zakharova
Institut für Theoretische Physik
Technische Universität Berlin
Berlin, Germany

ISSN 1860-0832 ISSN 1860-0840 (electronic)
Understanding Complex Systems
ISBN 978-3-030-21713-6 ISBN 978-3-030-21714-3 (eBook)
https://doi.org/10.1007/978-3-030-21714-3

This Springer imprint is published by the registered company Springer Nature Switzerland AG
The registered company address is: Gewerbestrasse 11, 6330 Cham, Switzerland

Preface

Synchronization phenomena represent one of the central issues in the modern research, in particular, on nonlinear dynamics, network science, and statistical physics with applications in almost all areas of science and engineering. Synchronization is ubiquitous in natural and man-made systems since their organization and functioning mostly depends on how different components are able to synchronize. This book provides an overview of state-of-the-art research on a peculiar type of partial synchronization pattern called *chimera states*. Discovered in the early 2000s, these hybrid states are made up of spatially separated domains of synchronized and desynchronized behavior. They arise surprisingly in networks of completely identical units and symmetric coupling topologies. Chimera states have been investigated both theoretically and experimentally and have wide-range applications in physics, biology, chemistry, and engineering. These intriguing patterns are named after a fabulous fire-breathing creature from Greek mythology that has a lion's, a goat's, and a snake's head. Like the counterintuitive dynamical state, this monster is also composed of incongruous parts.

This book describes recent developments on the formation of chimera states in various dynamical systems and networks with different topologies. Moreover, it discusses the role of time delay and stochasticity that arise naturally in real-world systems. We explore how the interplay of nonlinearity with network topology, time delay, and noise leads to a plethora of complex phenomena. In particular, we focus on the new dynamical behavior induced by noise and time delay. Furthermore, we address the question of robustness and control of chimera states that is especially relevant from the point of view of experiments and applications.

After introducing the reader to partial synchronization patterns and to the existing definitions of chimera states and explaining their main features, we provide a systematic overview of the established results on the subject (Chap. 1). In particular, we discuss the measures to detect chimera patterns and give a detailed overview of the systems in which chimera states have been found. Further, we systematically describe the network topologies for which chimera patterns have been reported. Then, the types of chimera states are considered and the existing methods of controlling chimeras are described. Moreover, we give examples of experiments on chimera states and discuss their main applications.

Further, we discuss special types of chimera states: amplitude chimeras and chimera death (Chap. 2). Chimera states have been initially found for the phase oscillator model and later extended to systems which involve not only phase but also amplitude dynamics and are named amplitude-mediated chimeras in the case when both amplitude and phase are characterized by chimera behavior. *Amplitude chimeras* are more complicated patterns in which chimera structures are formed with respect to the amplitudes only while the phases remain correlated for the whole network. *Chimera death* represents another recently discovered type of chimera states which, through death of the oscillations, generalizes the chimera feature of coexistence of spatially coherent and incoherent domains to steady states. In Chap. 2, we also discuss efficient control mechanisms for chimera states based on time delay and noise. In particular, we address the question of how time delay and noise influence the behavior of amplitude chimera states occurring in ring networks of Stuart-Landau oscillators.

Further, we focus on *coherence-resonance chimeras* in neural networks, an effect which combines coherence resonance and chimera states (Chap. 3). This noise-induced state is characterized by the coexistence of two different domains separated in space, where one part of the network is spiking coherently in space while the other exhibits incoherent spiking, i.e., the spiking of neighboring nodes is uncorrelated. In Chap. 3, we discuss time-delayed feedback control of the chimera patterns. Specifically, we focus on the role of noise and time delay for the chimera states occurring in ring networks of FitzHugh-Nagumo neurons in excitable regime. Finally, we describe the occurrence of chimera states in complex network structures beyond a nonlocally coupled ring (Chap. 4). Those include multiplex networks, networks with power-law coupling kernel, and networks with fractal connectivities. For multiplex networks, we first consider coherence resonance and further discuss chimera states and solitary states. Moreover, we describe coherence-incoherence patterns in networks with power-law coupling kernel and networks with fractal connectivities.

This book is aimed at researchers with a background in physics, applied mathematics, and engineering. It will not only be of great interest to the specialists working on chimera states and partial synchronization patterns, but also provides a valuable resource for the broader audience dealing with time-delayed and stochastic systems, complex networks and control.

Berlin, Germany
January 2020

Anna Zakharova

Acknowledgements

First and foremost, I thank my mother Larisa and my father Sergei for their selfless love and unconditioned support. I deeply appreciate their belief in me and the encouragement I have always been feeling in spite of the thousands of kilometers between Saratov and Berlin.

I thank my husband Richard for his endless love and enormous support. This book would not be possible without his encouragement, help, and patience. The permanent exchange of ideas and experiences with Richard allowed me to open up new perspectives on life, work, and ambition. I am still fascinated how highly Richard, not being a scientist, is interested in my research field and how strongly he is involved in my academic life, always having an open ear for me and offering his advice.

I wish to express my deepest gratitude to Prof. Eckehard Schöll. He has always inspired me with his enthusiasm, hard work, and dedication to science. I thank him very much for his encouragement, confidence, and supporting me on my way to scientific independence. He offered his help and advice, whenever I needed it. It was Prof. Eckehard Schöll who inspired me to write this book. Without his guidance and endless support, this book would not be possible.

I am very grateful to Prof. Vadim Anishchenko who sparked my interest in nonlinear dynamics as an undergraduate student and has been providing guidance and support ever since.

My first steps into the world of research were made under the careful and caring supervision of Prof. Tatiana Vadivasova. I am very thankful to her for fruitful and inspiring scientific discussions that I still have a pleasure to enjoy.

I would like to express my gratefulness to Prof. Jürgen Kurths who guided me through my first experience of international research in Germany and who has been supporting me ever since.

I am deeply thankful to the collaboration partners all over the world with whom I had the privilege to work and—in many cases—still enjoy working together (in alphabetical order): R. G. Andrzejak, V. Anishchenko, T. Banerjee, V. M. Bastidas, T. Brandes, K. Blyuss, J. C. Claussen, B. Fiedler, A. L. Fradkov, N. Fujiwara, A. Gjurchinovski, J. Hlinka, P. Hövel, S. Jalan, P. Jiruska, W. Just,

A. Koseska, J. Kurths, Y. N. Kyrychko, K. Lüdge, Y. Maistrenko, N. Malik, O. Omel'chenko, E. Panteley, E. Schöll, G. Strelkova, T. Vadivasova, M. Wolfrum.

I would like to thank the students on all levels from B.Sc. to Ph.D. and postdoctoral researchers, whom I had a pleasure to work with, as well as the members of the Schöll group for providing an inspiring working environment and friendly atmosphere (in alphabetical order): R. Aust, R. Berner, A. Blechschmidt, F. Böhm, A. zur Bonsen, A. Bukh, T. Chouzouris, A. Feoktistov, P. M. Geffert, M. Gerster, S. Ghosh, N. Heym, P. Kalle, M. Kapeller, S. Krishnagopal, J. Lehnert, S. Loos, M. Masoliver, M. Mikhaylenko, D. Nikitin, I. Omelchenko, J. Peters, W. Poel, L. Ramlow, G. Ruzzene, E. Rybalova, J. Sawicki, I. Schneider, L. Schülen, J. Siebert, A. Stuke, N. Semenova, V. Semenov, I. Shepelev, M. Thiele, L. Tumash, S. Ulonska, A. Vüllings, M. Winkler. In particular, I would like to acknowledge I. Omelchenko and M. Gerster for their assistance in the preparation of the overview and the tables in Chap. 1.

I acknowledge the support provided by the funding organizations including the Deutsche Forschungsgemeinschaft (DFG, German Research Foundation)—Projektnummer 163436311—SFB 910 and German Academic Exchange Service (DAAD) via exchange programs IKYDA and DAAD-DST.

Berlin, Germany
January 2020

Anna Zakharova

Contents

Chapter 1
Chimera Patterns in Complex Networks

Abstract This chapter provides a systematic overview of the established results on chimera states. We begin with a short historical note on how chimera states have been discovered. Having formulated the existing definitions of chimera states, we explain their main features. Moreover, we discuss the measures to detect chimera patterns and give a detailed overview of the systems in which chimera states have been found. Further, we systematically describe the network topologies for which chimera patterns have been reported. Then, the types of chimera states are considered and the existing methods of controlling chimeras are described. Finally, we give examples of experiments on chimera states and discuss their main applications.

1.1 Introduction

Synchronization phenomena in coupled nonlinear dynamical systems and networks [1–8] are of great importance in many areas ranging from physics and chemistry to biology, neuroscience, socio-economic systems, and engineering. Synchronization is ubiquitous in natural and man-made systems since their organization and functioning mostly depends on how different components are able to synchronize. For example, the synchronization of neurons is believed to play a crucial role in the brain under normal conditions, for instance in the context of cognition and learning [9], and under pathological conditions such as Parkinson's disease [10]. Fireflies are known to synchronize their flashing [11]. Synchronization of power grids is essential for their operation [12]. Chaos synchronization of lasers may lead to new secure communication schemes [3, 13, 14].

Synchronization is commonly understood as a collective behavior of coupled systems correlated in time. During synchronization, the coupled oscillatory units adjust their rhythms. There are various types of synchronization, for example, phase synchronization [15–20] and generalized synchronization [21–23]. In phase synchronization, only the suitably defined phases of the coupled systems but not their amplitudes are synchronized. This is particularly interesting in neural applications, where only the times when neurons spike are important but not the exact amplitudes. In generalized synchronization, coupled systems obey some—potentially nonlinear—

A. Zakharova, *Chimera Patterns in Networks*, Understanding Complex Systems,
https://doi.org/10.1007/978-3-030-21714-3_1

functional dependence. This functional dependence is generally not easily disclosed by just looking at the dynamics of the nodes and by standard time-series analysis.

A state where all the oscillating units undergo the same dynamics without a phase shift is called zero-lag synchronization. In coupled networks, it is often the case that not all the elements of the ensemble are synchronized. The examples are provided by group and cluster synchronization that have been investigated both theoretically [24–33] and experimentally [34–39]. In the case of group synchrony, the network consists of several groups where the nodes within one group are in zero-lag synchrony. Between the groups, no synchrony exists, or in the case of clusters states, synchrony with a constant phase shift is observed. Group and cluster synchronization can occur, for instance, in dynamics of nephrons [4], central pattern generation in animal locomotion [40], or population dynamics [15]. It is important to note that group synchronization can take place in networks where the local dynamics in different clusters or the coupling between clusters is non-identical [41].

More complex partial synchronization states occur when some parts of the network stop oscillating while the rest of the network is demonstrating synchronized dynamics. Two types of oscillation quenching mechanisms, i.e., amplitude death (AD) and oscillation death (OD), have been intensely studied both theoretically and experimentally [42, 43]. Amplitude death is associated with the stabilization of an already existing trivial steady state [44] while oscillation death is characterized by a newly born inhomogeneous steady state [45, 46]. Oscillation quenching has been observed across many man-made and natural systems ranging from lasers and electronic circuits [47, 48], chemical and biological networks including neurons [49–51] to climate systems [52]. Applications of AD are mainly in controlling physical and chemical systems (e.g., coupled lasers [53]) and suppressing neuronal oscillations [54, 55], while OD has been suggested as a mechanism to generate heterogeneity in homogeneous systems (e.g., stem cell differentiation [56] in morphogenesis). The investigation of partial amplitude and oscillation death in complex networks [33, 43, 57, 57, 58] is important for a better understanding of the emergence of local mesoscale structures on networks and their influence on the global dynamics [30, 32, 59, 60].

In the present book, we focus on a very prominent example of partial synchronization pattern called *chimera states*, which has recently gained much attention [61, 62]. These intriguing spatiotemporal patterns arise surprisingly in networks of symmetrically coupled identical units. In a chimera state, a network spontaneously splits into two parts with different dynamics separated in space: one demonstrating synchronization or coherent behavior and the other exhibiting desynchronization or incoherence. These symmetry-breaking states of partially coherent and partially incoherent behavior have wide-range applications in different fields of science and have been investigated both theoretically and experimentally.

The first review article on chimera states by Mark J. Panaggio and Daniel M. Abrams appeared in 2015 [63], followed by the review by Eckehard Schöll in 2016 [64]. Analytical study of chimera states turned out to be a challenging issue and current achievements on the mathematics behind chimeras are summarized in the review by Oleh Omel'chenko [65]. The findings on chimera states in neural systems are reported in the review by Soumen Majhi et al. [66]. A recent book on synchronization by Stefano Boccaletti et al. includes a chapter on chimera states [8]. A recent article by Anna Zakharova and Eckehard Schöll accessible to a broad audience discusses the aspect of control and the importance of chimeras for the study of brain networks [67]. Finally, recent developments and future perspectives of chimera states in complex networks are addressed in the Research Topic [68].

In this chapter, we provide a systematic overview of the established results on chimera states. We begin with a short historical note on how chimera states have been discovered (Sect. 1.2). Further, we formulate the existing definitions of chimera states and explain their main features (Sect. 1.3). Moreover, we discuss the measures to detect chimera patterns (Sect. 1.4). In Sect. 1.5, we give a detailed overview of the systems in which chimera states have been found. Furthermore, we systematically describe the network topologies for which chimera patterns have been reported (Sect. 1.6). The types of chimera states are considered in Sect. 1.7 and the existing methods of controlling chimeras are described in Sect. 1.8. Finally, we give examples of experiments on chimera states (Sect. 1.9) and discuss their main applications (Sect. 1.10).

1.2 Historical Note

The story of chimera in Greek mythology begins with the chimera from Carchemish, dated to 850-750 BC, which is considered to be a basis of the myth of the monster called *the chimera*. The Greek legend describes a fire-breathing monster that is composed of incongruous parts: a lion's, a goat's, and a snake's head (Fig. 1.1).

The *chimera state* in nonlinear dynamics appeared only recently in the beginning of the twenty-first century [61] and was named after the Greek beast by the American mathematician Steven Strogatz [62]. This counterintuitive spatiotemporal pattern also consists of incompatible parts: one demonstrating synchronization or coherent behavior and the other exhibiting desynchronization or incoherence [63, 64, 66]. These hybrid states surprisingly occur in networks of identical oscillators coupled in a symmetric way (Figs. 1.2 and 1.3).

Before the discovery of chimera states, it was assumed that ensembles of symmetrically coupled identical oscillatory elements can demonstrate either in-phase synchronization or complete incoherence. In networks of non-identical units, on the contrary, complex dynamical regimes such as partial synchronization were found. In this case, the splitting of the network into domains with different dynamics can be intuitively explained and originates from the inhomogeneity of the oscillators. The ground-breaking discovery was made by the Japanese physicist Yoshiki Kuramoto and his collaborator Dorjsuren Battogtokh in 2002. They considered a network of phase oscillators with a simple symmetric nonlocal coupling scheme and came up with an idea of using specially prepared initial conditions. Although all the elements were initially chosen to have the same natural frequency, the network spontaneously

Fig. 1.1 The chimera on a red-figure Apulian plate, c. 350-340 BC (Musée du Louvre)

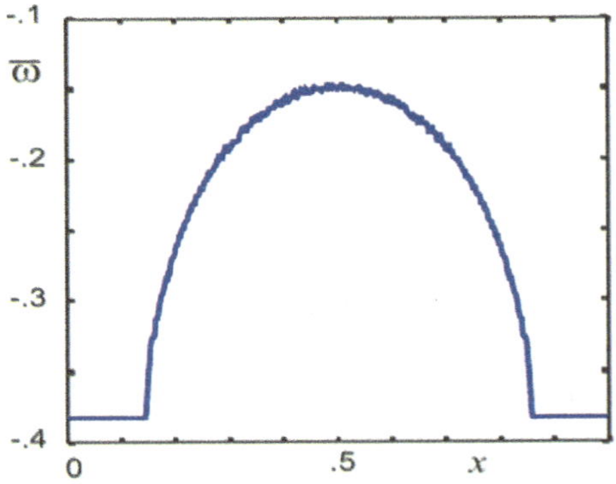

Fig. 1.2 Mean phase velocity profile of a chimera state in a network of coupled phase oscillators studied in [61]

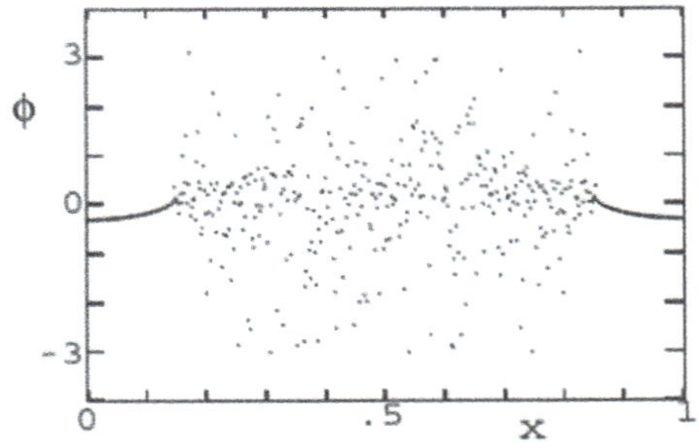

Fig. 1.3 Snapshot of a chimera state in a network of coupled phase oscillators studied in [61]

split into two spatially separated parts: a coherent domain of periodic in-phase oscillations and an incoherent domain exhibiting chaotic behavior in time and space. These *classical* chimera states are not a simple short-living trace of initial conditions, but a complex persistent pattern.

1.3 Definition and Main Features

The discovery of chimera states initiated a broad wave of further investigations including theoretical, numerical, and experimental studies. Besides the original model of coupled Kuramoto phase oscillators, chimera states have been found in various other systems including time-discrete maps [69–72] and time-continuous chaotic models [73], Boolean networks [74], Stuart-Landau oscillators [75, 76], Kuramoto model with inertia [77], Van der Pol oscillators [78], FitzHugh-Nagumo neural systems [79, 80], Hindmarsh-Rose oscillators [81], systems of type-I excitability [82], population dynamics [83, 84], quantum oscillator systems [85, 86], chemical oscillators [87–90], mechanical oscillators [91], laser systems [92–94], electronic circuits [95], optoelectronic networks [96], and many others. For a detailed overview of the systems explored for chimera states, see Sect. 1.5 and Tables 1.1 and 1.2 therein.

It is important to note that often the features of chimera states are model specific. In some cases, a particular model under study defines the type and properties of the observed pattern. Nevertheless, it is possible to formulate some general trends for chimera states. A relevant question is related to the definition of a chimera state. The original definition given by Abrams and Strogatz [62] is as follows: *a chimera state is a spatiotemporal pattern in which a system of identical oscillators is split into coexisting regions of coherent (phase and frequency locked) and incoherent (drifting) oscillation.* The more general version of the original definition is widely used in the nonlinear dynamics community: *chimera states correspond to the situation when an ensemble of identical elements self-organizes into two coexisting and spatially separated domains with different behavior, i.e., spatially coherent (synchronized) and incoherent (desynchronized).* Therefore, chimera states are possible not only for oscillatory regimes. As we will show in Chap. 2, steady-state solutions can also form chimera patterns called *chimera death*, where no temporal dynamics throughout the network is observed [75].

A mathematically strict definition of so-called *weak chimeras* has been suggested by Ashwin and Burylko [97] for a system of N identical coupled phase oscillators, described as an ordinary differential equation on the torus $(\varphi_1(t), \ldots, \varphi_N(t)) \in \mathbb{T}^N = [0, 2\pi)^N$:

$$\dot{\varphi}_i = \omega_i + \sum_{j=1}^{N} K_{ij} g(\varphi_i - \varphi_j), \tag{1.1}$$

where $\varphi_i(t)$ is the phase of the ith oscillator at time t, K_{ij} is the strength of the coupling, ω_i is the natural frequency of the ith oscillator, and $g(\varphi)$ is a smooth 2π-

periodic coupling function. The definition is as follows: $A \subset \mathbb{T}^N$ is a *weak chimera* state for a coupled phase oscillator system, if it is a connected chain-recurrent [98] flow-invariant set such that on each trajectory within A there exist indices i, j, and k such that $\Omega_{ij} \neq 0$ and $\Omega_{ik} = 0$, where

$$\Omega_{ij} := \lim_{t\to\infty} \tfrac{1}{t}(\varphi_i(t) - \varphi_j(t)) = 0. \tag{1.2}$$

It is important to note that for every particular case, one should carefully define the notion of "domain", "incoherence", and "coherence". For example, in the case of *classical* chimeras found in phase oscillator systems, incoherence refers to the phases of oscillating units, while so-called *amplitude chimeras* are characterized by coexisting domains with coherent and incoherent amplitudes, while the phases are correlated throughout the whole network (Chap. 2). The generalization towards networks of elements which have more complicated local dynamics has been a significant achievement for the study on chimera states [63, 78, 93]. In particular, nodes which not only involve phase but also amplitude dynamics have been considered. As described in [99, 100], in such systems, amplitude-mediated chimeras can be found, which show a chimera behavior with respect to the phases as well as with respect to the amplitudes. And the *amplitude chimeras* mentioned above represent, therefore, another type of chimera states. They are characterized by strictly correlated phase dynamics throughout the whole network, but coexisting domains of coherent and incoherent amplitude dynamics. These *amplitude chimeras* were first described in [75]. A crucial difference to classical phase chimeras is that the spatial incoherence does not imply chaotic behavior in time. In fact, all nodes of an amplitude chimera perform periodic oscillations, but in the incoherent domain, the spatial sequence of the positions of the centers of oscillation is completely random [75, 76, 101–104]. Amplitude chimeras were also described in different models: networks of ecological oscillators (Rosenzweig-MacArthur system) with nonlocal [105] and power-law coupling [84] and nonlocally coupled ring networks of Rayleigh oscillators [106]. Moreover, transitions between amplitude-mediated phase chimeras and amplitude chimeras were found [106]. Recently, it has been shown that filtering suppresses amplitude chimeras [107]. A systematic overview of the types of chimera states is given in Sect. 1.7 (see also Tables 1.6 and 1.7).

For the studies on chimera states, the topology of the underlying network and coupling parameters play a crucial role. Initially it was believed that to induce chimeras, a nonlocal coupling is significant (Fig. 1.4). In such networks, every node has P nearest neighbors in each direction on a ring. Additionally, a normalized number of nearest neighbors $r = P/N$ (also called coupling range), where N is the total number of elements in the network, is used to compare networks of different sizes. By tuning P, one can interpolate between local (nearest neighbor, $P = 1$) coupling and global coupling (all-to-all, $P = N/2$). In general, decreasing the coupling range leads to the formation of multicluster chimeras, i.e., chimera states with more than one incoherent domain.

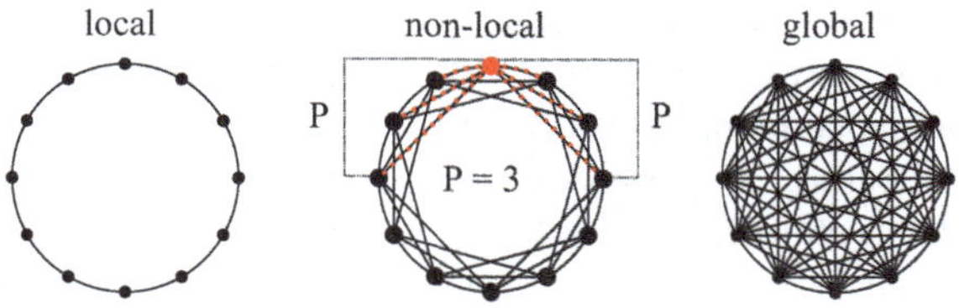

Fig. 1.4 Schematic illustration of local, nonlocal, and global coupling schemes

The notion of space and consequently spatially separated domains is clear for networks with nonlocal coupling, where the neighborhood of a particular node is well defined. The global coupling topology, on the contrary, does not provide a clear notion of space. Nevertheless, the occurrence of chimeras has now been accepted not only for nonlocal coupling, but also for global coupling [93, 99, 100, 108–110] and even local coupling [105, 111, 112]. Moreover, chimeras have been found for irregular topologies [113–117] and hierarchical, quasi-fractal connectivities [80, 83, 118–123]. Here, we will consider nonlocally coupled ring networks (Chaps. 2 and 3) as well as multilayer networks [124–126] (Chap. 4). An overview of the network topologies for which chimera states have been investigated is provided in Sect. 1.6 (see also Tables 1.3, 1.4, and 1.5 therein).

In the nonlocal coupling, one has two control parameters: coupling range and coupling strength. The former is controlled by the kernel of the nonlocal coupling function: in previous studies, generally a trigonometrically (sinusoidal kernel in [62]) or exponentially decaying function (in the original work by Kuramoto [61]) or a rectangular kernel have been used. We will study rectangular (Chaps. 2 and 3) and power-law coupling kernels (Chap. 4).

Another important question is how to distinguish chimera states from other partial synchronization patterns such as, for example, cluster states. Chimera states are often seen as a dynamical scenario of transition from completely synchronized to completely irregular behavior. Alternatively, such a transition can occur via another partial synchronization pattern, namely, *solitary state*. Its mechanism is different from that of a chimera state. The term solitary comes from the Latin solitarius and can be understood as alone, lonely, or isolated. In the case of chimera states, the system spontaneously splits into coexisting domains of coherent (e.g., synchronized) and incoherent behavior, which are localized in space. On the contrary, for solitary states, it is typical that individual "solitary" oscillators start leaving the synchronous cluster at random positions in space [127, 128]. Solitary states represent a soft transition to incoherence since the oscillators are leaving the coherent cluster gradually. Chimera states are characteristic of a sharp transition due to the fact that the occurring incoherent domain has initially a certain finite (typically large) size. In the present book, solitary states will be investigated in networks with power-law coupling kernel and in multiplex networks [126] (Chap. 4).

One of the most important questions is related to *stability* of the observed regimes. It is known that chimera states are long-living transients for finite-size systems and their lifetime for phase oscillators grows exponentially with the system size [74, 129]. As the number of oscillators in the network $N \to \infty$, chimera states become stable.

Often chimeras are transients towards the in-phase synchronized oscillatory regime or these two states can coexist in the parameter space. Typically, the basin of attraction of the chimera states is relatively small compared to that of the synchronized state. Therefore, in many situations, specially prepared initial conditions are required to detect chimera patterns [130]. Also this is the reason why these hybrid states were not detected for a long time. Nevertheless, chimera states have been shown to be robust with respect to various perturbations ranging from noise [102] to inhomogeneities introduced into the local dynamics or the network topology [80]. Moreover, it has been shown that noise can play a constructive role for chimera states, i.e., induce these hybrid patterns. The noise-induced chimeras have been discovered recently and called *coherence-resonance chimeras* [131–133] (Chap. 3). These patterns combine features of coherence resonance [134, 135] and chimera states.

Originally found in excitable systems, the counterintuitive effect of *coherence resonance* describes a non-monotonic behavior of the regularity of noise-induced oscillations in the excitable regime, leading to an optimum response in terms of regularity of the excited oscillations for an intermediate noise strength. Excitability means that the systems rests in a stable steady state, but can be excited by a sufficiently strong perturbation (external input or noise) to emit a spike before returning to the stable steady state. Later, it has been shown that coherence resonance can also occur in non-excitable systems [136]. For a generalized Van der Pol model with a subcritical Hopf bifurcation, it has been demonstrated that its mechanism is different from conventional coherence resonance in excitable systems [137]. The possibility of time-delayed feedback control for this type of coherence resonance has been reported in experiments on an electronic circuit combined with numerical simulations, and an analytical approach [138]. Further, analytical tools to estimate the stationary distribution and the time correlations in nonlinear stochastic delay differential equations have been developed [139, 140]. These tools allowed to introduce a new quantity to measure coherence resonance induced by a saddle-node bifurcation of periodic orbits. Moreover, the phenomena of coherence resonance have been systematically investigated in time-delay-coupled networks of FitzHugh-Nagumo elements in the excitable regime for different coupling topologies [141]. Using numerical simulations, the interplay of noise, time-delayed coupling, and network structure has been examined. Very recently it has been shown that coherence resonance can be induced in multilayer structures by weak multiplexing [142].

Additionally, the noise-based control mechanisms developed in [131, 143] reveal an alternative direction for chimera control complementary to recent deterministic control schemes [126, 144–148]. Furthermore, it has been demonstrated that applying time-delayed feedback allows to control coherence-resonance chimeras by adjusting delay time and feedback strength [133]. Chimera relay synchronization can be controlled by time delay in the inter-layer coupling [149] as well as by the interplay of time delays in the intra- and inter-layer coupling [150]. The existing methods of controlling chimeras are summarized in Sect. 1.8 (see also Table 1.8).

The first experimental works on chimera states appeared one decade after their theoretical discovery in optical light modulators [151], chemical [87, 88], mechanical [91, 152], electronic or optoelectronic oscillators [92, 94, 153], electrochemical systems [89, 110, 154], electronic circuits [74, 95], and optical combs [155]. The examples of experiments on chimera states are provided in Sect. 1.9 (see also Table 1.9).

In real-world systems, chimera states might play a role, e.g., in the unihemispheric sleep of birds and dolphins [156], in first-night effect in humans [157], in epileptic seizures [122, 158–161] or even perceptual organization [162] and behavioral sensitization [163, 164], in power grids [12], or in social systems [165]. For a detailed overview of the main applications of chimera states, see Sect. 1.10 and Table 1.10 therein.

1.4 Quantitative Measures

The first works on chimera states investigate the phase oscillator model [61, 62]:

$$\dot{\varphi}(x,t) = \omega - \int_{-\pi}^{\pi} G(x-x')\sin[\varphi(x,t) - \varphi(x',t) + \alpha]dx', \tag{1.3}$$

where $\varphi(x,t)$ is the phase of the oscillator at position x at time t, ω is the frequency, α is the phase lag parameter, and $G(x-x')$ is the coupling kernel. The phases are assumed to be continuously distributed over the interval $[-\pi, \pi]$ with periodic boundary conditions. It was shown that the value of the phase lag parameter close to $\pi/2$ is essential for the detection of chimera states [166].

To detect the coexistence of coherence and incoherence in space, the visual inspection of the snapshot (Fig. 1.3) and space-time plot for the state variable of the system $\varphi(x,t)$ is helpful. For the quantitative characterization of chimera states in the continuum limit, it is common to use the local order parameter R that measures the amount of coherence in the system:

$$R(x,t)e^{i\Theta(x,t)} = \int_{-\pi}^{\pi} G(x-x')e^{i\Theta(x',t)}dx'. \tag{1.4}$$

For a one-dimensional ring network consisting of a finite number N of elements with phases $\Theta_1, \ldots, \Theta_N$, the spatial coherence and incoherence of the chimera state can be characterized by a real-valued local order parameter [69, 167]:

$$Z_k(t) = \left| \frac{1}{2\delta_Z} \sum_{|j-k|\le\delta_Z} e^{i\Theta_j} \right|, \quad k = 1, \ldots N, \tag{1.5}$$

where δ_Z is a spatial window size, i.e., the summation is performed for the neighbors of the k-th element belonging to this window. The local order parameter $Z_k = 1$ indicates that the k-th unit belongs to the coherent part of the chimera state, and $Z_k < 1$ for incoherent parts. Therefore, coexisting regions with Z_k close to unity (coherence) and $Z_k < 1$ (incoherence) serve as an indicator of chimera states.

Another quantitative measure of chimera states is based on the mean phase velocity:

$$\omega_k = \langle \frac{d\varphi_k(t)}{dt} \rangle_{\Delta t}, \tag{1.6}$$

where Δt is the time interval for averaging. In general, one has to define the phase and the mean phase velocity individually for every particular model to quantify chimera states. The mean phase velocity profile is a clear indication of chimera states: it is flat for coherent domain of chimera (synchronized oscillators) and arc-shaped for the incoherent domain (desynchronization) (Fig. 1.2). It is often the case that a particular type of chimera state requires a specific measure to indicate the existence of chimeras in the system. For example, in the *amplitude chimera* state, considered in Chap. 2, all the network elements are characterized by the same average frequency. Therefore, the mean phase velocity profile does not provide any information on the hybrid states and alternative measures have to be developed (see Chap. 2).

Further indicators of chimeras are based on correlation measures [168, 169]. To characterize chimera states, Kemeth et al. suggested measures for spatial g_0 and temporal h_0 coherence that are applicable to arbitrary data sets [168]. Their measure for spatial coherence is based on the local curvature of the observable that is quantified by the second derivative in one-dimension. In the general case for any number of spatial dimensions, the curvature is characterized by the Laplacian. The local curvature at each point in space is obtained by applying the discrete Laplacian $\mathbf{D}$ on each snapshot containing the spatial data f. In the case of one spatial dimension for one snapshot at time t, we obtain

$$\hat{\mathbf{D}}f = \Delta x^2 \mathbf{D} f = f(x+\Delta x, t) - 2f(x,t) + f(x-\Delta x, t). \tag{1.7}$$

Next, the normalized probability density function g of $|\hat{\mathbf{D}}|$ is calculated, where $g(|\hat{\mathbf{D}}| = 0)$ measures the relative size of spatially coherent regions in each temporal realization, D_m indicates the maximal value of $|\hat{\mathbf{D}}|$. Consequently, for a fully synchronized state $g(|\hat{\mathbf{D}}| = 0) = 1$, for the incoherent state $g(|\hat{\mathbf{D}}| = 0) = 0$. The coexistence of synchrony and incoherence is characterized by a value between 0 and 1 of $g(|\hat{\mathbf{D}}| = 0)$. Further, it is assumed that for spatially extended systems, a point for which the absolute local curvature is less than one percent of the maximum curvature present in the system should be considered as coherent, and as incoherent otherwise. Finally, with the threshold $\delta = 0.01 D_m$, the spatial correlation measure is defined as follows:

$$g_0(t) := \int_0^{\delta} g(t, |\hat{D}|) d|\hat{D}|. \tag{1.8}$$

1.5 Chimera States in Different Systems

Chimera states have been discovered for nonlocally coupled ring networks of Kuramoto phase oscillators [61, 62]. Since the first observation of chimera patterns, this model has received growing interest of researchers [129, 166, 167]. Besides nonlocally coupled rings, phase oscillators have been investigated for the case of two interacting identical populations [170], where chimera states correspond to the regime when one population is synchronized and the other is desynchronized. Later, it has been shown that the network elements do not have to be identical and chimera states can be observed in heterogeneous networks of Kuramoto phase oscillators [171, 172]. For a system of coupled identical phase oscillators, a mathematically strict definition of chimera states (see Sect. 1.3) has been formulated [97].

Besides the original model of coupled Kuramoto phase oscillators, chimera states have been found in various other systems. These include, for example, time-discrete logistic maps in nonlocally coupled ring networks [69, 73]. Further works on logistic maps discuss specific types of chimeras in nonlocally coupled rings (see Sect. 1.7 for more details) [173], and networks with fractal topologies [121]. In [174], the possibility of observing chimeras in the networks of non-identical logistic maps with nonlocal coupling has been demonstrated. For chaotic systems, both time-discrete (Henon map, Lozi map) and time-continuous (Lorenz system), it has been shown that the hyperbolicity of the chaotic attractor impedes the emergence of chimera states [70]. Temporal intermittency and liftetime of chimera states have been investigated in nonlocally coupled ring networks of Henon maps [71]. For a 2D square lattice of nonlocally coupled cubic maps, the so-called double-well chimeras have been uncovered [72]. Moreover, chimera patterns have been detected in nonlocally coupled rings of time-continuous chaotic models, namely Rössler and Lorenz oscillators [73]. In Boolean networks, chimera states have been found both in numerical simulations and experiments on nonlocally coupled electronic oscillators [74].

An important step for the investigation of chimera state is their generalization towards systems that involve not only phase but also amplitude dynamics. For example, networks of Stuart-Landau oscillators with global coupling have been studied in the presence of delayed feedback [175, 176]. Networks of nonlocally and nonlinearly coupled Stuart-Landau oscillators have been analyzed in [177].

Chimera patterns occurring in generic oscillators (Stuart-Landau oscillators) are discussed in detail in the present book (Chap. 2). Interestingly, in networks with nonlocal topology and symmetry-breaking coupling, the formation of amplitude chimera states and chimera death patterns is possible [75, 76, 101]. The stability of amplitude chimeras has been investigated using Floquet theory and it has been shown that they represent saddle orbits in the underlying phase space of the network [103]. Further, the question of robustness and control of amplitude chimera states in the presence of noise and time delay has been addressed. It has been found that the lifetime of amplitude chimeras can be controlled by varying the noise intensity [102] and the value of time delay [104]. Moreover, it has been shown that long-range interaction with distance-dependent power-law coupling, which is very common in physics and biology, can induce amplitude chimeras and chimera death [84]. In more detail, the investigation of a network of coupled ecological oscillators has revealed that the variation of the power-law exponent associated with the coupling can induce transitions between spatial synchrony, amplitude chimeras, and various chimera patterns like chimera death and chimera-like coexistence of synchronized oscillation and death. Therefore, by tuning the power-law exponent, one can induce or control chimera patterns. Amplitude chimeras have also been found in nonlocally coupled ecological oscillators [105] and Rayleigh oscillators [106]. Moreover, transitions between amplitude-mediated phase chimeras and amplitude chimeras have been found [106]. Chimera death pattern has been shown to exist in networks with local [105] and global coupling [178].

Furthermore, the observation of chimera patterns has been extended to a broad range of oscillatory systems including planar oscillators [179] and Kuramoto model with inertia [77, 180]. The pendulum-like oscillators have been shown to demonstrate chimera patterns [181] and in [152], imperfect chimera states have been found in experiments on mechanical oscillators. Further, nonlocally coupled networks of Van der Pol oscillators have been studied. In more detail, the nonlinearity parameter of the individual units has been shown to influence the multiplicity of the incoherent domains in the chimera patterns [78]. Chains of Duffing oscillators have been analyzed in [182].

FitzHugh-Nagumo oscillators, which are widely used for modeling neuronal dynamics, demonstrate chimera states in nonlocally coupled rings, where the increase of the coupling strength leads to the formation of the multichimera states with several incoherent domains [79]. Chimera states have been shown to be robust with respect to inhomogeneities in both local dynamics (heterogeneous elements) and network topology (non-symmetrical links) [80]. Surprisingly, chimera states can also occur in the excitable FitzHugh-Nagumo neurons in the presence of noise [131, 132]. These patterns noise-induced patterns are called coherence-resonance chimeras. They occur in networks of nonlocally coupled neurons and can be controlled by time-delayed feedback [133]. Moreover, time delay induces a specific type of chimera called period-two coherence-resonance chimera. The formation of coherence-resonance chimeras and their control is discussed in detail in Chap. 3.

Moreover, chimera and multichimera states have been found for nonlocally coupled Hindmarsh-Rose neurons [81], as well as SNIPER model (type-I excitability) [82]. Other neuronal models in which chimera states have been observed are leaky integrate-and-fire neurons [183], Morris-Lecar neurons [184], and Hodgkin-Huxley neurons [185]. Furthermore, realistic C. elegans brain networks can demonstrate chimera patterns [186, 187]. Networks of pulse-coupled neural networks also show chimera behavior [188, 189]. Multicluster and traveling chimera states can be observed in nonlocally pulse-coupled oscillators [190, 191]. For a recent review on chimeras in neuronal systems, see [66].

As a result of growing interest to chimera states, more and more fields of research got involved in the study on these peculiar patterns. They have been found in networks describing population dynamics [83, 84], in epidemic models [192], chemical [87–90] and mechanical oscillators [91], electronic and optoelectronic oscillators [92, 94, 153]. In particular, it has been shown that even very small globally coupled network consisting of four lasers can demonstrate chimera states [93]. Especially interesting are the systems where both numerical and experimental observation of chimera patterns is possible. These include, for example, optical networks with nonlocal couplings [151], electronic circuits [95], optoelectronic networks [96]. Furthermore, chimera states have been detected for quantum oscillator systems [85, 86] and

Table 1.1 Systems in which chimeras have been found (part 1)

System type	Topology	Comment	References
Kuramoto phase oscillator	Ring with nonlocal coupling	First observation	[61]
Kuramoto phase oscillator	Ring with nonlocal coupling	Irregular motion in time as finite-size effect	[166]
Kuramoto phase oscillator	Ring with nonlocal coupling	Chaotic transients, lifetime grows exponentially with system size	[129]
Kuramoto phase oscillator	Ring with nonlocal coupling	Spectral properties of chimera states	[167]
Kuramoto phase oscillator	Ring with nonlocal coupling	Two-population model	[170]
Kuramoto phase oscillator	Ring with nonlocal coupling	Heterogeneous networks	[171, 172]
Kuramoto phase oscillator	Kuramoto-Sakaguchi coupling	Weak chimeras	[97]
Kuramoto phase oscillator	Ring with nonlocal coupling	Coining the name *chimera states*	[62]
Kuramoto phase oscillator	Ring with nonlocal phase-coupling	Multicluster and traveling chimeras	[190]
Kuramoto phase oscillator	Nonlocal non-identical phase-coupling	Multicluster and traveling chimeras	[191]
Kuramoto phase oscillator	Ring with nonlocal coupling	Continuum limit	[195]
Kuramoto phase oscillator	Ring with nonlocal coupling	Partially coherent twisted states	[196]
Kuramoto phase oscillator	Ring with nonlocal coupling	Impact of small phase lag	[197]
Kuramoto phase oscillator	Ring with nonlocal coupling	Universal relations of local order parameters	[198]
Kuramoto phase oscillator	Ring with nonlocal coupling	Self-localized excitation	[199]
Modified Ikeda time-delayed equation	Single oscillator with time-delayed feedback	Virtual chimeras	[153]
Kuramoto model with inertia	Two globally coupled populations	Intermittent chaotic chimeras	[77]
Kuramoto model with inertia	Two globally coupled populations	Dynamics for varying inertia	[180]
Chemical oscillators	Nonlocally coupled ring and square lattice	Experiment	[87]
Pendulum-like oscillators	Two globally coupled populations	Two-population network	[181]
Pendulum-like oscillators	Local and nonlocal coupling	Imperfect chimera states	[152]

(continued)

Table 1.1 (continued)

System type	Topology	Comment	References
Damped mechanical pendula	Nonlocal hierarchical coupling of two subpopulations	Mechanical oscillators	[91]
Stuart-Landau oscillator	Global coupling	Delayed feedback stimulation	[175]
Stuart-Landau oscillator	Global coupling	Spatially modulated delayed feedback	[176]
Stuart-Landau oscillator	Symmetry-breaking nonlocal coupling	Amplitude chimera and chimera death	[75, 76, 101]
Stuart-Landau oscillator	Symmetry-breaking nonlocal coupling	Stability of amplitude chimera	[103]
Stuart-Landau oscillator	Symmetry-breaking nonlocal coupling	Role of delay for amplitude chimera	[104]
Stuart-Landau oscillator	Symmetry-breaking nonlocal coupling	Role of noise for amplitude chimera	[102]
Stuart-Landau oscillator	Nonlocal and nonlinear coupling	Turbulent chimeras	[177]
Stuart-Landau oscillator	Two populations with global coupling	Network of planar oscillators	[179]
Stuart-Landau oscillator	Symmetry-breaking mean-field diffusive coupling	Chimera death	[178]
Stuart-Landau oscillator	Symmetry-breaking nonlocal coupling	Amplitude chimera (with filtering)	[107]
Ginzburg-Landau oscillator	2D local coupling	Continuous media	[194]

SQUID metamaterials with local interactions [193]. Recently, chimera states have been observed in continuous media [194]. The systems in which chimeras have been found are summarized in Tables 1.1 and 1.2, where the corresponding references are given.

1.6 Chimera States in Networks with Various Topologies

Chimera states have been initially studied in two types of network topologies. The first model investigated in the original works represents a *ring with nonlocal coupling*, where each network element interacts with a certain number of its neighbors. By tuning the number of nearest neighbors, this coupling allows to interpolate between two limit cases: local (nearest neighbor) coupling and global (all-to-all) coupling. Nonlocal topology can be described by different coupling kernels: exponential [61], cosine-like [62, 200], piecewise-constant [69, 79, 129, 166], or power-law-like [171,

Table 1.2 Systems in which chimeras have been found (part 2)

System type	Topology	Comment	References
Ecological oscillator (Rosenzweig-MacArthur)	Power-law coupling	Amplitude chimera, chimera death	[84]
Ecological oscillator (Rosenzweig-MacArthur)	Power-law coupling	Chimera-like coexistence of synchronized oscillation and death	[84]
Ecological oscillator (Rosenzweig-MacArthur)	Ring with nonlocal coupling	Chimera-like coexistence of synchronized oscillation and death	[105]
Ecological oscillator (Rosenzweig-MacArthur)	Local coupling	Chimera death	[105]
Rayleigh oscillators	Ring with nonlocal coupling	Amplitude-mediated and amplitude chimeras	[106]
Van der Pol oscillator	Ring with nonlocal coupling	Multichimera states	[78]
Van der Pol oscillator	Ring with nonlocal coupling	Quantum signatures of chimeras	[85, 86]
FitzHugh-Nagumo oscillator	Ring with nonlocal coupling	Multichimera states	[79]
FitzHugh-Nagumo oscillator	Ring with nonlocal coupling	Robustness of chimeras	[80]
FitzHugh-Nagumo oscillator	Local coupling	Experiments on electronic circuit	[95]
FitzHugh-Nagumo excitable regime with noise	Ring with nonlocal coupling	Coherence-resonance chimeras	[131, 132]
FitzHugh-Nagumo excitable regime with noise	Ring with nonlocal coupling	Coherence-resonance chimeras, control by delay	[133]
Logistic maps, Rössler/Lorenz system	Ring with nonlocal coupling	Coherence-incoherence transition	[73]
Logistic maps	Ring with nonlocal coupling	Amplitude and phase chimeras	[173]
Logistic maps	Fractal (hierarchical) connectivity	Nested chimeras	[121]
Non-identical logistic maps	Ring with nonlocal coupling	Robustness of chimeras	[174]
Henon/Lozi map, Lorenz system	Ring with nonlocal coupling	Hyperbolicity of chaotic attractors	[70]
Henon maps	Ring with nonlocal coupling	Temporal intermittency and liftetime of chimeras	[71]

(continued)

Table 1.2 (continued)

System type	Topology	Comment	References
Chaotic/periodic maps, Rössler system	Ring with nonlocal coupling	Coherence-incoherence transition	[69]
Cubic maps	2D lattice	Double-well chimeras	[72]
Hindmarsh-Rose oscillators	Ring with nonlocal coupling	Multichimera states	[81]
Hindmarsh-Rose oscillators	C. elegans network	Chimeras in C. elegans	[186, 187]
Photonic nonlinear system	Semiconductor laser with two feedback delays	Experiment	[94]
Semiconductor laser	Global coupling	Lang-Kobayashi equations	[93]
Leaky integrate-and-fire	2D fractal toroidal geometry	Stripe-and-grid chimeras	[183]
Leaky integrate-and-fire	Two pulse-coupled populations	Influence of dilution and noise	[189]
Type-I Morris-Lecar neurons	Ring with nonlocal coupling	Multichimera states	[184]
SNIPER model	Ring with nonlocal coupling	Multichimera states	[82]
Hodgkin-Huxley neurons	Abrams-Strogatz/Kuramoto coupling	Thermally sensitive neurons	[185]
Boolean phase oscillators	Ring with nonlocal coupling	Resurgence of chimeras	[74]
Duffing oscillators	Ring with local coupling	Lyapunov spectra	[182]
Lattice Limit Cycle model	Nonlocal coupling/hierarchical coupling	Population dynamics	[83]
Belousov-Zhabotinsky model	Global coupling of two populations	Experiment, chemical osc.	[88]
Chemical oscillator	Various coupling/max. 20 units	Experiment	[89]
Chemical oscillator	Exponentially decaying nonlocal coupling	Experiment, spiral wave chimeras	[90]
Liquid-crystal spatial light modulator	Nonlocal coupling/coupled-map lattice	Experiment	[151]
Fiber-coupled laser diodes	Global coupling	Experiment	[96]
SQUID oscillator	Ring with local coupling	SQUID metamaterials	[193]

201]. In [202], the authors observe persistent chimera states in nonlocally coupled phase oscillators, where coupling function includes higher harmonics. Moreover, the occurrence of chimeras has been explained by the interplay between initial conditions and nonlocal coupling [130].

The second topology is a *two-population model*, i.e., a network that consists of two identical globally coupled subpopulations characterized by weaker global (all-to-all) coupling between the populations [170]. In [203], two small populations have been considered. Chimera states have been observed for two subpopulations of pendulum-like elements [181] and for two interacting populations of Stuart-Landau oscillators [204]. As an extension, chimera states have been found in triangular networks of phase oscillators [205] and networks of three subpopulations [206].

In both cases, there is a spatial order of the oscillators the network: in the first case, the oscillators have their fixed positions on the ring. In the second case, coherence and incoherence are observed within a certain population: one of the subpopulations is synchronized, while the other one is desynchronized. The global (all-to-all) coupling topology, on the contrary, does not provide a clear notion of space. Nevertheless, the occurrence of chimeras has now been accepted also for global coupling [93, 99, 100, 108, 109, 207–209]. Similarly, the locally coupled rings or chains with only nearest-neighbor connections started to attract the interest in the context of chimera states [105, 111, 112]. For example, it has been shown that chimeras can be observed in ring networks with unidirectional local nonlinear interaction [112]. In [182], the chain of Duffing oscillators has been studied.

Therefore, the local, nonlocal, and global coupling schemes have been extensively studied. However, the structures of real-world networks are organized in a more complex way. Recent results in the area of neuroscience increased the interest in irregular coupling topologies and disclosed an intricate architecture in the neuron interconnectivity of the human and mammalian brain: the connectivity of the neuron axons network represents a quasi-fractal geometry. These studies initiated the investigation of different networks with quasi-fractal connectivity and the mechanisms of formation of chimera states in such networks have been systematically explored. Hierarchical topologies can be generated using a classical Cantor construction algorithm for a fractal set [210, 211]. This iterative hierarchical procedure starts from a *base pattern* or initiation string b_{init} of length b, where each element represents either a link ('1') or a gap ('0'). In each iterative step, each link is replaced by the initial base pattern, while each gap is replaced by b gaps. Thus, each iteration increases the size of the final bit pattern, such that after n iterations, the total length is $N = b^n$. Since the hierarchy is truncated at a finite n, we called the resulting pattern quasi-fractal. Using the resulting string as the first row of the adjacency matrix $\boldsymbol{G}$, and constructing a circulant adjacency matrix $\boldsymbol{G}$ by applying this string to each element of the ring, a ring network of $N = b^n$ nodes with hierarchical connectivity is generated. For networks of FitzHugh-Nagumo oscillators, it has been shown that this particular coupling results in chimera patterns with nested incoherent parts characterized by complex fragmented structure. Their specific structure depends on the complex interplay between the link geometry and the initial conditions [33, 80]. In networks of logistic maps with hierarchical connectivities, it has been shown that the symmetry of the base pattern promotes chimera states, and asymmetric connectivities result in complex nested chimera patterns [121]. Further, the existence of chimera states in a population dynamics model has been demonstrated and it turned out that quasi-fractal coupling topologies induce traveling multichimera states for this model [83]. Multi-chimera states have also been detected in networks of leaky integrate-and-fire model with excitatory coupling and hierarchical connectivity [119]. For networks of Van der Pol oscillators, the stepwise transition from a nonlocal to a quasi-fractal connectivities in ring networks has been investigated [118]. Using the clustering coefficient and the symmetry properties to classify different topologies with respect to the occurrence of chimera states, it has been shown that symmetric connectivities with large clustering coefficients promote the emergence of chimera states, while they are suppressed by slight topological asymmetries or small clustering coefficient. Moreover, the interplay of fractal topology and time-delayed coupling has been considered for Van der Pol oscillator [120] and FitzHugh-Nagumo model [123]. In the parameter plane of coupling strength and delay time tongue-like regions of existence of chimera states alternating with regions of coherent dynamics have been detected. It has been shown that by varying the time delay, one can deliberately stabilize desired spatiotemporal patterns in the system. Very recently a mathematically constructed network with modular fractal connectivity has been compared to an empirical structural neural connectivity derived from diffusion weighted magnetic resonance imaging [122]. In more detail, for networks of FitzHugh-Nagumo neurons, the properties of chimeras and partially synchronized states have been analyzed and the regions of their stability have been investigated.

Moreover, chimera states have been demonstrated for networks with strongly irregular topologies [113–117], such as random non-complete networks of phase oscillators [115] or small-world networks of pulse-coupled oscillators [160]. In [212], the authors have analyzed the synchronizable clusters and stable chimeras in complex networks of optoelectronic systems. Further complex topologies that have been investigated for chimera states include long-range interaction with distance-dependent power-law coupling, which is very common in physics and biology [84]. In particular for networks of ecological oscillators (Rosenzweig-MacArthur system), it has been shown that the variation of the power-law exponent associated with the coupling can induce transitions between spatial synchrony, amplitude chimeras, and various other chimera patterns. In networks of FitzHugh-Nagumo oscillators, power-law coupling also allows observation of chimera states [201]. Additionally, it leads to formation of solitary states and other complex spatiotemporal patterns such as tree-like perturbations (see Chap. 4).

Recently, multilayer networks have been suggested to offer a better representation of the topology and dynamics of real-world systems in comparison with isolated one-layer structures [213–215]. In multilayer networks, the nodes are distributed in different layers according to the type of relation they share. For example, in the case of a neuronal network, the neurons can form different layers depending on their connectivity through a chemical link or by an ionic channel. In brain networks, different regions can be seen connected by functional and structural neural networks. From the point of view of applications of chimeras to neuronal and brain networks, it is especially important to understand their formation mechanisms in multilayer networks. Multiplex networks represent a special class of multilayer networks where the layers contain the same number of nodes and the inter-layer links are allowed only for replica nodes, i.e., there is one-to-one correspondence between the nodes in different layers (no cross-links between the nodes in different layers). Chimera states have been reported in multilayer networks with nonlocal intra-layer topology for various models: logistic maps [124, 125], FitzHugh-Nagumo oscillators [126], and phase oscillators [20, 216]. In more detail, the interplay of time delay and multiplexing reveals a novel type of a chimera state, a layer chimera, where one layer of the network demonstrates coherent dynamics and the other layer is characterized by incoherent behavior [124]. Real-world multilayer networks can be in general non-identical, i.e., the layers can have distinct structure or the local dynamics across the layers can be different. In [125], it has been shown that non-identical multiplexing, i.e., when the layers have different topology, promotes chimera states. Moreover, chimeras have been demonstrated in multilayer networks of coupled and uncoupled neurons [217]. A two-layer network, where one layer is represented by a ring of nonlocally coupled Henon maps and the other layer is a ring of nonlocally coupled Lozi maps, reveals a new type of spatiotemporal structure, a solitary state chimera [218]. In a two-layer network of phase oscillators with diffusive drive-response coupling between the layers, the generalized synchronization of chimera states has been investigated [23]. Further, the occurrence of chimeras has been studied for a two-layer network of phase oscillators, where the number of elements and their natural frequencies is different across the layers [20]. Interestingly, the required coupling strength to achieve the mean-field phase synchronization in such a network increases with the mismatches in the network sizes and the oscillators' natural frequencies. A multilayer heterogeneous network of chaotic maps in the chimera state regime has been analyzed in [219]. In more detail, transmitting the master chimera structure of the first driving layer to the subsequent 19 layers is considered, and the peculiarities of forced synchronization are studied depending on the nature and degree of heterogeneity of the network, as well as on the degree of asymmetry of the inter-layer coupling. The synchronization scenarios of chimeras in three-layer multiplex networks have been investigated in [149, 150, 220, 221]. In particular, an intriguing phenomenon of relay (or remote) synchronization between the layers which are not directly connected is studied for logistic maps [221] and FitzHugh-Nagumo oscillators [149, 150, 220]. Therefore, the notion of relay synchronization has been extended to chimera states. A special type of multiplex network, where the coupling between the layers is much smaller than that within the layers, has been shown to have a significant impact on the formation of chimera states in a two-layer network of FitzHugh-Nagumo neurons [126]. In particular, strategies of controlling chimeras based on weak multiplexing have been developed: the desired state in one layer can be achieved without manipulating its parameters, but only by adjusting the other layer. A further method of chimera control is reported for two-layer networks, where each layer is a small nonlocally coupled ring of Van der Pol oscillators [222]. In more detail, the tweezer control, applied to only one layer, successfully stabilizes chimera patterns in the other, uncontrolled layer, even in the case of non-identical layers.

In [223], chimera states are discussed for local, nonlocal, global coupling topologies, as well as for modular, temporal, and multilayer networks. In the present book, we investigate in detail the formation of chimera states in networks with power-law coupling kernel, in multiplex networks, and in networks with fractal topologies (Chap. 4).

While previous studies have been mainly focused on stationary network topologies, recent works have demonstrated that chimera patterns are possible in time-varying complex networks [224], as well as in dynamical networks with adaptive couplings [225–227]. The types of interactions can be also different, and in [228], the authors study the impact of repulsive coupling on the chimera patterns.

There occurs a question whether it is possible to observe coexistence of coherence and incoherence in the networks of higher dimensions. Indeed, two-dimensional chimera patterns have been found in networks with 2D-lattice structures, and can appear as incoherent spots, stripes, or incoherent spots in spiral waves. In [229], the authors demonstrate chimera states for phase oscillators on a flat torus, further works consider 2D lattices of leaky integrate-and-fire neurons [183], 2D networks of locally coupled oscillators [230], and chimera patterns in 2D networks of coupled neurons [231].

Despite extremely high numerical effort, chimera patterns have been demonstrated numerically even in 3D networks: on a surface of a sphere [232], and in 3D lattice of phase oscillators [233]. The topologies for which chimera states have been investigated as well as the corresponding references are given in Tables 1.3, 1.4 and 1.5.

1.7 Types of Chimera States

Since the discovery of classical chimeras in network of coupled Kuramoto phase oscillators in 2002 [61], a large variety of chimera states has been detected. The classical or phase chimeras can be characterized by several spatial domains where the phases are synchronized or desynchronized. Moreover, the oscillators in the coherent domains are frequency locked, while the frequencies of the elements in the incoherent domains are different. Further, the systems involving not only phase but also amplitude dynamics have been investigated and amplitude-mediated chimeras have been discovered [99, 100]. Purely amplitude chimeras have been observed in networks of coupled Stuart-Landau oscillators with symmetry-breaking coupling [75, 76]. An important difference of these patterns from classical phase chimeras is that the spatial incoherence does not imply chaotic behavior in time. In fact, all nodes in the amplitude chimera regime perform periodic oscillations and in the incoherent domain, the spatial sequence of the positions of the centers of oscillation is completely random [75, 76, 101, 103]. Amplitude chimeras have been shown to exist in the presence of noise [102] and time delay [104]. Moreover, they have been described in different models: networks of ecological oscillators (Rosenzweig-MacArthur system) with nonlocal [105] and power-law coupling [84] and nonlocally coupled ring

Table 1.3 Topologies for which chimeras have been found (part 1)

Topology	System type	Comment	References
Ring with nonlocal coupling (exponential)	Kuramoto phase oscillator	First observation	[61]
Ring with nonlocal coupling (cosine-like)	Kuramoto phase oscillator	Naming of *chimera states*	[62]
Ring with nonlocal coupling (piecewise-constant)	Kuramoto phase oscillator	Irregular motion in time as finite-size effect	[166]
Ring with nonlocal coupling (piecewise-constant)	Kuramoto phase oscillator	Chaotic transients	[129]
Ring with nonlocal coupling (piecewise-constant)	Chaotic/periodic maps, Rössler system	Coherence-incoherence transition	[69]
Ring with nonlocal coupling (piecewise-constant)	FitzHugh-Nagumo oscillator	Multichimera states	[79]
Ring with nonlocal coupling (power law)	Kuramoto phase oscillator	Heterogeneous networks	[171]
Ring with nonlocal coupling (power law)	FitzHugh-Nagumo oscillator	Chimeras and solitary states	[201]
Ring with nonlocal coupling	Kuramoto phase oscillator	Persistent chimera states	[202]
Ring with nonlocal coupling (symmetry-breaking)	Stuart-Landau oscillator	Amplitude chimera, chimera death	[75, 76, 101]
Ring with nonlocal coupling (symmetry-breaking)	Stuart-Landau oscillator	Amplitude chimera (stability)	[103]
Ring with nonlocal coupling (symmetry-breaking)	Stuart-Landau oscillator	Amplitude chimera (with noise)	[102]
Ring with nonlocal coupling (symmetry-breaking)	Stuart-Landau oscillator	Amplitude chimera (with delay)	[104]
Ring with nonlocal coupling (symmetry-breaking)	Stuart-Landau oscillator	Amplitude chimera (with filtering)	[107]
Ring with nonlocal coupling	Rayleigh oscillators	Amplitude-mediated and amplitude chimeras	[106]
Ring with nonlocal coupling	Ecological oscillator (Rosenzweig-MacArthur system)	Chimera death, chimera-like coexistence of synchronized oscillation and death	[105]
Ring with nonlocal coupling	Kuramoto phase oscillator	Clustered chimera states	[208]
Ring with nonlocal coupling	Complex Ginzburg-Landau equation	Amplitude-mediated chimeras	[99]

Table 1.4 Topologies for which chimeras have been found (part 2)

Topology	System type	Comment	References
Power-law coupling	Ecological oscillator (Rosenzweig-MacArthur)	Amplitude chimera, chimera death	[84]
Power-law coupling	Ecological oscillator (Rosenzweig-MacArthur)	Chimera-like coexistence of synchronized oscillation and death	[84]
Scale-free distribution of coupling strength	Phase oscillator	Inhomogeneous coupling	[113]
Ring with local nonlinear unidirectional coupling	Linear dissipative oscillators	Impact of additional diffusive coupling	[112]
Ring with local coupling	Duffing oscillators	Lyapunov spectra	[182]
Local coupling	Ecological oscillator (Rosenzweig-MacArthur)	Chimera death	[105]
Local coupling	Phase/Stuart-Landau/Winfree oscillator	Diffusive coupling	[111]
Two identical globally coupled subpopulations	Kuramoto phase oscillator	Nonlocal coupling decreases with distance	[170]
Two rings of Kuramoto-Sakaguchi type coupling	Kuramoto phase oscillator	Two populations	[203]
Two globally coupled populations	Pendulum-like oscillators	Two-population network	[181]
Two rings with nonlocal coupling	Stuart-Landau oscillators	Two populations	[204]
Two rings with nonlocal coupling	Kuramoto phase oscillator	Alternating chimera states	[207]
Three rings with nonlocal coupling connected as a ring	Kuramoto phase oscillator	Bistable chimera attractors	[205]
Three rings with nonlocal coupling, connected as chain/ring	Kuramoto phase oscillator	Varying network topology	[206]
Three-layer network	FitzHugh-Nagumo oscillator	Inter-layer delay, relay synchronization of chimeras	[149, 220]

(continued)

Table 1.4 (continued)

Topology	System type	Comment	References
Three-layer network	FitzHugh-Nagumo oscillator	Inter- and intra-layer delay, relay synchronization of chimeras	[150]
Three-layer network	Logistic maps	Relay synchronization of chimeras	[221]
Community-structured network	Kuramoto phase oscillator	Global coupling within community	[114]
Two-population network	Kuramoto phase oscillator	Random non-complete networks	[115]
Ring with nonlocal coupling	Kuramoto phase oscillator	Random-link removal	[116]
Ring with nonlocal coupling	Kuramoto phase oscillator	Scale-free and Erdös-Renyi networks	[117]
Global coupling (symmetry-breaking)	Stuart-Landau oscillator	Chimera death	[178]
Global coupling	Complex Ginzburg-Landau equation	Amplitude-mediated chimeras	[100]
Global coupling	Kuramoto- Sakaguchi model with internal delayed feedback	Dynamically sustained bistability	[108]
Global coupling	Bistable Stuart-Landau-type oscillator (no delay)	Dynamically sustained bistability	[108]
Global coupling	Stuart-Landau oscillators	Clustering as a prerequisite for chimera states	[109]
Global coupling	Ginzburg-Landau equation	Two types of cluster states	[209]
Global coupling	Lang-Kobayashi equations	Semiconductor laser	[93]
Small-world network	Integrate-and-fire-like oscillators	Pulse-coupled oscillators	[160]
Undirected and unweighted networks	Electro-optic system	Network-structural conditions for chimeras	[212]

networks of Rayleigh oscillators [106]. Moreover, transitions between amplitude-mediated phase chimeras and amplitude chimeras have been found [106]. Recently, it has been shown that filtering suppresses amplitude chimeras [107]. The generalization of chimera patterns to steady-state solutions has been originally detected in nonlocally coupled network of Stuart-Landau oscillators and called chimera death

Table 1.5 Topologies for which chimeras have been found (part 3)

Topology	System type	Comment	References
Ring with nonlocal coupling	FitzHugh-Nagumo oscillator	Inhomogeneous elements	[80]
Irregular coupling topology	FitzHugh-Nagumo oscillator	Identical elements	[80]
Ring with nonlocal hierarchical connectivity	Lattice limit cycle model	Population dynamics	[83]
Ring with nonlocal hierarchical connectivity	Van der Pol oscillator	Transition from nonlocal to hierarchical topology	[118]
Ring with nonlocal hierarchical connectivity	Leaky integrate-and-fire	Multichimera states	[119]
Ring with nonlocal hierarchical connectivity	Van der Pol oscillator	Interplay of fractal topology and delay	[120]
Ring with nonlocal hierarchical connectivity	FitzHugh-Nagumo oscillator	Interplay of fractal topology and delay	[123]
Quasi-fractal (hierarchical) connectivity on a ring	Logistic maps	Nested chimeras	[121]
Empirical DTI topology/modular fractal topology	FitzHugh-Nagumo neuron	Empirical neural versus modular fractal connectivity	[122]
Multiplex network of two nonlocally coupled rings	Logistic map	Interplay of delay and multiplexing	[124]
Multiplex network of two nonlocally coupled rings	Logistic map	Non-identical multiplexing	[125]
Multiplex network of two nonlocally coupled rings	FitzHugh-Nagumo neurons	Weak multiplexing	[126]
Multilayer network	Hindmarsh-Rose neurons	Nonlocally coupled layer and uncoupled layer	[217]
Two-layer network of two nonlocally coupled rings	Van der Pol oscillators	Tweezer control	[222]
Multilayer network (20 layers)	Logistic maps	Heterogeneous layers, forced sync	[219]
Two-layer network	Phase oscillators	Non-identical layers, generalized sync of chimeras	[23]

(continued)

Table 1.5 (continued)

Topology	System type	Comment	References
Two-layer network	Phase oscillators	Heterogeneous layers, mean-field phase sync of chimeras	[20]
Time-varying network of two coupled populations	Kuramoto phase oscillator	Stable/breathing/alternating chimeras	[224]
Adaptive coupling	Kuramoto phase oscillator	Emergence of hierarchical multilayer structures	[225]
Two-layer adaptive multiplex network	Kuramoto phase oscillator	Synchronization of chimera states	[226]
Global coupling	Pulse-coupled phase oscillators	Itinerant chimeras	[227]
Two rings with local coupling	Kuramoto phase oscillator	Multiplex network	[228]
Nonlocal coupling on a torus	Kuramoto phase oscillator	Conditions for 2D chimeras	[229]
Two-dimensional arrays, nonlocal coupling	Kuramoto phase oscillator	Stripes, spots, spirals	[234]
2D fractal toroidal geometry	Leaky integrate-and-fire	Stripe-and-grid chimeras	[183]
Two-dimensional locally coupled networks	Stuart-Landau oscillator/Hindmarsh-Rose neuron/Rulkov map	Identical oscillators	[230]
Two-dimensional toroidal geometry	FitzHugh-Nagumo/leaky integrate-and-fire	Spots, rings, stripes and grids	[231]
Surface of sphere, nonlocal coupling	Kuramoto phase oscillator	Spot and spiral chimeras	[232]
3D grid topology with periodic boundary conditions	Kuramoto phase oscillator	3D chimeras	[233]

[75]. Later, chimera death patterns have been extended to local [105] and global coupling [178] as well as long-range interactions with distance-dependent power-law coupling [84].

It is important to distinguish the amplitude chimeras found in time-periodic systems [75, 76] from those detected in chaotic systems [173]. In more detail, the incoherent domain of the amplitude chimera state found for chaotic systems (for example, in logistic maps [173]) is characterized by completely asynchronous chaotic dynamics. In contrast to that, all elements of an amplitude chimera in a network of Stuart-Landau systems demonstrate periodic dynamics in time. A special kind of phase chimera states reported in [173] has been shown to be typical for chaotic systems. In this phase, chimera regime coherent domains of in-phase oscillations coexist with incoherent domains where the sequence of nodes performing in-phase and antiphase oscillations is random in space.

A peculiar *solitary state chimera* pattern that combines features of chimera and solitary states has been detected in a two-layer network, where one layer is represented by a ring of nonlocally coupled Henon maps and the other layer is a ring of nonlocally coupled Lozi maps [218]. Further, the possibility of obtaining the solitary state chimera in the one-dimensional ring of nonlocally coupled Henon maps has been reported [235]. In more detail, it has been shown that both solitary states and the solitary state chimera can be induced by noise which multiplicatively modulates the coupling strength. Moreover, it has been demonstrated that these states can also arise due to multistability in the network. Recently, *solitary state chimera* has been found in one-layer neural networks [236].

It is often the case that the nature of the model under study defines a particular chimera type. Moreover, certain network topologies reveal and favor specific chimera patterns. Also the interaction of local dynamics with the network topology can determine the resulting chimera regime. For example, a special kind of phase chimera states as mentioned above is characteristic of chaotic systems [173]. Nested chimeras, where incoherent domains have complex fragmented structure, are observed in networks with fractal connectivities [80, 83, 118–121, 123]. The interplay of multiplex network topology and time delay leads to the formation of a layer chimera, where one layer exhibits coherent dynamics and the other layer demonstrates incoherent behavior [124]. A 2D square lattice of nonlocally coupled cubic maps shows double-well chimeras [72].

In two-dimensional networks, the spiral wave chimera states (spiral chimeras) are observed [237–239]. In this state, an ordered spiral wave rotates around a core consisting of asynchronous oscillators. In particular, they have been demonstrated for phase oscillators [240], Rössler systems [241], chemical oscillators [87] and recently a stability of spiral chimera states on a torus has been discussed [242].

The possibility of observing chimeras in a single system has been shown for an oscillator with delayed feedback. These states have been called virtual chimera states [153] and demonstrated both numerically and experimentally. Further types of chimera patterns include self-organized alternating chimera states in oscillatory media [243]. In [97], a mathematical definition of weak chimeras has been introduced, and later they have been studied in the networks with broken symmetry [244].

Various further features of chimera states have been analyzed. For instance, it has been shown that noise can not only decrease the lifetime of chimeras [102], but also play a constructive role, i.e., induce these hybrid patterns. The noise-induced chimeras have been discovered recently and called coherence-resonance chimeras [131, 132]. Time delay has been shown to control these patterns and induce novel patterns called period-two coherence-resonance chimeras [133]. A special attention has been paid to the study of basins of attractions for chimera states and the so-called

Table 1.6 Types of chimeras (part 1)

Chimera type	System type	Topology	References
Classical chimera state	Phase oscillator (Kuramoto model)	Ring with nonlocal coupling	[61]
Amplitude-mediated chimeras	Complex Ginzburg-Landau equation	Ring with nonlocal coupling	[99]
Amplitude-mediated chimeras	Complex Ginzburg-Landau equation	Global coupling	[100]
Amplitude chimera	Stuart-Landau oscillator	Symmetry-breaking nonlocal coupling	[75, 76, 101]
Chimera death	Stuart-Landau oscillator	Symmetry-breaking nonlocal coupling	[75, 76, 101]
Chimera death	Stuart-Landau oscillator	Symmetry-breaking mean-field diffusive coupling	[178]
Amplitude chimera (stability)	Stuart-Landau oscillator	Symmetry-breaking nonlocal coupling	[103]
Amplitude chimera (with noise)	Stuart-Landau oscillator	Symmetry-breaking nonlocal coupling	[102]
Amplitude chimera (with delay)	Stuart-Landau oscillator	Symmetry-breaking nonlocal coupling	[104]
Amplitude chimera (with filtering)	Stuart-Landau oscillator	Symmetry-breaking nonlocal coupling	[107]
Amplitude chimera	Ecological oscillator (Rosenzweig-MacArthur)	Power-law coupling	[84]
Chimera death	Ecological oscillator (Rosenzweig-MacArthur)	P1ower-law coupling	[84]
Chimera-like coexistence of synchronized oscillation and death	Ecological oscillator (Rosenzweig-MacArthur)	Power-law coupling	[84]
Chimera-like coexistence of synchronized oscillation and death	Ecological oscillator (Rosenzweig-MacArthur)	Ring with nonlocal coupling	[105]
Chimera death	Ecological oscillator (Rosenzweig-MacArthur)	Local coupling	[105]
Amplitude-mediated chimeras	Rayleigh oscillators	Ring with nonlocal coupling	[106]
Amplitude chimeras	Rayleigh oscillators	Ring with nonlocal coupling	[106]
Amplitude and phase chimeras	Logistic maps	Ring with nonlocal coupling	[173]

(continued)

Table 1.6 (continued)

Chimera type	System type	Topology	References
Double-well chimeras	Cubic maps	2D lattice	[72]
Weak chimeras	Phase oscillator (Kuramoto model)	Kuramoto-Sakaguchi coupling	[97]
Weak chimeras	Phase oscillators	Coupling function without symmetry	[244]
Spiral wave chimeras	Phase oscillator	2D structure	[240]
Spiral wave chimeras	Rössler system	2D structure (square lattice)	[241]
Spiral wave chimeras	Phase oscillators	2D structure (torus)	[242]
Spiral wave chimeras	Chemical oscillators	2D structure (square lattice)	[87]
Multichimera states	FitzHugh-Nagumo oscillator	Ring with nonlocal coupling	[80]

stable, breathing and unstable chimera patterns have been identified [245]. Finally, the heteroclinic switching between chimeras has been investigated in [246] where the existence of the so-called saddle weak chimeras has been demonstrated. The classification scheme for chimera states based on the spatial and temporal correlations has been proposed in [168]. Tables 1.6 and 1.7 provide the systematic summary of the network topologies for which chimera states have been found.

1.8 Control of Chimera States

One of the most challenging questions is related to *control* of chimera states. For example, the observation of these peculiar spatiotemporal patterns in small-size systems is a nontrivial task since there are two principal obstacles which are difficult to overcome. First, it is known that chimera states are usually chaotic transients that eventually collapse to the uniformly synchronized state [129]. Second, the position of the incoherent domain is not stationary but rather moves erratically along the oscillator array [166]. Therefore, the necessity occurs to develop control techniques. In [144], the authors have proposed a feedback control scheme for chimera states, which allows to stabilize chimera states and extend their lifetime. The authors of [145] introduced an asymmetry induced spatial position control for phase oscillators, which allows to fix preferable position of coherent and incoherent domains. To solve both problems, the tweezer control method has been developed [146, 147]. This feedback control scheme, like a tweezer, is especially useful in experiments, where usually only small networks can be realized. In particular, the action of the tweezer control has been analyzed in small nonlocally coupled networks of Van der Pol and FitzHugh-Nagumo oscillators, and the ranges of optimal control parameters

Table 1.7 Types of chimeras (part 2)

Chimera type	System type	Topology	References
Multichimeras	Lattice limit cycle model	Ring with nonlocal coupling/hierarchical coupling	[83]
Multichimeras	Van der Pol oscillators	Ring with hierarchical coupling	[118]
Multichimeras	Leaky integrate-and-fire	Ring with nonlocal hierarchical connectivity	[119]
Traveling/nontraveling/nested chimeras	Van der Pol oscillators	Ring with nonlocal hierarchical connectivity	[120]
Nested chimeras	FitzHugh-Nagumo oscillator	Ring with nonlocal hierarchical connectivity	[123]
Nested chimeras	Logistic maps	Fractal (hierarchical) connectivity	[121]
Virtual chimeras	Single oscillator with time-delayed feedback	Strongly asymmetric nonlinear delayed feedback	[153]
Chimera death	Stuart-Landau oscillators	Mean-field diffusive coupling	[247]
Alternating chimeras	Modified complex Ginzburg-Landau equation	Nonlinear global coupling	[243]
Solitary state chimera	Henon maps	Ring with nonlocal coupling	[235]
Solitary state chimera	Henon and Lozi maps	Two-layer network	[218]
Solitary state chimera	FitzHugh-Nagumo oscillator	Ring with nonlocal coupling	[236]
Layer chimera	Logistic maps	Multiplex network of two nonlocally coupled rings	[124]
Coherence-resonance chimera	FitzHugh-Nagumo (excitable) with noise	Nonlocally coupled ring	[131, 132]
Period-two coherence-resonance chimera	FitzHugh-Nagumo (excitable) with noise and delay	Nonlocally coupled ring	[133]
Stable/breathing/unstable saddle chimeras	Kuramoto-Sakaguchi model	Two populations	[245]
Saddle weak chimeras	Phase oscillator	Coupled populations	[246]
Stationary/turbulent/breathing chimeras	Classification scheme for various systems	Different topologies	[168]

have been determined [147]. The tweezer control scheme allows for stabilization of chimera states with different shapes, and can be used as an instrument for controlling the coherent domains size, as well as the maximum average frequency difference of the oscillators. To fix the position of chimera states, it has been also proposed to include a block of excitable units into the oscillatory population [248] or apply pinning control technique [249]. Recently, a method for controlling chimera states via minimal coupling modification has been proposed for a ring-shaped network of nonlocally coupled phase oscillators [148]. In more detail, the links of one element in the network (called pacemaker) are made unidirectional. This control scheme can be easily implemented and allows to induce chimeras for parameters and initial conditions for which they do not form spontaneously. Furthermore, the pacemaker attracts the incoherent part of the chimera state, thus controlling its position.

Multilayer network topologies open up new possibilities for chimera control. Recently, it has been shown that excitation and suppression of chimera states can be achieved in multiplex networks [250]. The interplay of time delay and multiplexing can be used for controlling chimeras and leads to the formation of a so-called, layer chimera, where one layer of the network demonstrates coherent dynamics and the other layer is characterized by incoherent behavior [124]. In [125], it has been shown that non-identical multiplexing, i.e., when the layers have different topology, promotes chimera states. Chimera control by weak multiplexing has been proposed in [126] for a two-layer network of FitzHugh-Nagumo oscillators. Weak multiplexing means that the coupling between the layers is much smaller than that within the layers. The advantage of this control scheme is that it allows to achieve the desired state in a certain layer without manipulating its parameters, and it works for weak coupling between the layers [126]. A further method of chimera control is reported for two-layer networks, where each layer is a small nonlocally coupled ring of Van der Pol oscillators [222]. In more detail, the tweezer control, applied to only one layer, successfully stabilizes chimera patterns in the other, uncontrolled layer, even in the case of non-identical layers. For multiplex networks of coupled logistic and Henon maps, it has been shown that chimera states can be induced by introducing heterogeneous delays in a fraction of inter-layer links, referred to as multiplexing delays [251]. Moreover, the incoherent domain of the chimera state can be regulated by the appropriate choice of both inter- and intra-layer coupling strengths, whereas the extent and the position of the incoherence regime can be adjusted by appropriate placing and strength of the multiplexing delays.

The noise-based control mechanisms have been developed in [131, 143]. In more detail, chimera control has been studied in a single element, a nonlinear oscillator model with negative time-delayed feedback under external deterministic and stochastic forcing [143]. It turned out that one-cluster and multicluster chimeras can be achieved by adjusting the external forcing frequency to appropriate resonance conditions. If a stochastic component is superimposed to the deterministic external forcing, chimera states can be induced in a way similar to stochastic resonance, they appear, therefore, in regimes where they do not exist without noise. In [131], it has been shown that noise on one hand can induce chimera patterns and on the other hand it allows to control it. In more detail, by properly choosing the noise intensity, one can achieve the desired regime of the network: steady state, coherence-resonance chimera, or other patterns. Moreover, by fine-tuning the noise intensity, it is possible to adjust the size of the incoherent domain of the chimera pattern. The robustness and control of amplitude chimera states in the presence of noise has been considered in [102]. It has been demonstrated that the lifetime of amplitude chimeras can be controlled by varying the noise intensity. Furthermore, it has been demonstrated that applying time-delayed feedback allows to control coherence-resonance chimeras by adjusting delay time and feedback strength [133]. Delay has been shown to control of amplitude chimeras [104], and control of chimera relay synchronization in multiplex networks by delay [149, 150, 220] has been demonstrated recently. Moreover, for networks of Van der Pol oscillators [120] and FitzHugh-Nagumo neurons [123] with fractal connectivity, it has been shown that by varying the time delay in the coupling, one can deliberately stabilize desired spatiotemporal patterns in the system. For long-range interaction with distance-dependent power-law coupling, it has been shown that chimeras can be induced and controlled by tuning the power-law exponent [84]. Table 1.8 gives a systematic overview of the methods used to control chimera states.

1.9 Chimera States in Experiments

The experimental observation of chimera states in networks of coupled oscillatory units is a nontrivial task, especially taking into account their transient nature and the fact that laboratory setups are usually restricted to a small number of coupled elements. Here we provide some examples of experiments on chimera states. One decade after their theoretical discovery, the first experimental works on chimeras appeared in optical [151] and chemical systems [87, 88]. Further experiments involved mechanical systems and demonstrated chimeras for coupled metronomes [91, 152]. Pendulum-like oscillators with local and nonlocal coupling investigated in [152] revealed the so-called imperfect chimera states. Electronic or optoelectronic oscillators show chimera states [92, 94, 153] and electrochemical systems have demonstrated chimera patterns in two dimensions [89, 110, 154]. Chimeras have also been shown experimentally in electronic circuits [95], optical combs [155] as well as in Boolean networks [74]. In more detail, in [74], networks of nonlocally coupled electronic oscillators that can be described approximately by a Kuramoto-like model have been investigated. The fast timescale of these oscillators (on the order of 100 ns) allows to study the scaling of the transient time of large networks of more than a hundred nodes, which has not been previously confirmed in an experiment and could potentially be important in many natural networks. The experimental networks showed long complex transients from random initial conditions on the route to network synchronization. The transients displayed complex behaviors, including resurgence of chimera states. The spatial domain of the chimera state has been shown to move around the network and alternate with desynchronized dynamics. It has been shown that the average transient time increases exponentially with the network size

Table 1.8 Control of chimeras

Control	Topology and system type	Comment	References
Feedback control	Phase oscillator with nonlocal coupling	Stabilizing chimeras	[144]
Gradient dynamics	Phase oscillator with nonlocal coupling	Controlling drift of chimera	[145]
Tweezer control	Nonlocally coupled Van der Pol oscillators	Small networks	[146]
Tweezer control	Nonlocally coupled Van der Pol/FitzHugh-Nagumo oscillators	Optimal control parameters	[147]
Control by noise	Nonlocally coupled Stuart-Landau oscillator	Amplitude chimera lifetime	[102]
Control by time delay	Nonlocally coupled Stuart-Landau oscillator	Amplitude chimera and chimera death	[104]
Control by means of filtering	Nonlocally coupled Stuart-Landau oscillator	Suppression of amplitude chimera	[107]
Tuning power-law exponent	Ecological oscillators (Rosenzweig-MacArthur) with power-law coupling	Inducing and controlling of chimeras	[84]
Pinning control	FitzHugh-Nagumo/Kuramoto nonlocally coupled	Inducing chimeras, controlling the position	[249]
Block of excitable units	Nonlocally coupled ring of FitzHugh-Nagumo oscillators	Controlling the position of chimeras	[248]
Pacemaker control	Nonlocally coupled ring of phase oscillators	Inducing chimeras, controlling the position	[148]
Multiplexing	Multiplex network of Kuramoto/Hindmarsh-Rose oscillators	Inducing/suppressing chimeras	[250]
Interplay of delay and multiplexing	Two nonlocally coupled rings of logistic map	Layer chimera	[124]
Weak multiplexing	Multiplex FitzHugh-Nagumo network (oscillatory)	Inducing/suppressing chimeras	[126]

(continued)

Table 1.8 (continued)

Control	Topology and system type	Comment	References
Multiplexing delays	Multiplex network, logistic and Henon maps	Heterogeneous delays in inter-layer links	[251]
External periodic/stochastic forcing	Single oscillator with time-delayed feedback	Multiclusters; stochastic resonance	[143]
Noise	Nonlocally coupled FitzHugh-Nagumo network (excitable)	Coherence-resonance chimeras	[131]
Time-delayed feedback	Nonlocally coupled FitzHugh-Nagumo neurons (excitable)	Coherence-resonance chimeras	[133]
Inter-layer time delay	Nonlocally coupled FitzHugh-Nagumo neurons	Multiplex network	[149, 220]
Inter- and intra-layer time delay	Nonlocally coupled FitzHugh-Nagumo neurons	Multiplex network	[150]
Time delay	Ring with nonlocal hierarchical connectivity, Van der Pol oscillator	Interplay of fractal topology and delayed coupling	[120]
Time delay	Ring with nonlocal hierarchical connectivity, FitzHugh-Nagumo oscillator	Interplay of fractal topology and delayed coupling	[123]

and can be modeled as a Poisson process in experiment and simulation [74]. Small chimera states have been observed experimentally in optical globally coupled networks containing only four elements [96]. The possibility of observing chimeras has been even shown for a single system. These states have been called virtual chimera states [153] and demonstrated both numerically and experimentally for an oscillator with delayed feedback. More recently, spiral chimera states have been for the first time observed experimentally in chemical oscillator systems [90]. The examples of experimental works on chimera states are given in Table 1.9.

Table 1.9 Experimental works on chimera states

Experiment	System type	References
Liquid-crystal spatial light modulator	Nonlocal coupling/coupled-map lattice	[151]
Chemical oscillators	Belousov-Zhabotinsky model, global coupling of two populations	[88]
Chemical oscillators	Networks of up to 20 units with various coupling	[89]
Photosensitive chemical oscillators	Nonlocally coupled ring and square lattice	[87]
Chemical oscillators	Exponentially decaying nonlocal coupling	[90]
Damped mechanical pendula	Nonlocal coupling of two populations	[91]
Pendulum-like oscillators	Local and nonlocal coupling	[152]
Electronic oscillator	Electronic frequency modulation nonlinear delay oscillator	[153]
Optoelectronic system	Delayed feedback tunable semiconductor laser	[92]
Photonic nonlinear system	Semiconductor laser with two feedback delays	[94]
Network of electrochemical reactions	Ring of nonlocally coupled elements	[154]
Oscillatory electrochemical system	Stuart-Landau model, nonlinear global coupling	[110]
Coupled electronic circuits	FHN model, local coupling	[95]
Electronic oscillators	Boolean phase oscillators, nonlocal coupling	[74]
Optical comb	Passively mode-locked quantum dot laser	[155]
Optoelectronic oscillators	Fiber-coupled laser diodes, global coupling	[96]

1.10 Applications of Chimera States

Chimera states can be linked to various processes occurring in real-world systems. Here we provide some examples. One of the most promising fields of application is neuroscience, where chimera states can be associated with unihemispheric sleep of mammals (seals and dolphins) [156], birds (great frigatebirds) [252] or with the first-night effect in humans [157]. During this peculiar unihemispheric sleep, one hemisphere of the brain is synchronized (sleeping), while the other is characterized by erratic dynamics (awake). Further applications include epileptic seizures, where the electrical activity in the brain is excessive and synchronous. Partial synchronization patterns as a model for unihemispheric sleep have been recently analyzed in a network of FitzHugh-Nagumo oscillators with empirical structural connectivity measured in healthy human subjects [253].

Studying chimera states can give further insights into the underlying mechanisms of the initiation and, in particular, termination of epileptic seizures [122, 158–161]. Chimera states have also been discussed in the context of so-called bump states in neural networks, where localized coherent regions are surrounded by incoherence [254]. Perceptual organization [162] and behavioral sensitization [163, 164] are also considered as possible applications of chimera states. Recent studies have shown that chimera-like states can be observed in a neuronal network model of the cat brain [255].

In power distribution networks, the blackout problem can be associated with partial desynchronization and chimera patterns [12]. Therefore, investigating the possibility of chimera states occurrence in power grids could be useful for maintaining stable synchronization. Moreover, chimera states can be applied to laser networks

Table 1.10 Applications of chimera states

Application	Comment	References
Unihemispheric sleep of seals, dolphins	General neuroscience review	[156]
Unihemispheric sleep of birds	Great frigatebirds	[252]
Unihemispheric sleep of humans	Neurobiological experiment	[157]
Epilepsy	Review on epilepsy	[158]
Epilepsy	Experiment (human/mouse/rat/zebrafish)	[159]
Epilepsy	Pulse-coupled oscillators	[160]
Epilepsy	Chimeras in human brains	[161]
Epilepsy	Empirical DTI brain network	[122]
Bump states	Neural networks	[254]
Human visual experience	Neurobiological experiment	[162]
Human brain rhythms	Neurobiological experiment	[163]
Rat brain rhythms	Neurobiological experiment	[164]
Cat brain	Numerical simulation	[255]
Power-grid networks	Optimization of stability	[12]
Semiconductor laser	Global coupling	[93]
Four delay-coupled lasers	Small chimera states	[256]
Social systems	Two interacting populations	[165]

[93, 256]. And finally, social systems are often characterized by coexistence of coherent and incoherent groups, where one population reaches consensus, while the other population remains disordered [165]. The application examples of chimera states are summarized in Table 1.10.

Chapter 2
Amplitude Chimeras and Chimera Death in Ring Networks

Abstract In the present chapter, we discuss special types of chimera states: amplitude chimeras and chimera death. In the intriguing amplitude chimera regime, coherence-incoherence patterns are formed with respect to the amplitudes only. Chimera death generalizes chimera patterns to steady states through the death of oscillations. We first describe amplitude chimeras and chimera death in the deterministic case without time delay. Further, efficient control mechanisms based on time delay and noise are discussed. In particular, we address the question of how time delay and noise influence the behavior of amplitude chimera states in ring networks of Stuart-Landau oscillators.

2.1 Introduction

Chimera states have been initially found for the phase oscillator model [61, 62]. Later they have been extended to systems which involve not only phase but also amplitude dynamics and named amplitude-mediated chimeras in the case when both amplitude and phase are characterized by chimera behavior [99, 100]. More complicated patterns in which chimera structures are formed with respect to the amplitudes while the phases remain correlated for the whole network have been first reported in [75]. This particular type of chimera states, *amplitude chimeras*, is investigated in this chapter [75, 76, 101–104]. Additionally, we study another recently discovered type of chimera states, *chimera death* [75], which, through the death of the oscillations, generalizes the chimera feature of coexistence of spatially coherent and incoherent domains to steady states.

While modeling real-world systems, it is important to take stochasticity and time delay into account. Arising naturally, these two factors lead to a plethora of complex phenomena with applications to various fields. Moreover, both may result in opposite effects and can be exploited for control purposes. In this chapter, we investigate the dynamics in the presence of time delay and noise. The sections are organized as follows. First, we introduce the model (Sect. 2.2). Then, we describe amplitude chimeras and chimera death in the deterministic case without delay (Sect. 2.3). Further, efficient control mechanisms based on time delay (Sect. 2.4) and noise (Sect. 2.5) are

A. Zakharova, *Chimera Patterns in Networks*, Understanding Complex Systems,
https://doi.org/10.1007/978-3-030-21714-3_2

discussed. In particular, we address the question of how time delay and noise influence the behavior of amplitude chimera states in ring networks of Stuart-Landau oscillators. Section 2.6 provides the conclusions.

2.2 Stuart-Landau Model

For our studies, we consider the paradigmatic model of Stuart-Landau (SL) oscillator [61, 110, 257–264]. We investigate a network of N coupled elements, where the local deterministic dynamics of each node $j \in \{1, ..., N\}$ is given by $\dot{z}_j = f(z_j)$, with the normal form of a supercritical Hopf bifurcation

$$f(z_j) = (\lambda + i\omega - |z_j|^2)z_j, \tag{2.1}$$

where $z_j = x_j + i\, y_j = r_j e^{i\phi_j} \in \mathbb{C}$, with $x_j, y_j, r_j, \phi_j \in \mathbb{R}$, and $\lambda, \omega > 0$. At $\lambda = 0$, a Hopf bifurcation occurs, so that for $\lambda > 0$, the single Stuart-Landau oscillator exhibits self-sustained oscillations with frequency ω and radius $r_j = \sqrt{\lambda}$, and the unique fixed point $x_j = 0$, $y_j = 0$ is unstable.

Here we focus on a ring network of N nonlocally coupled Stuart-Landau oscillators. In the deterministic case without delay (see Sect. 2.3), we use the following model:

$$\dot{z}_j = f(z_j) + \frac{\sigma}{2P} \sum_{k=j-P}^{j+P} (\mathrm{Re} z_k - \mathrm{Re} z_j). \tag{2.2}$$

The coupling parameters, which are identical for all links, are the coupling strength $\sigma \in \mathbb{R}$ and the coupling range P/N, where P corresponds to the number of nearest neighbors in each direction on a ring. Here we consider coupling only in the real parts, since this breaks the rotational S^1 symmetry of the system. With respect to applications, this means that the oscillators are coupled only through a single real variable x.

Further, we study nonlocally coupled ring networks of N oscillators with different types of delay in the coupling introduced via the delay operator $\mathscr{D}(\cdot)$ (see Sect. 2.4):

$$\dot{z_j} = f(z_j) + \frac{\sigma}{2P} \sum_{k=j-P}^{j+P} \left[\mathrm{Re}(\mathscr{D}[z_k(t)]) - \mathrm{Re}(z_j(t))\right]. \tag{2.3}$$

Depending on the type of the delay in the coupling, the delay operator $\mathscr{D}$ could act upon the state function $z(t)$ in different forms, e.g., $\mathscr{D}_1[z(t)] = z(t - \tau)$ in the case of *constant* time delay τ, $\mathscr{D}_2[z(t)] = z(t - \tau(t))$ for *time-varying* delay, with time dependence given by the function $\tau(t)$, or $\mathscr{D}_3[z(t)] = \int_0^\infty G(t')z(t - t')\mathrm{d}t'$ for *distributed* delay, where $G(t)$ is a kernel characterizing the delay distribution.

Finally, we investigate the role of noise of intensity $D > 0$ using the following model (see Sect. 2.5):

$$\dot{z_j} = f(z_j) + \frac{\sigma}{2P} \sum_{k=j-P}^{j+P} (\text{Re}z_k - \text{Re}z_j) + \sqrt{2D}\xi_j(t), \tag{2.4}$$

where the coupling and the noise are only applied to the real parts, and $\xi_j(t) \in \mathbb{R}$ is additive Gaussian white noise [265], i.e., $\langle \xi_j(t) \rangle = 0,\ \forall j$, and $\langle \xi_i(t)\xi_j(t') \rangle = \delta_{ij}\delta(t - t'),\ \forall i, j$, where δ_{ij} denotes the Kronecker-Delta and $\delta(t - t')$ denotes the Delta-distribution. Hence the noise is spatially uncorrelated.

2.3 Deterministic Dynamics Without Delay

We begin the description of hybrid patterns with the consideration of deterministic dynamics without delay [75, 76, 103]. As discussed above, chimera states are related to the symmetry breaking of the system and correspond to the situation when an ensemble of identical elements self-organizes into two coexisting and spatially separated domains with dramatically different behavior, i.e., spatially coherent and incoherent oscillations [61, 62]. Another fascinating effect which requires the break-up of the system's symmetry as a crucial ingredient is *oscillation death* (OD) which refers to stable inhomogeneous steady states (IHSS) which are created through the coupling of self-sustained oscillators. This regime occurs when a homogeneous steady state splits into at least two distinct branches—upper and lower—which represent a newly created IHSS [266–269]. For a network of coupled elements, oscillation death implies that its nodes occupy different branches of the IHSS. Oscillation death is inherent in various systems and its existence has been confirmed experimentally, e.g., in chemical reactors [50], chemical oscillators [270], chemical droplets [271], electronic circuits [48], and thermokinetic oscillators [272]. Moreover, it is especially significant in the light of applications to biological systems like neuronal networks [51], genetic oscillators [273], calcium oscillators [267], and it has been proposed as a basic mechanism for morphogenesis and cellular differentiation, for instance, stem cell differentiation [56]. Oscillation death has been shown to exist for special time-delayed [46] and repulsive [274] types of coupling as well as coupling through conjugate [275, 276] or dissimilar variables [277]. Furthermore, in the context of network topologies, it has been studied for the two limit cases of global (all-to-all) and local coupling [278]. Here we focus on nonlocal coupling, which interpolates between these limit cases. An analytical explanation for IHSS occurring in a network of nonlocally coupled Stuart-Landau oscillators with symmetry-breaking coupling is provided in [45]. The impact of symmetry-preserved coupling and global network topology on amplitude chimeras and chimera death is discussed in [279]. The role of mean-field diffusive coupling [178] and distance-dependent power-law coupling [84] has also been considered. Moreover, it has been shown that synchronized oscillation and stable steady state can coexist forming a particular spatiotemporal pattern in networks with nonlocal coupling topology [105]. In this state, a subset of oscillators populate spatially synchronized oscillation and stable steady state in a random manner, and the rest of the elements oscillate in synchrony. This spatiotemporal state is unlike the chimera state (where coherent and incoherent oscillations coexist) and the chimera death state (where neighboring oscillators populate two branches of OD in coherent and incoherent manner).

Therefore, chimera states and oscillation death possess a relevant common feature, both implying the break-up of symmetry in a dynamical network. Here we provide systematically the bridging between these two phenomena that have been previously studied independently. This introduces more insight into the concept of symmetry breaking in general, and allows to fill the existing gaps in understanding chimeras and oscillation death in particular. The chimera death pattern considered in this section represents a connection between chimera states and oscillation death. In this regime, the population of oscillators splits into distinct coexisting domains of (i) spatially coherent oscillation death (where neighboring oscillators populate essentially the same branch of the inhomogeneous steady state) and (ii) spatially incoherent oscillation death (where the sequence of populated branches of neighboring nodes is completely random in the inhomogeneous steady state). Moreover, we analyze the chimera behavior with respect to the amplitude, rather than the phase, i.e, amplitude chimeras. In this state, one sub-population is oscillating with spatially coherent amplitude, while the other displays oscillate with spatially incoherent amplitudes.

In this section, we first define and describe the main properties of amplitude chimeras and chimera death (Sect. 2.3.1). Next, we discuss the transient behavior of amplitude chimeras (Sect. 2.3.2) and introduce a global measure which can distinguish between chimera patterns and completely coherent dynamics like in-phase synchronization or traveling waves (Sect. 2.3.3). Further, we consider the role of initial conditions for amplitude chimera state (Sect. 2.3.4). Then we investigate the relative size of its incoherent domains depending on coupling parameters (Sect. 2.3.5). Section 2.3.6 discusses the impact of system size on the lifetime of amplitude chimeras. Finally, we analyze the stability of amplitude chimeras in Sect. 2.3.7. And Sect. 2.3.8 provides a summary on the deterministic dynamics.

2.3.1 Amplitude Chimeras and Chimera Death

First, we fix the strength of the interaction between the networks elements σ and vary the number of nearest neighbors P in (2.2). The resulting dynamical states are shown in Fig. 2.1 where the left column displays space-time plots, color coded by the variable y and the right column shows the corresponding snapshots. We observe a pattern characterized by chimera behavior with respect to amplitude dynamics, i.e., amplitude chimeras [75]. In more detail, the network splits into two domains with different dynamics: one group of elements demonstrates oscillations with spatially coherent amplitude, while the other is oscillating with spatially incoherent amplitudes (Fig. 2.1a, f). For every oscillator in the network, one can calculate the center of mass $z^{c.m.} = \int_0^T z_j(t)dt/T$, where $T = 2\pi/\omega$ is the period of oscillations. It turns out that the nodes belonging to the coherent cluster of the amplitude chimera perform oscillations around the origin, while the centers of mass for the elements from the incoherent part are shifted away from the origin. To measure this shift for each oscillator, we calculate the distance between the center of mass and the origin $r^{c.m.}$ (Fig. 2.2a). The resulting curve strikingly resembles the mean phase velocity profile observed for the classical phase chimera states. Interestingly, the mean phase velocity profile $\bar{\omega}_j$ for amplitude chimera regime is flat (Fig. 2.2b). It means that the averaged phase velocity is the same for all oscillators, independently of the fact whether they belong to coherent or incoherent part. This property is a characteristic signature of amplitude chimeras. Figure 2.2c displays the phase portraits of all network elements in the complex $z = x + iy$ plane. One can clearly see limit cycles with different amplitudes and centers of mass. The elements from the incoherent domain are characterized by smaller amplitudes. Moreover, they oscillate around their individual centers of mass that are shifted away from the origin. The nodes from the coherent cluster perform oscillations with larger amplitudes around the origin and move faster along the limit cycle if compared to the elements from the incoherent domain. Therefore, the patterns we describe here are pure amplitude chimera states. In contrast to amplitude-mediated chimeras, where both phase and amplitude are in a chimera regime [99, 110], here we observe chimera behavior exclusively with respect to amplitude dynamics.

Next, for increasing the number of nearest neighbors P, amplitude chimeras disappear and in-phase synchronization takes place throughout the network (Fig. 2.1b, g). Further increasing the number of nearest neighbors P leads to the intriguing regime called *chimera death* that has features of both oscillation death and chimera state [75] (Fig. 2.1c–e, h–j). In the chimera death state, the oscillations are quenched in a particular way. In more detail, the network splits into two different parts. In one of them, the neighboring nodes are correlated since they populate the same branch of the inhomogeneous steady state (IHSS). Therefore, this domain of the network forms spatially coherent oscillation death. In the other one, the sequence of branches of the IHSS populated by neighboring nodes is completely random in space. In other words, this domain forms spatially incoherent oscillation death. The term "coherent/incoherent" refers here to the coherence/incoherence in space. Temporal coherence/incoherence does not apply here since there is no dynamics in time. It is important to note that due to symmetry reasons, for a node on the upper branch $y^{*1} \approx +1$ of the IHSS in the left half of the system, there always exists a mirror state shifted by phase π (antiphase) in the complex plane, i.e., located on the lower branch $y^{*2} \approx -1$, in the right half of the system. By changing the number of nearest neighbors P in the network, one can observe the formation of clusters within the coherent domains of the chimera death pattern. The elements within such a cluster populate one and the same branch of the IHSS. In particular, the increase of the coupling range P/N reduces the number of clusters formed in the coherent spatial

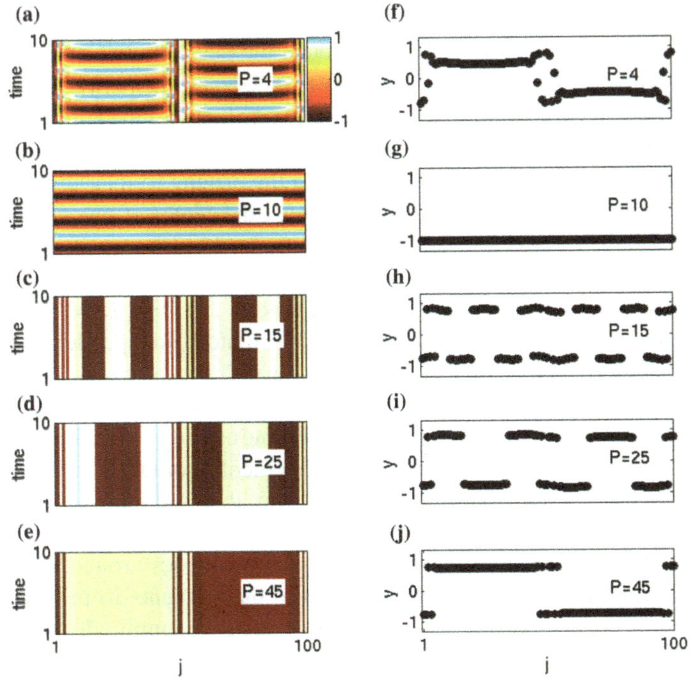

Fig. 2.1 Space-time plots (left column) and snapshots for the variable $y_j(t)$ (right column) in a network of Stuart-Landau oscillators for coupling strength $\sigma = 10$ and different values of nearest neighbors number P. **a, f** $P = 4$: amplitude chimera; **b, g** $P = 10$: in-phase synchronized regime; **c, h** $P = 15$: multicluster (>3) chimera death; **d, i** $P = 25$: 3-cluster chimera death; **e, j** $P = 45$: 1-cluster chimera death. Other parameters: $N = 100$, $\lambda = 1$, $\omega = 2$ [76]

domain. This can be seen from Fig. 2.1 where the number of clusters decreases from (c, h) (5 clusters) via (d, i) (3 clusters) to (e, j) (1 cluster). Figure 2.3a–c displays chimera death state with the maximum number of clusters in the coherent domain. Figure 2.3c shows the phase portrait of chimera death regime: it typically consists of two fixed points shifted by a phase π, corresponding to the two branches y^{*1}, y^{*1} of the IHSS. Interestingly, these two fixed points are located near the centers of mass of those oscillations in Fig. 2.2c with the smallest amplitude, i.e., in the center of the incoherent cluster of the amplitude chimera state.

Further, we investigate the behavior of the network depending on coupling parameters: coupling range P/N and coupling strength σ. The resulting map of dynamical regimes is shown in Fig. 2.4. One can see that steady-state solutions represented by chimera death states are dominating. In Fig. 2.4, these states correspond to red regions with different hatching. In more detail, for large coupling range values, we observe chimera death with two coherent parts, where the elements from one part populate the upper branch and the nodes from the other part populate the lower branch of IHSS (see 1-CD region (light and dark red stripes) in Fig. 2.4). For smaller

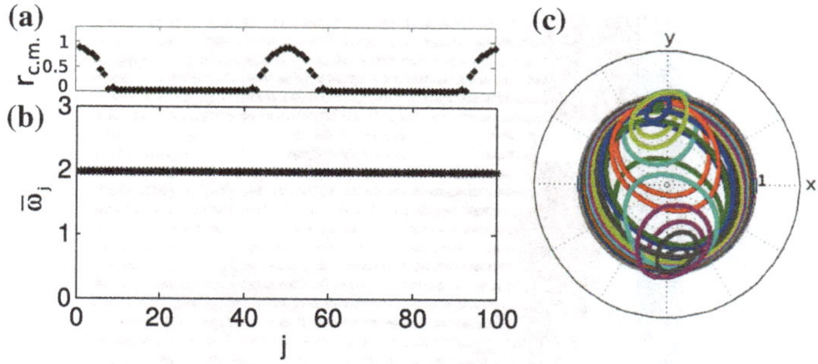

Fig. 2.2 Amplitude chimera: **a** snapshot at time $t = 1000$ for $r^{c.m.}$ (the distance between the center of mass for every oscillator and the origin); **b** mean phase velocity profile; **c** phase portraits of all oscillators in the complex $z = x + iy$ plane. Parameters: $N = 100$, $P = 4$, $\sigma = 14$, $\lambda = 1$, $\omega = 2$ [76]

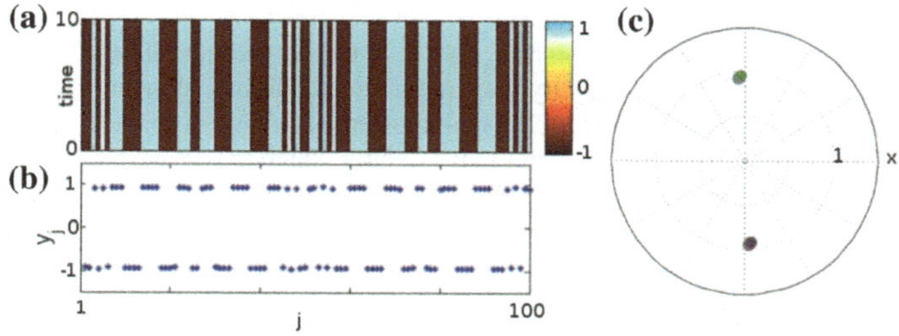

Fig. 2.3 Multicluster (>3) chimera death: **a** space-time plot for the variable y_j; **b** snapshot for the variable y_j; **c** phase portrait of all oscillators in the complex $z = x + iy$ plane. Parameters: $N = 100$, $P = 5$, $\sigma = 26$, $\lambda = 1$, $\omega = 2$ [76]

coupling range values, the clusters are formed within each coherent part. For example, coherent domains may consist of three clusters each (see 3-CD region (yellow and dark red stripes) in Fig. 2.4). If we decrease the coupling range even further, more than three clusters are formed (see N-CD region (dark red region) in Fig. 2.4). Figure 2.1 displays chimera death patterns with different number of clusters in the coherent domain [see space-time plots in panels (c, d, e) and snapshots in panels (h, i, j)]. The empty diamonds in the map of regimes (Fig. 2.4) mark the corresponding values of coupling range and coupling strength. There are two regions corresponding to oscillatory dynamics in the network: amplitude chimera region (blue in Fig. 2.4) and in-phase synchronization region (light green in Fig. 2.4). Amplitude chimeras are detected for rather small number of nearest neighbors and for the whole range of coupling strength values considered in the diagram (Fig. 2.4). Interestingly, chimera death region is highly multistable: chimera death states with different number of clusters in the coherent domain coexist. Initial conditions have a strong impact on the phase diagram, especially on the borders separating chimera death states with different number of clusters. However, the boundary between the steady-state solutions and the oscillatory solutions is less affected by initial conditions (see Sect. 2.3.4). In

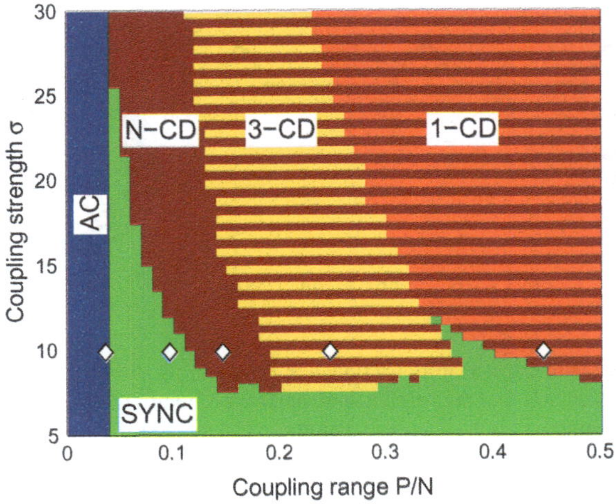

Fig. 2.4 Map of dynamical regimes for $N = 100$, $\lambda = 1$, $\omega = 2$ in the plane of coupling range P/N and coupling strength σ for specially prepared initial conditions, showing 1-cluster chimera death (1-CD); 3-cluster chimera death (3-CD); multicluster (>3) chimera death (N-CD); amplitude chimera (AC); in-phase synchronized oscillations (SYNC). Diamonds mark the parameter values chosen in Fig. 2.1. Other parameters: $N = 100$, $\lambda = 1$, $\omega = 2$ [76]

the considered network, the particular type of transition to steady state depends on the strength of interaction (see Fig. 2.4). In the case of weak interaction ($\sigma = 10$), the transition from amplitude chimera to chimera death occurs via in-phase synchronized regime for increasing coupling range (see diamonds in Fig. 2.4). Strong interaction between the network elements (for example, $\sigma = 26$) makes a direct transition from amplitude chimera to chimera death possible when the coupling range is increased from $P/N = 0.04$ to $P/N = 0.05$.

2.3.2 *Transient Behavior of Amplitude Chimeras*

It is known that classical chimera patterns are transient states. Moreover, they strongly depend on initial conditions. The amplitude chimera states considered here are long-living transients and can exist for thousands of intrinsic oscillation periods for specially prepared initial conditions. For completely random initial conditions, they disappear very fast and in-phase synchronization is observed.

Next, we investigate temporal evolution of the network for the fixed values of system's parameters. Therefore, we calculate snapshots for the variable y at different values of time t (Fig. 2.5a–d). The initial condition is chimera death pattern ($t = 0$). Its trace can still be seen at $t = 4$ (Fig. 2.5d). Later amplitude chimera state is observed (see Fig. 2.5c for $t = 65$). The deformation of the amplitude chimera pattern starts at approximately $t = 125$ (Fig. 2.5b). Finally, it transforms into an in-phase

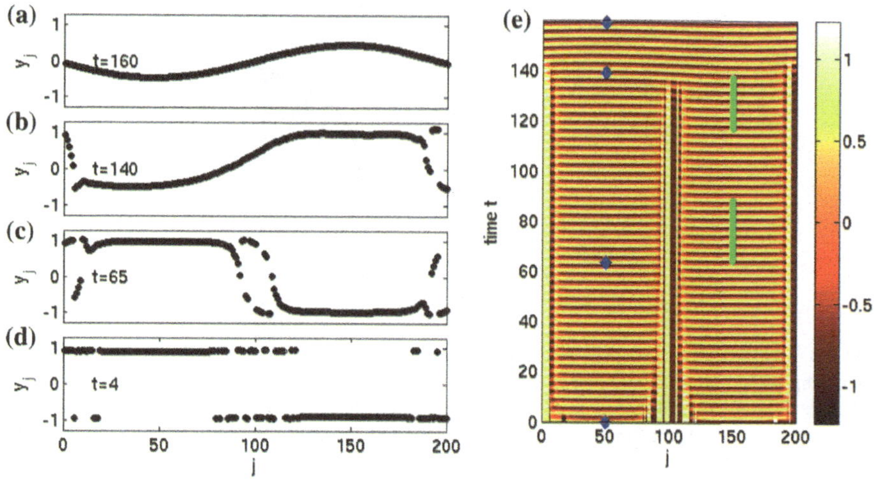

Fig. 2.5 Transient amplitude chimera. Snapshots for the variable y_j (left panel): **a** $t = 160$: in-phase synchronized oscillations; **b** $t = 140$: deformed amplitude chimera at the transition point; **c** $t = 65$: amplitude chimera; **d** $t = 4$: trace of initial condition. The right panel shows the space-time plot for the variable $y_j(t)$. Blue diamonds mark the time values chosen for the snapshots in the left panel. The green lines mark the time ranges chosen for the phase portraits in Fig. 2.6. Parameters: $N = 200$, $\lambda = 1$, $\omega = 2$, $P = 5$ and $\sigma = 20$ [76]

synchronized regime (Fig. 2.5a). The transition described above is also displayed in the space-time diagram (Fig. 2.5e).

To gain more insight into the transition scenario, we inspect phase portraits in the amplitude chimera regime (Fig. 2.6a) and during the transition (Fig. 2.6b). In more detail, we select three network elements: one from the coherent domain ($j = 50$) and the other two from the incoherent part ($j = 99, 102$). Two vertical green lines in the space-time diagram in Fig. 2.5e. Mark the time intervals used for Fig. 2.6a and b, respectively. In the amplitude chimera regime, the element from the coherent domain ($j = 50$) performs oscillatory motion with a large amplitude around the origin. The two nodes from the incoherent part ($j = 99, 102$) are characterized by smaller amplitudes and oscillate around their individual centers of mass which are shifted away from the origin in two opposite directions (Fig. 2.6a). During the transition the centers of mass for elements from the incoherent part move towards the origin (Fig. 2.6b) while the amplitudes of the incoherent oscillations increase until the value of the coherent oscillations is reached. From Fig. 2.6b, one can see that the phase trajectories for the elements $j = 99, 102$ have spiral shapes during the transition.

It is worth noting that initial conditions have an impact on the transition scenario. For instance, asymmetric initial conditions can result in a different transition mechanism involving an asymmetric state: the elements from the incoherent domain join one by one their neighboring synchronized state.

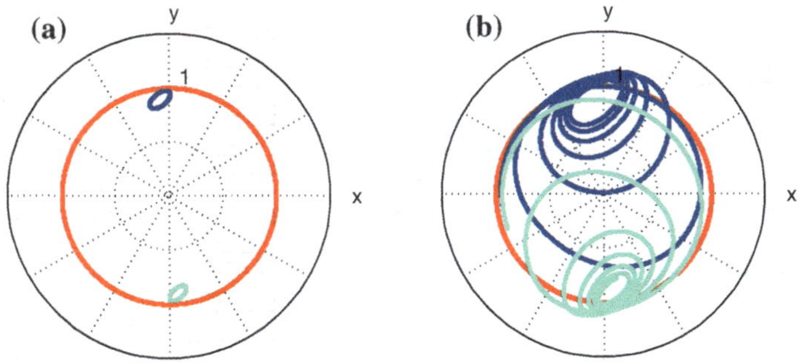

Fig. 2.6 Phase portraits of three selected nodes of the network in the complex $z = x + iy$ plane for **a** amplitude chimera state and **b** during the transition to in-phase synchronization. Red color indicates the trajectory of node $j = 50$ (from the coherent domain). Dark and light blue colors mark trajectories of nodes from the incoherent part $j = 99$ and $j = 102$ correspondingly. Time intervals as indicated in green in Fig. 2.5. Parameters: $N = 200$, $\lambda = 1$, $\omega = 2$, $P = 5$, $\sigma = 20$ [76]

2.3.3 *Detection of Transient Time of Amplitude Chimeras*

To detect whether the system is in the in-phase synchronized oscillatory state, the global Kuramoto order parameter [61, 280] can be used:

$$R(t) = \frac{1}{N}\sum_{i=1}^{N} z_j(t), \tag{2.5}$$

where the value $R(t) = 1$ indicates in-phase synchronization and $R(t) < 1$ corresponds to all other configurations. However, this order parameter cannot be usefully applied to determine the transient time of an amplitude chimera. We note that the in-phase synchronized oscillation is not the only asymptotically stable solution the system can approach; traveling waves are also possible, but $R(t)$ cannot distinguish between amplitude chimeras and coherent traveling waves since both give $R(t) \approx 0$.

In order to define an appropriate order parameter, we introduce the center of mass $(x_j^{c.m.}, y_j^{c.m.})$ of each oscillator j [75]:

$$y_j^{c.m.} = \frac{1}{T}\int_t^{t+T} y_j(t')\,\mathrm{d}t', \quad \text{with } T = \frac{2\pi}{\omega}, \tag{2.6}$$

and an analogous definition of $x_j^{c.m.}$. The shift of the center of mass from the origin is given by

$$r_j^{c.m.} = \sqrt{(y_j^{c.m.})^2 + (x_j^{c.m.})^2}. \tag{2.7}$$

These measures allow to distinguish between coherent and incoherent domains: $x_j^{c.m.}$, $y_j^{c.m.}$, and hence also $r_j^{c.m.}$, vanish for nodes within the coherent domains of the amplitude chimera. In contrast, all nodes within incoherent domains have finite values. Here, the spatial sequence of positive and negative signs of $y_j^{c.m.}$ is completely random. Therefore, these quantities can serve as local order parameters.

Here we use global order parameter based on their second moments [102]

$$c_2(t) = \sqrt{\frac{1}{N}\sum_{i=1}^{N}\left(y_i^{c.m.}(t)\right)^2} \tag{2.8}$$

to detect the transition from an amplitude chimera to a coherent oscillating state. Figure 2.7 shows two examples with different initial conditions for such a transition, one where the approached stable asymptotic solution is an in-phase synchronized oscillation, and one where it is a coherent traveling wave. Both figures depict a space-time plot (upper panel), and the temporal evolution of the global order parameters $R(t)$ and $c_2(t)$ for a time series around the transition. As is clearly visible, in contrast to the global Kuramoto order parameter, c_2 is capable to detect both types of transition. Another advantage of this order parameter is that its value solely depends on the relative size of the incoherent domains of the amplitude chimera, which in turn depends only on the system parameters. In particular, it does not depend on the pattern of disorder within the incoherent domains, and hence not on the specific realization of the initial condition. This means that the value c_2 is a characteristic quantity of a chimera, which distinguishes it from other patterns. We define the relative number of nodes within the incoherent domains as the total number of nodes within both incoherent domains normalized by the overall number of nodes within the whole network. To calculate the relative number of nodes within the incoherent domains, we use here a modified version [102]:

$$\bar{c}_2(t) = \frac{1}{N}\sum_{j=1}^{N}\Theta\left(\left(y_j^{c.m.}(t)\right)^2 - \alpha\right), \tag{2.9}$$

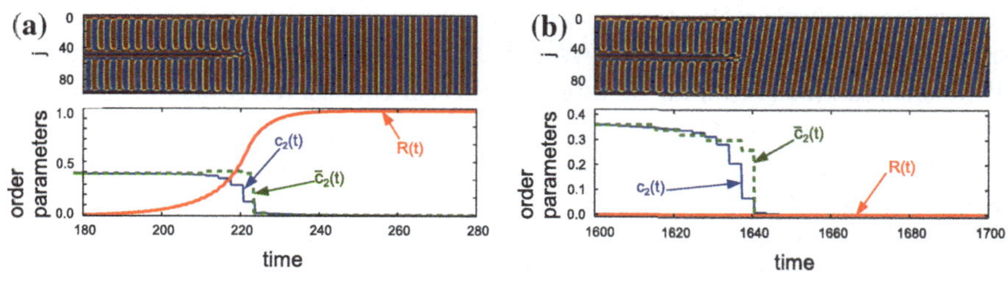

Fig. 2.7 Transition from an amplitude chimera state to **a** in-phase synchronized oscillation, **b** coherent traveling wave. Upper panels: space-time plots, lower panels: time series of the order parameters: global Kuramoto order parameter $R(t)$, second moments c_2, $\bar{c}_2$ with $\alpha = 0.0361$. Parameters $N = 100$, $P/N = 0.04$, $\sigma = 14$, $\lambda = 1$, $\omega = 2$. In **a** and **b** different realizations of the initial conditions shown in Fig. 2.8b with $Q = N/4$ are used [102]

with the Heaviside function $\Theta(x)$, and some fixed threshold $\alpha > 0$. The Heaviside function is non-zero only for those nodes where $(y_i^{c.m.}(t))^2$ is above this threshold. For the choice $\alpha = 0.0361$, the predicted number agrees very well with visual determination of the incoherent domain size, which we have checked with numerous amplitude chimera configurations in a wide range of system parameters.

2.3.4 Role of Initial Conditions

The deterministic system (2.2) is known to demonstrate multistable behavior [75]. Both types of chimera states appear in coupling parameter regimes, where other oscillation death patterns and coherent oscillatory states can be found as well. In order to increase the probability of finding chimera states, we use specially prepared initial conditions. A very simple initial condition that produces transient amplitude chimeras in a certain parameter regime (of about $0.01 < P/N < 0.05$, $\sigma < 33$), is when all nodes of one half of the network ($1 \leq j \leq \frac{N}{2}$) are set to the same value $(x_j, y_j) = (x_0^1, y_0^1)$ (excluding the choice $(0, 0)$), and the rest is set to $(x_0^2, y_0^2) = (-x_0^1, -y_0^1)$. Hence, amplitude chimera states can evolve out of an initial configuration that only consists of two completely coherent parts. We choose the values $(x_0^1, y_0^1) = (\sqrt{0.5}, -\sqrt{0.5})$, so that the nodes start on the limit cycle with $r = \sqrt{\lambda} = 1$, which is the solution for the in-phase synchronized oscillation. The amplitude chimera lifetime nevertheless appears to be of the same order for other values (e. g., $(x_0^1, y_0^1) = (1, -1)$).

By adding random numbers to y_j, we construct a more general class of specially prepared random initial conditions for amplitude chimeras. In particular, we add a random number drawn from a Gaussian distribution with variance V to y_j of the Q nodes on the left and on the right side of the borders between both halves (at $j = \frac{N}{2}$ and $j = N$), as indicated in Fig. 2.8a, with $Q \in \mathbb{N}$ and $0 < Q \leq \frac{N}{4}$. Besides the range Q of incoherence, we also vary $V \geq 0$. For a proper choice of the two initial condition parameters (Q and V), we obtain amplitude chimeras. Using the achieved amplitude chimera lifetime as a quality measure for the initial condition, we compare multiple realizations of the specially prepared random initial conditions for the deterministic system with $P/N = 0.04$ and $N = 100$. We observe that among all considered kinds of initial conditions (different choices for Q and V, symmetry conditions, x_j randomized as well, a different underlying distribution for the random numbers), the applied symmetry of the initial condition has the greatest effect upon the transient time.

Figure 2.9 shows the transient times and their mean value (solid lines) for multiple realizations of the initial conditions following three different symmetry schemes. For the particular choice ($Q = 10$, $V = 2$), all symmetry types lead on average to shorter lifetimes than an initial condition without random component (black dashed line). For the initial configurations without symmetry, a random number is chosen independently for each node within the four incoherent intervals. These configurations clearly create the shortest amplitude chimera lifetimes, lasting at most for a couple of

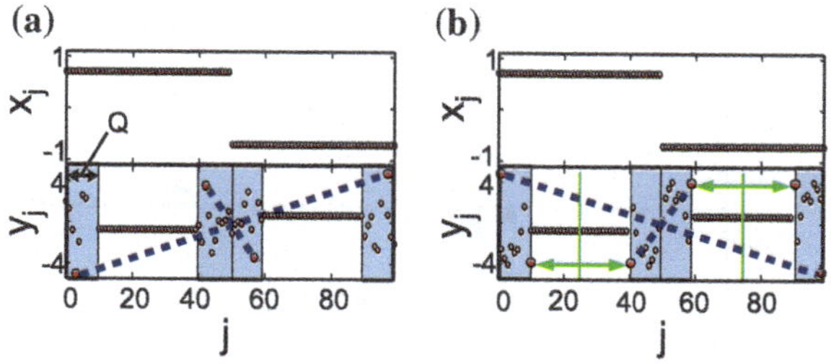

Fig. 2.8 Specially prepared initial conditions for amplitude chimeras: **a** point-symmetric type, **b** fully symmetric type. Top panels: x_j, bottom panels: y_j. Dashed lines indicate the point symmetry about the center, vertical solid lines and arrows (green) indicate the axial symmetry within both network halves. System size: $N = 100$ [102]

oscillation periods. This symmetry type also leads to the shortest transients in other regimes of Q, and V (not shown here). In contrast, for the point-symmetric initial conditions, we mirror the random numbers used for $j \in \{1, ..., \frac{N}{2}\}$ with respect to the center $j = 0$, $y_j = 0$, and use their negative counterparts for the second half. We hence only generate $2Q$ random numbers in total. The initial configurations are point-symmetric with respect to the center, see Fig. 2.8a. The lifetimes of the occurring amplitude chimeras are much longer than in the non-symmetric case. However, the symmetry type which leads to the longest lifetimes is the one referred to as full symmetry; the initial conditions fulfill two symmetries: The randomly chosen values of the positions of the nodes within the first incoherent interval $1 \leq j \leq Q$ are mirrored to the nodes $\frac{N}{2} - Q \leq j \leq \frac{N}{2}$, by setting $z_j = z_{\frac{N}{2}+1-j}$. To obtain the positions of the second network half, then a phase shift of π is applied, such that the "antiphase partner" condition is fulfilled ($z_j = -z_{j+\frac{N}{2}}$ and $j \mod N$). Thus, we only generate Q different random numbers in total. The configurations are again point-symmetric with respect to the center, and have an additional axial symmetry with orthogonal axes through $j = \frac{N}{4}$ and $j = \frac{3N}{4}$, as indicated in Fig. 2.8b. Of course, the simple initial condition with no randomization also fulfills these symmetry conditions and can therefore be regarded as one special type of the fully symmetric specially prepared initial conditions (with $V = 0$).

We have further tested another type of initial condition that solely fulfills the antiphase partner condition: $z_j = -z_{j+\frac{N}{2}}$, but has no other symmetries. This type of initial condition also certainly leads to transient amplitude chimeras, but only within very narrow ranges of Q and V. For $Q = 10$, $V = 2$, the mean lifetime (of about $t_{tr} \approx 49$), is only slightly increased compared to the non-symmetric initial condition (not shown here).

Since the symmetry which is applied to the initial conditions remains preserved during the dynamic evolution, this observation means that the fully symmetric amplitude chimeras are most stable and have the longest lifetimes.

By decreasing the variance in the interval $0.1 \leq V \leq 2$, the mean amplitude chimera lifetimes increase. In the range of small variances of about $V < 0.5$, amplitude chimeras occur for all choices of the incoherence range Q, and the particular choice of Q does not influence the transient time much. For $Q = \frac{N}{4}$, all nodes are

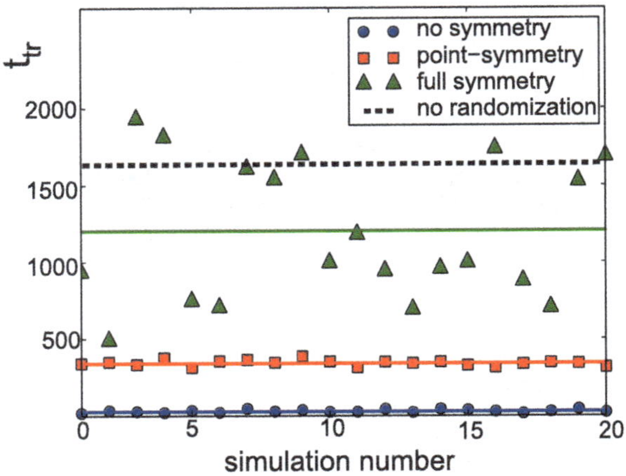

Fig. 2.9 Transient times of amplitude chimera states t_{tr} for 20 realizations of specially prepared random initial conditions (no symmetry, point symmetry, full symmetry, depicted by different symbols) with $Q = 10$, $V = 2$. Horizontal solid lines: mean values. Dashed line: t_{tr} for the initial condition with $V = 0$ (no randomization). System parameters: $N = 100$, $\sigma = 14$, $P/N = 0.04$, $\lambda = 1$, $\omega = 2$ [102]

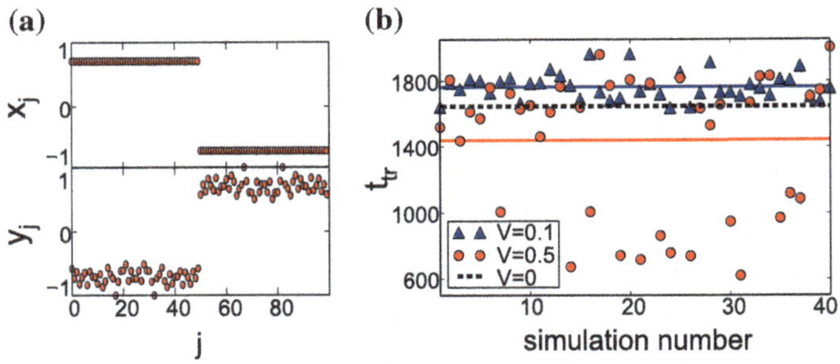

Fig. 2.10 **a** Fully symmetric random initial condition with range of incoherence $Q = \frac{N}{4}$ and variance $V = 0.1$. **b** Transient times of amplitude chimeras t_{tr} for 40 realizations of the initial condition shown in **a**: circles: $V = 0.5$, triangles: $V = 0.1$. Horizontal solid lines: mean values, dashed line: t_{tr} for $V = 0$. System parameters: $N = 100, \sigma = 14, P/N = 0.04, \lambda = 1, \omega = 2$ [102]

randomized, see Fig. 2.10a, which appears to be a natural choice. Figure 2.10b shows the corresponding transient times belonging to a set of 40 realizations of the specially prepared random initial condition with $Q = \frac{N}{4}$ and $V = 0.5$, and for a set with $V = 0.1$. The mean transient times are much longer than for $V = 2$ (cf. Figure 2.9). They are at least of the same order (and can be larger) as the transient time for the simple initial condition with no randomization, $V = 0$ (dashed black line). For the choice $V = 0.1$, the transient times are increased as compared to $V = 0.5$.

Besides oscillatory states, oscillation death states can occur in a large variety of different spatial patterns. Our numerical results suggest that in the appropriate parameter regime, every amplitude chimera snapshot can be used as initial condition to certainly produce a chimera death state. How many clusters in the coherent domain

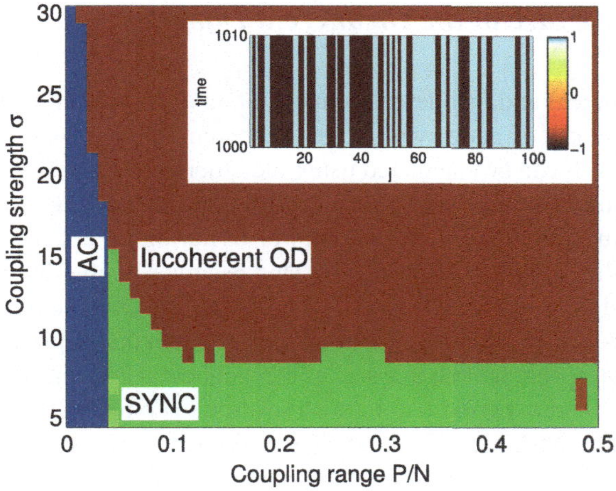

Fig. 2.11 Map of dynamical regimes for $N = 100$, $\lambda = 1$, $\omega = 2$ in the plane of coupling range P/N and coupling strength σ for random initial conditions with symmetries: amplitude chimera (AC); in-phase synchronized oscillations (SYNC); incoherent oscillation death (Incoherent OD). The inset shows a space-time plot for the variable $y_j(t)$ for the coupling strength $\sigma = 30$ and coupling range $P/N = 0.5$ [76]

of the chimera death pattern occur depends on the initial condition as well as on the parameter choice (see Sect. 2.5.4).

To illustrate the role of initial conditions for the dynamic regimes in the system, we compare the map of regimes calculated for a specially prepared initial set (Fig. 2.4) and for random initial conditions with symmetries (Fig. 2.11). For the diagram shown in Fig. 2.4, we use as initial condition an amplitude chimera profile at a fixed time (similar to Fig. 2.5c), which is obtained in the following way: For a fixed set of parameters in the amplitude chimera regime, the system is divided into two equal domains; half of the nodes are located on the upper branch and the other half is on the lower branch of the inhomogeneous steady state; this initial condition then evolves into an amplitude chimera state which is used as initial condition for all other parameter values (σ, P/N). The random initial distribution with symmetries leads to a significantly different phase diagram (Fig. 2.11) in comparison with the specially prepared initial conditions (Fig. 2.4). The domain of the amplitude chimera is decreased for strong coupling if compared with Fig. 2.4. Moreover, the regime of chimera death obtained from carefully chosen initial conditions (Fig. 2.4) is replaced by the regime of incoherent oscillation death (inset in Fig. 2.11) which results from random initial conditions. It should be noted that the map of regimes shown in Fig. 2.11 is obtained for one single realization of random initial distribution and no averaging of initial conditions has been provided. Therefore, it is not statistically representative that chimera death is not observed in Fig. 2.11. This is merely due to the fact that the probability to achieve chimera patterns starting from one random realization of the initial condition is very low. However, the occurrence of chimeras for random initial conditions, in general, is not excluded.

2.3.5 Relative Size of the Incoherent Domains

The ratio of coherent and incoherent domains of an amplitude chimera can be measured by the number of nodes within the incoherent domains divided by the total number of nodes. It can be calculated using the order parameter $\overline{c}_2(t)$. Interestingly, in the time evolution of the system, this quantity typically fluctuates by less than one order of magnitude, during the chimera, as well as during the coherent synchronized oscillation, or during the traveling wave. This becomes visible by considering the temporal evolution of $\overline{c}_2(t)$, as exemplarily shown in Fig. 2.7. In contrast to $R(t)$ and $c_2(t)$, the values of $\overline{c}_2(t)$ show a sharp drop to zero when the amplitude chimera disappears, illustrating that the relative size of the incoherent domains remains constant until the last oscillation period before the transition to complete coherence.

Note that the relative size of the incoherent domains is not related to the choice of Q, but is completely independent of the initial condition. Even if the incoherent domain is chosen larger or smaller initially, it evolves into a domain with a characteristic size corresponding to the given set of parameters. This is an important general distinguishing factor of chimeras which emphasizes that they are self-organized patterns.

In contrast, the relative size of the incoherent domains does depend on the choice of the coupling parameters. This is shown in Fig. 2.12 for coupling parameters varied in the range $1 \leq P \leq 20$ and $5 \leq \sigma \leq 30$. For each choice or the coupling parameters, the same snapshot of an amplitude chimera is used as initial condition. The maximum simulation time is $t = 5000$. The relative size of the incoherent domains varies over a wide range from about 0.14 to about 0.6, i.e., we observe amplitude chimeras where

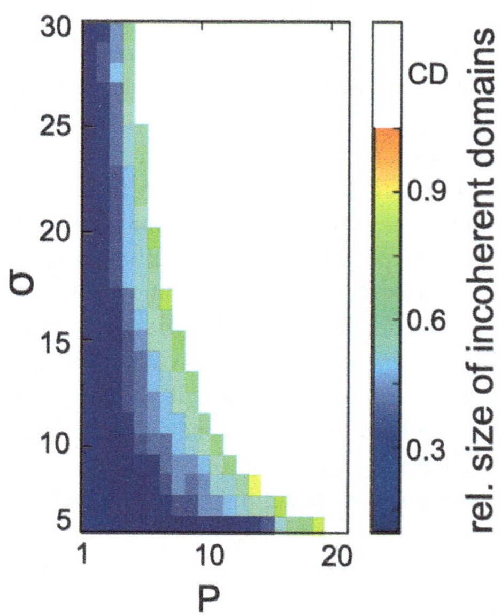

Fig. 2.12 Relative size of the incoherent domains in the plane of coupling strength σ and nearest neighbors number P. Initial condition: a fixed snapshot of an amplitude chimera. The light gray area denotes chimera death (CD). Other parameters: $N = 100$, $\lambda = 1$, $\omega = 2$ [102]

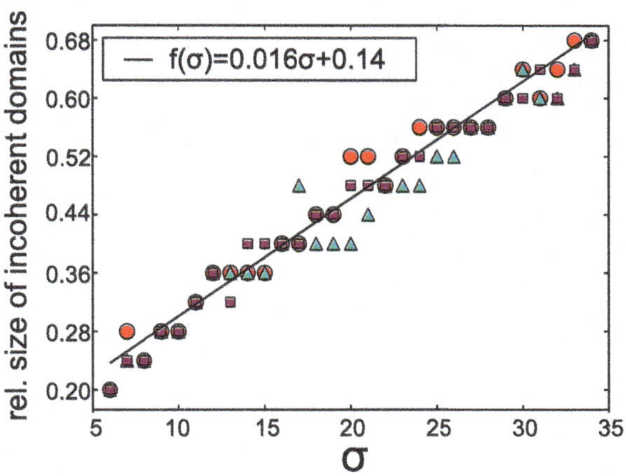

Fig. 2.13 Relative size of the incoherent domains of amplitude chimeras versus coupling strength σ, measured by $\overline{c}_2(t = 100)$. Squares, triangles, and disks correspond to three different initial conditions. Solid line: linear fit to disks. System parameters: $N = 100$, $P/N = 0.04$, $\lambda = 1$, $\omega = 2$ [102]

just 14% of the nodes show a spatially incoherent behavior, up to configurations with 60%. At the border between the oscillatory regime and the chimera death regime, we detect the amplitude chimeras with the largest incoherent domains. The width of the incoherent domains appears to increase linearly with P, as well as with σ. In Fig. 2.13, the relative size of the incoherent domains in dependence upon the coupling strength is depicted for three different initial conditions. The actual spread of the values is due to the fact that, firstly, $\overline{c}_2(t)$ can assume only multiples of $\frac{1}{N}$, and, secondly, the relative number of nodes can change only in multiples of $\pm\frac{4}{N}$, due to the symmetry in the initial conditions (see Sect. 2.3.4).

As opposed to the linear increase with σ and P, the relative size of the incoherent domains is roughly constant for all system sizes. For an exemplary choice of the coupling parameters, Fig. 2.14 shows (a) snapshots and (b) the corresponding shifts from the origin of the centers of mass, for five different system sizes N.

2.3.6 *Impact of System Size*

For classical phase chimeras, it has been established that the transient nature of the chimera states is a finite-size effect [63, 129]. In the limit of an infinitely large system, the phase chimera is stable. This is opposite to the system size dependence of the amplitude chimera states, where we find that their lifetime decreases with N.

For various system sizes N, Fig. 2.15 shows the median of the transient times for a constant coupling strength $\sigma = 14$ and a constant coupling range $P/N = 0.04$, averaged over 20 initial conditions (black disks). The data for an additional single initial condition are also shown by cyan triangles. Both data sets can be well described by the following dependence of the amplitude chimera lifetime on the system size:

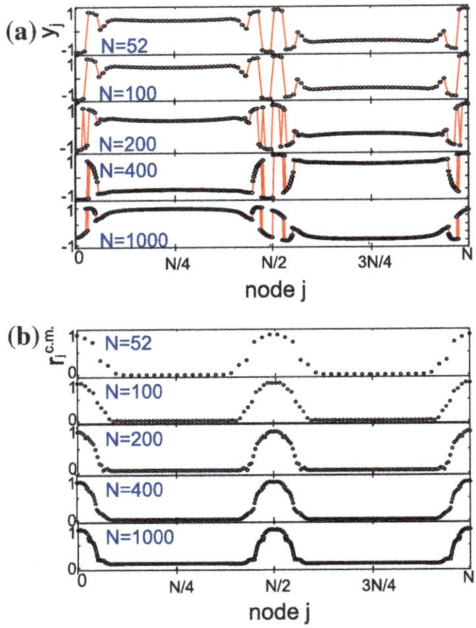

Fig. 2.14 Snapshots of amplitude chimeras for different system sizes N, but constant coupling range $P/N = 0.04$: **a** y_j, **b** shift of the center of mass $r_j^{c.m.}$ from the origin. Parameters: $t = 140$, $\sigma = 14$, $D = 0$, $\lambda = 1$, $\omega = 2$. Initial conditions: see Sect. 2.3.4 [102]

$$t_{tr}(N) = \left(\frac{\alpha_1}{N}\right)^2 + \alpha_2. \tag{2.10}$$

The black curve in Fig. 2.15 is a fit with $\alpha_1 \approx 3802.29$ and $\alpha_2 \approx 596.23$. This indicates that amplitude chimeras should exist, albeit with shorter lifetime, in systems of every size. That means that they would also appear in the "thermodynamic limit" of infinitely large systems with a finite lifetime (for instance, $t_{tr} \approx 600$ for $N \to \infty$, for the type of initial condition used in Fig. 2.15). These results show that the mechanism of the transient is different from phase chimeras, where the transition is triggered by a sweeping front which travels through the whole system. Rather, the incoherent oscillators of the amplitude chimeras catch up locally with the coherent ones to synchronize, which is independent of the global system size as long as this size is large enough [76]. This reasoning is in line with our observations in Sect. 2.3.5 that the relative size of the incoherent domains firstly, is roughly constant for all N (Fig. 2.14), and secondly, stays constant during the temporal evolution of a chimera until the last period before the transition to the asymptotic solution and then drops immediately to zero (Fig. 2.7).

Amplitude chimeras also occur in small networks. However, in this case, it becomes more difficult to clearly distinguish domains of coherence and incoherence. A minimum example is shown in Fig. 2.16 for $N = 12$ nodes. Here amplitude chimeras are found only for $P = 1$ ($r = 1/12$). Two coherent domains with elements $j = 3, 4$ and $j = 9, 10$, respectively, can be seen. Between these domains, there are incoherent domains where the oscillation amplitude decreases, and the center of mass

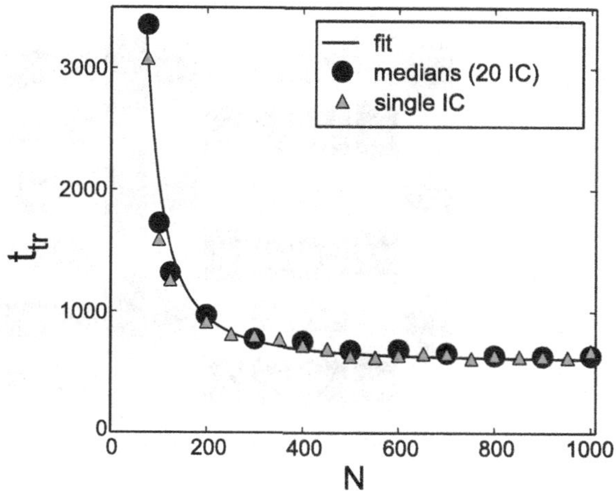

Fig. 2.15 Transient times of amplitude chimeras t_{tr} versus system size N, for the constant coupling range $P/N = 0.04$. Disks: medians of t_{tr} over 20 initial conditions (IC), solid line: fit to (2.10) with $\alpha_1 \approx= 3802.29$, and $\alpha_2 \approx 596.23$. Triangles: t_{tr} for a single initial condition. Other parameters: $\sigma = 14$, $\lambda = 1$, $\omega = 2$ [102]

is shifted from the origin, see the phase portraits in Fig. 2.16b. Note that the antiphase symmetry between elements $j = 1, ..., 6$ and $j = 7, ..., 12$ is observed.

2.3.7 Stability Analysis of Amplitude Chimeras

It is important to note that in the amplitude chimera state, individual oscillators never become chaotic: they remain periodic and thus in the incoherent domain, all the oscillators are temporally periodic but spatially chaotic. According to the classification provided in [168], amplitude chimeras are categorized as transient chimeras. Indeed amplitude chimeras are transients towards the completely synchronized state. Their lifetime strongly depends on the initial conditions in the deterministic case. In contrast to classical chimeras [74, 129] or transient chaos in spatially extended systems [281, 282], where the transient time exponentially increases with the system size, for amplitude chimeras the transient time decreases and saturates for large system size [102]. Thus, the transient nature of amplitude chimeras cannot be related to a finite-size effect.

In the present chapter, we investigate the phase-space structure of amplitude chimeras.[1] We propose that they represent saddles in the high-dimensional phase space of the corresponding network, i.e., they have both stable and unstable manifolds, which can explain their transient nature. Here we continue investigating amplitude chimeras in a ring of Stuart-Landau oscillators with nonlocal coupling and focus on their stability using Floquet analysis. In particular, we calculate the real parts of the Floquet exponents for a wide range of system parameters (coupling strength and range) in the regime where amplitude chimeras exist. Positive real parts of Flo-

[1] Portions of the following text have been re-printed from [103] with the permission of IOP Publishing.

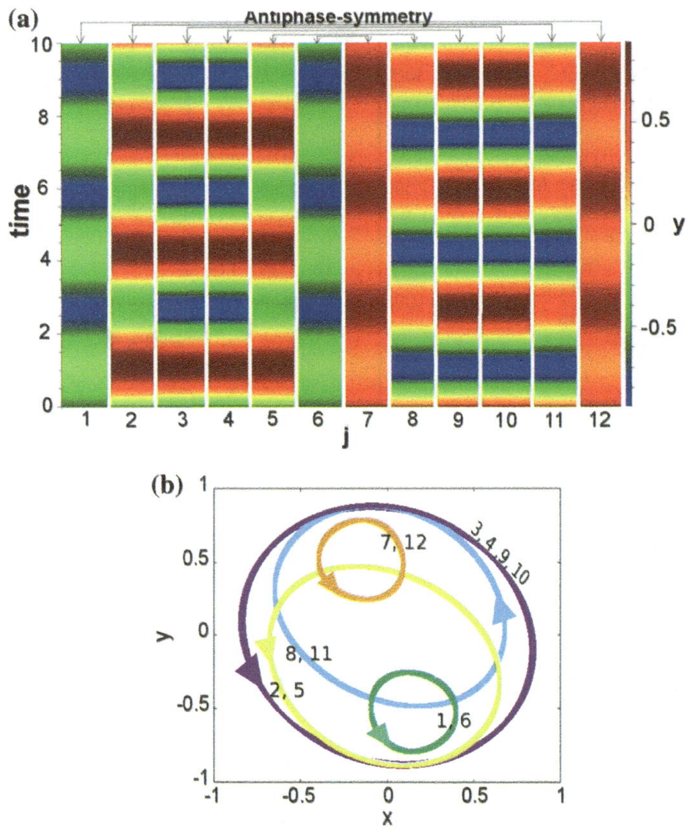

Fig. 2.16 **a** Space-time plot of an amplitude chimera state $y_j(t)$ in a very small network. **b** Phase portrait. Parameters: $N = 12$, $\lambda = 1$, $\omega = 2$, $r = \frac{1}{12}$, $\sigma = 5$ [103]

quet exponents correspond to the unstable manifold of the saddle and negative ones characterize its stable manifold. All investigations are performed for a deterministic system without noise and time delay. Using the eigenvectors spanning the stable and unstable manifolds of the saddle cycle, we relate the amplitude chimera lifetime to the geometric structure of the phase space and investigate the structural change with increasing coupling range, and its effect upon the lifetime.

Method

The Floquet theory is a mathematical tool to study the local stability of periodic solutions, i.e., limit cycles. For a system of differential equations

$$\dot{\mathbf{x}} = \mathbf{f}(\mathbf{x}(t)), \tag{2.11}$$

with $\mathbf{x}(t) \in \mathbf{R}^n$ assume that there exists a periodic solution $\chi(t) = \chi(t + T)$. Writing (2.2) in terms of the real variables x_j and y_j, we have $n = 2N$. We analyze the stability of the periodic orbit by considering solutions in its vicinity:

$$\mathbf{x}(t) = \chi(t) + \delta\mathbf{x}(t). \tag{2.12}$$

The linearized equation

$$\delta\dot{\mathbf{x}}(t) = \mathbf{Df}(\chi(t))\delta\mathbf{x}(t), \tag{2.13}$$

where $\mathbf{Df}(\chi(t))$ is the Jacobian, evaluated at $\chi(t)$, has the solution

$$\delta\mathbf{x}(t) = \mathbf{U}(t)\delta\mathbf{x}(0) \tag{2.14}$$

with the initial condition $\delta\mathbf{x}(0)$, and the fundamental matrix $\mathbf{U}(t)$ satisfies

$$\dot{\mathbf{U}}(t) = D\mathbf{f}(\chi(t))\mathbf{U}(t), \qquad \mathbf{U}(0) = \mathbf{1}. \tag{2.15}$$

The time evolution operators obey

$$\mathbf{U}(t+T) = \mathbf{U}(t)\mathbf{U}(T), \tag{2.16}$$

and $\mathbf{U}(T)$ is called monodromy matrix.

In Fig. 2.17, we illustrate this by a simple geometric picture (which strictly holds only for a symmetric monodromy matrix): The initial conditions of fundamental solutions $\phi_k(0)(k = 1, 2, ..., n)$ for $t = 0$ are located on the unit sphere S. Each point of S is mapped to the ellipsoid E, which contains all the fundamental solutions $\phi_k(T)$ after one period T and whose principal axes correspond to the eigenvectors $\mathbf{p}_k$ and the eigenvalues μ_k (Floquet multipliers) of $\mathbf{U}(T)$:

$$\mathbf{U}(T)\mathbf{p}_k = \mu_k\mathbf{p}_k \tag{2.17}$$

provides N eigenvalues μ, numbered by index k, and corresponding eigenvectors $\mathbf{p}$. The eigenvalues μ_k of the monodromy matrix, the so-called *Floquet multipliers*, characterize the stability of a period orbit $\chi(t)$.

If all $|\mu_k| < 1$, then the ellipsoid E is located inside the sphere S, i.e., after one period T, the perturbations in all the directions decrease. If at least one $|\mu_k| > 1$, then the perturbations increase exponentially and the corresponding orbit is unstable. For

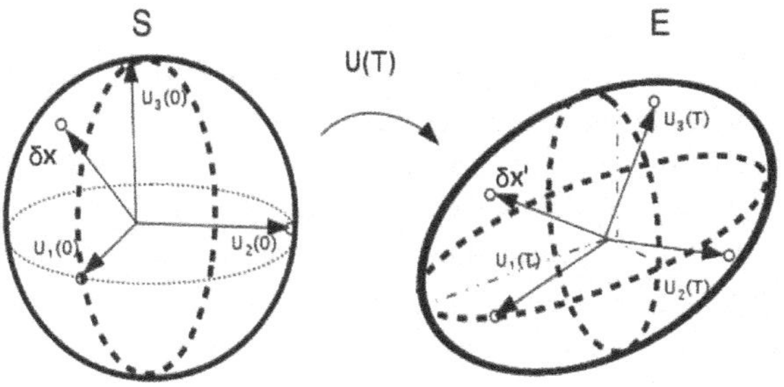

Fig. 2.17 Geometric interpretation of monodromy matrix $\mathbf{U}(T)$ for a three-dimensional Euclidean space (see text) [103]

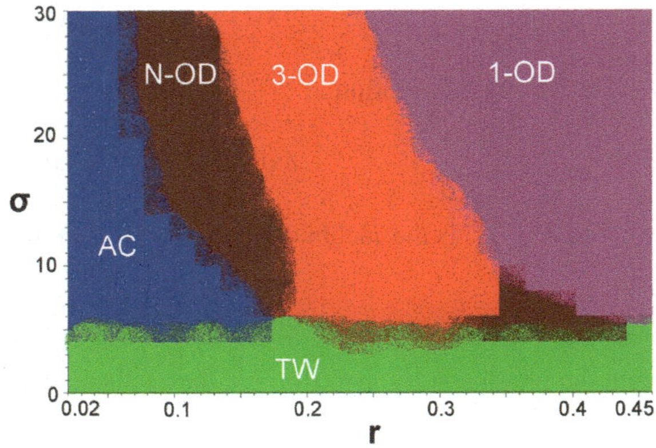

Fig. 2.18 Map of dynamic regimes for $N = 100$ in the plane of coupling strength σ and coupling range r. 1-OD: one-cluster oscillation death; 3-OD: three-cluster oscillation death; N-OD: multicluster (>3) oscillation death; AC: amplitude chimera; TW: traveling wave. Other parameters: $\lambda = 1$, $\omega = 2$ [103]

periodic orbits $\chi(t)$, there exists always one Floquet multiplier $|\mu_k| = 1$ (Goldstone mode), where the perturbation is along the orbit. The Floquet multipliers are related to the Floquet exponents $\Lambda_k + i\Omega_k$ by $\mu_k = \exp(\Lambda_k + i\Omega_k)T$. The periodic orbit is stable, if all $\Lambda_k < 0$ (except for the Goldstone mode), and unstable if at least one $\Lambda_k > 0$.

Numerical results

Amplitude chimera states, which are long-living transients, have been found for small values of the coupling range r and sufficiently strong coupling strength σ [75, 76, 102]. Weakly coupled systems exhibit coherent traveling waves or in-phase synchronized states. For large σ, an increase of coupling range r leads to oscillation death distinguished by different numbers of clusters (Fig. 2.18). The transient time t_{tr} of amplitude chimeras is depicted in Fig. 2.19a in dependence on r and σ. In our simulations, the transient time t_{tr} is defined as the time when no oscillations with shifted center of mass are observed any more [102]. Decreasing the coupling strength σ for a fixed value of coupling range r corresponds to an increase of the lifetime of the amplitude chimera. The same tendency is observed for a fixed value of σ when increasing r. Dark red color denotes $t_{tr} > 5000$, i.e., $t_{tr} > 1600T$. Thus, amplitude chimeras are particularly long-living in networks with weak coupling and large coupling range.

Figure 2.19b shows the maximum real part Λ of the Floquet exponents calculated from the monodromy matrix, which is averaged over at least 100 periods $T \approx \pi$; we do not show Λ for states with $t_{tr} < 320$. In Fig. 2.19c, we display the number of Floquet exponents with positive real parts Λ for each set of parameters σ and r. Most amplitude chimera states have one positive Λ, but some have two, corresponding to one or two unstable directions in phase space, respectively. For instance, increasing σ at fixed $r = 0.02$, several transitions between one and two Floquet exponents with positive real part occur; they are characterized by zero-eigenvalue bifurcations (transcritical or pitchfork bifurcations of limit cycles). It should be noted that the second positive exponent is approximately 10 times smaller than the first one, and

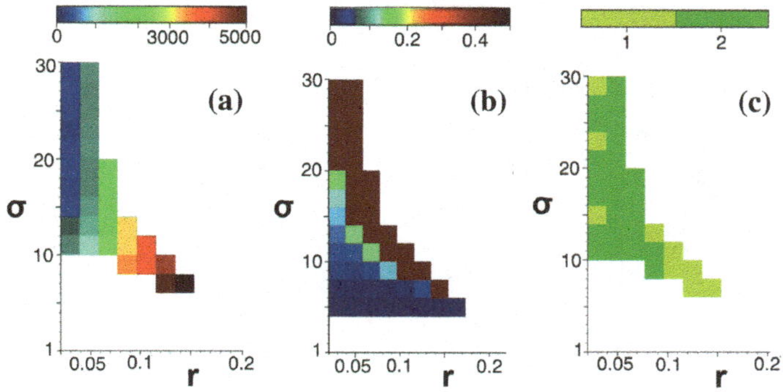

Fig. 2.19 **a** Transient time of amplitude chimera states in the (σ, r) plane. Maximum integration time $t = 5000$. **b** Largest positive real parts of Floquet exponents Λ in the (σ, r) plane. **c** Number of positive real parts of Floquet exponents Λ in the (σ, r) plane. Other parameters: $N = 100$, $\lambda = 1$, $\omega = 2$ [103]

sometimes difficult to distinguish from the Goldstone mode which corresponds to $\Lambda = \Omega = 0$, and is always present.

For a fixed coupling range r, we observe an increase of the largest positive real part Λ of the Floquet exponents with increasing σ (Fig. 2.19b). Similarly, when changing the network topology by increasing r at fixed σ, we also observe that Λ increases. Comparing Fig. 2.19a, b, we find that within the same network topology, i.e., fixed r, the transient times decrease while the positive real part of the dominant Floquet exponent increases. The escape rate Λ from the saddle along the unstable direction increases, which leads to a shorter transient time $t_{tr} \sim \frac{1}{\Lambda}$. However, for constant σ with increasing r, both the lifetime of the amplitude chimera and Λ increase. These two different cases are visualized by plotting t_{tr} versus Λ in Fig. 2.20 for constant r (solid red curve) and constant σ (dashed blue curve), respectively.

These results suggest that amplitude chimeras are saddle orbits in phase space with a small number (one or two) of unstable dimensions.

Eigenvectors

Essential information on the structure of the phase space near the saddle is given by the eigenvectors $\mathbf{p}_k$ of the monodromy matrix $\mathbf{U}(T)$ corresponding to the Floquet multipliers μ_k, see (2.17). The eigenvectors associated with positive real parts of Floquet exponents span the unstable manifold of the saddle. In Fig. 2.21b, c, we show the two eigenvectors associated with positive real parts Λ of the Floquet exponents, and in (d), the eigenvector related to the leading stable Floquet exponent, for an amplitude chimera state in a network with $N = 400$. For reference, a snapshot of the amplitude chimera is given in Fig. 2.21a; it illustrates the location of coherent ($j \in (65, ..., 135), \in (265, ..., 335)$, shaded blue) and incoherent domains. The coherence and incoherence shows up most prominently in the imaginary part (y-variable), i.e., the right column. Panels (b) and (c) show that the components of the eigenvector associated with the unstable manifold have much more distributed values in the

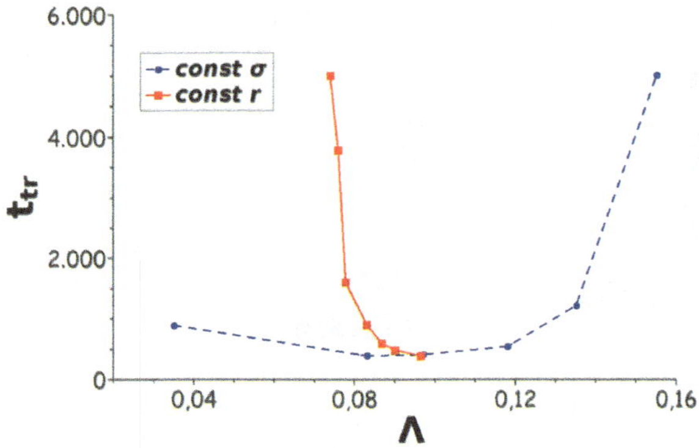

Fig. 2.20 Dependence of transient time t_{tr} upon real part Λ of Floquet exponents **a** for constant $r = 0.03$ and varying σ, **b** for constant $\sigma = 12$ and varying r. Parameters: $N = 100, \lambda = 1, \omega = 2$ [103]

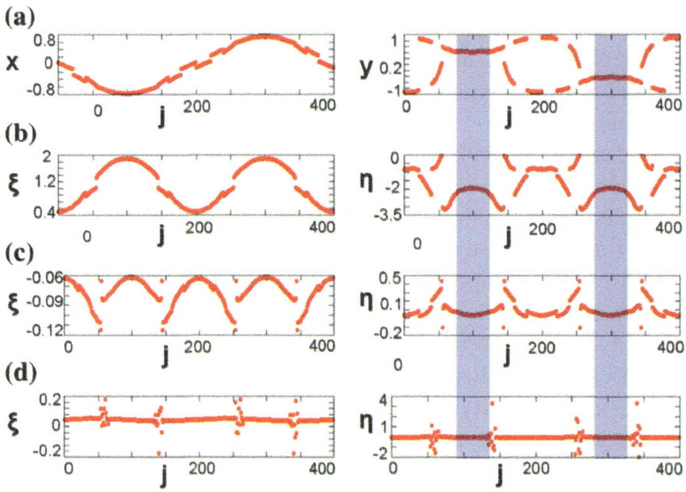

Fig. 2.21 Real (left column) and imaginary parts (right column) of **a** snapshot z_j, **b** eigenvectors associated with the largest positive $\Lambda = 0.135$, **c** with the second positive $\Lambda = 0.053$, **d** with the first negative $\Lambda = -0.133$. Parameters: $N = 400, \lambda = 1, \omega = 2, \sigma = 23, r = 0.06$ [103]

incoherent domain than the nodes in the coherent domain which are almost constant. This is due to the fact that incoherent domains represent sources of instability, which push our network away from the amplitude chimera state towards the completely synchronized state. In contrast, the components of the eigenvector associated with a negative real part of the Floquet exponents (stable manifold of the amplitude chimera) are almost zero everywhere except at the boundaries between coherent and incoherent domains.

Phase-space structure

Based upon the eigenvectors, we will now discuss the change in phase-space structure induced by varying the network topology, i.e., the coupling parameter r. The geometric features of phase space in the vicinity of the amplitude chimera (AC) are schematically represented in Fig. 2.22f. Two trajectories approaching the amplitude chimera saddle along the stable manifold (solid lines) and then escaping along the unstable manifold (dashed lines) with escape rate Λ are shown. If more neighbors are coupled to each element, i.e., for larger r, this leads to an increase in the width of the incoherent domain (Fig. 2.22a, b) because those synchronized elements which are located at the edge of the coherent domain experience more influence from the elements in the incoherent domain, and hence they also become desynchronized. A larger incoherent domain means that it takes longer time to reach the completely synchronized global attractor, hence the lifetime increases. Figure 2.22c, d shows the corresponding snapshots for larger $N = 400$. Again, the width of the incoherent domains increases with r, and additionally it can be seen that with increasing N, the width for fixed r shrinks, cf. panels (a) and (c), or (b) and (d), respectively. This explains why the lifetime decreases with increasing N, in contrast to classical chimeras [74, 129]. Figure 2.22e depicts the relative width of the two incoherent domains versus r for $N = 100$ which clearly follows a linear relation.

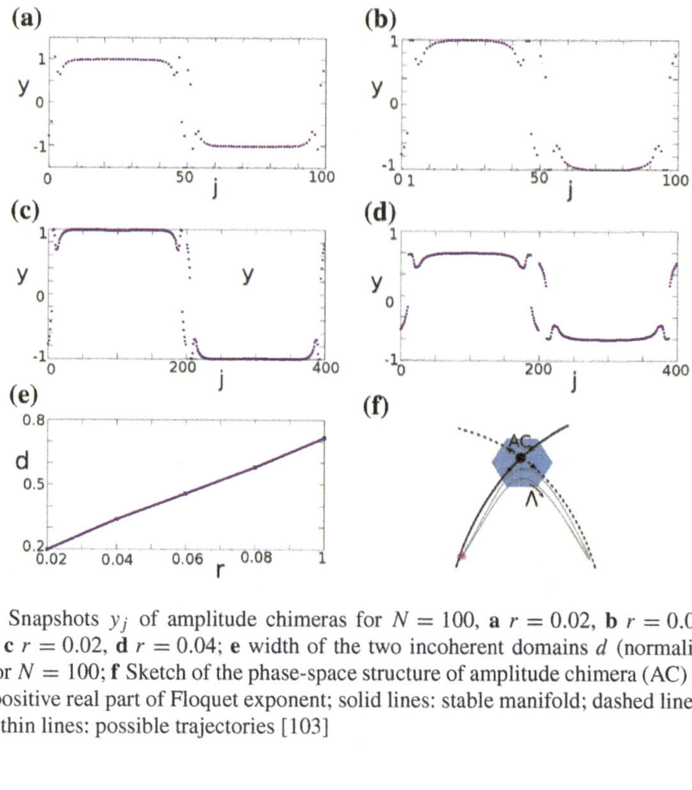

Fig. 2.22 Snapshots y_j of amplitude chimeras for $N = 100$, **a** $r = 0.02$, **b** $r = 0.04$, and for $N = 400$, **c** $r = 0.02$, **d** $r = 0.04$; **e** width of the two incoherent domains d (normalized by N) versus r for $N = 100$; **f** Sketch of the phase-space structure of amplitude chimera (AC) as a saddle orbit. Λ: positive real part of Floquet exponent; solid lines: stable manifold; dashed lines: unstable manifold; thin lines: possible trajectories [103]

For all parameter values here, we choose the same initial condition which is approximately equal to the amplitude chimera, and is constructed by simulating the system for 6500 time-steps ($t \approx 20T$) starting from a fully antisymmetric state as in 2.3.4: the first half of the nodes $j \in (1, ..., N/2)$ is set to $(x_j, y_j) = (-1, 1)$, whereas the second half $j \in (N/2 + 1, ..., N)$ is set to $(x_j, y_j) = (1, -1)$. The observation of amplitude chimera states is related with the symmetry of the initial condition. While the chimera itself is a saddle in phase space, the dominant unstable direction (given by the symmetric eigenvector corresponding to the eigenvalue with largest positive real part) is orthogonal to antisymmetric initial conditions. Normally long transients occur which are caused by tiny violations of the symmetry of the system during numerical integration. However, on changing the parameters as demonstrated above, a second small unstable eigenvalue may occur and the corresponding unstable direction lies within the subspace of antisymmetric initial conditions. This small subleading eigenvalue Λ_2 then causes transient behavior which is in agreement with the observed dependence of the transient time $t_{tr} \sim 1/\Lambda_2$ on the system parameters, since Λ_2 decreases with increasing width of the incoherent domain and hence with r.

For the small size network (see Fig. 2.16), the calculation of the three dominant Floquet multipliers gives $\Lambda = 0.363$, $\Lambda = -0.359$, and the Goldstone mode $\Lambda \approx 0.059$, in agreement with our previous analysis.

Therefore, the stability analysis of amplitude chimera states in Stuart-Landau oscillator networks using Floquet theory allows to establish the phase-space structure of these long-living transient states. Moreover, it shows that they are saddles with one or two repelling directions given by the Floquet exponents with positive real parts and the corresponding eigenvectors. Amplitude chimeras are particularly long-living (thousands of periods) in weakly coupled networks or in networks whose topology is characterized by a large number of nearest neighbors. We have explained the behavior of the transient times in dependence on the strength σ and the range r of the nonlocal coupling by the changes in phase-space structure. In particular, the increase of the chimera lifetime with increasing coupling range at fixed coupling strength results from the increase in the width of the incoherent domain. For antisymmetric initial conditions, the escape rate from the saddle is determined by numerical deviations from strict antisymmetry or by the second unstable eigenvector, whose eigenvalue has a real part Λ_2 which decreases with increasing r. At the same time, the positive real part Λ_1 of the leading Floquet exponent increases with r, but has no influence upon the chimera lifetime since its direction is orthogonal in phase space. If the coupling strength σ is increased at fixed coupling range r, the positive real part Λ of the Floquet exponents increases, and the chimera lifetime t_{tr} decreases inversely proportional to Λ. Thus, amplitude chimera states with small coupling strength σ and large coupling range r have the longest lifetimes. We have also presented simulations of amplitude chimera states in a minimum network of $N = 12$ elements, and showed that these are saddles with one unstable direction.

2.3.8 Summary

In conclusion, in this section, we have considered complex spatiotemporal patterns occurring in deterministic networks of nonlocally coupled ring networks of Stuart-Landau oscillators. We have related two different phenomena—chimera states and oscillation death—which have a common signature of symmetry breaking in dynamical networks, thus revealing two intriguing types of chimeras, i.e., coexisting incongruous spatially coherent and incoherent states: amplitude chimera and chimera death. These two findings generalize chimeras to amplitude dynamics and to inhomogeneous steady states, respectively. Moreover, we have uncovered the transition scenarios between these partially coherent spatiotemporal patterns. In particular, we have demonstrated that a network of identical elements with nonlocal symmetry-breaking coupling can be driven out of the oscillatory regime of amplitude chimera by two distinct mechanisms: either via passing through an in-phase synchronized state, a scenario which occurs at low coupling strength, or at sufficiently large coupling strength by an abrupt suppression of oscillations and therefore, the transition to chimera death, by increasing the coupling range. Moreover, transitions from in-phase synchronized oscillations to inhomogeneous stationary patterns with various numbers of coherent and incoherent clusters (chimera death) can be induced by increasing the coupling strength. These findings deepen our general understanding of partial synchronization patterns in complex networks. Additionally, these results represent an important step towards the systematic study of oscillation death in the context of complex network topologies with nonlocal interactions of variable range and symmetry-breaking coupling.

Further, we have shown that amplitude chimeras are long-living transients and systematically analyzed their transient dynamics and lifetime. To detect the transitions from amplitude chimeras to in-phase synchronized oscillations or coherent traveling waves, we have introduced a set of global order parameters. These can be also used as a measure for the width of the incoherent domains of amplitude chimeras. We have demonstrated that amplitude chimera lifetimes depend sensitively on the particular type of initial condition. We have introduced a class of specially prepared random initial conditions that produce long lasting amplitude chimeras. We have shown that initial configurations that fulfill symmetry condition that is also found in oscillation death patterns, result in the longest living amplitude chimeras. The width of the incoherent domains of amplitude chimera pattern relative to the overall system size is independent of the used initial condition and the total system size. It increases linearly with the coupling strength as well as with the coupling range. The amplitude chimeras appear as long lasting transients in systems of all sizes, in contrast to classical phase chimeras whose transient times increase exponentially with the network size [74, 129]. The amplitude chimera lifetime decreases with the network size, but approaches a finite value in the thermodynamic limit.

Our numerical findings give an idea on the structure of the phase space for amplitude chimeras. More specifically, we show that amplitude chimera states can be represented by saddle states in the phase space of the network. This fact is confirmed by the stability analysis of amplitude chimera states performed using Floquet theory. In more detail, we demonstrate that amplitude chimeras are saddles with one or two repelling directions.

Since we have studied a generic model of coupled self-sustained oscillators, these results can be applied to a wide class of systems ranging from laser models, communication networks, and power grids to biological networks. We believe that they are of particular importance for synthetic biology. For instance, while building synthetic circuits, one can initiate the transitions between different regimes of operation, e.g., from self-sustained oscillations (amplitude chimera) to oscillation suppression (chimera death) by adjusting the architecture of the network (tuning coupling range and coupling strength) without changing the local dynamics (tuning parameters of the individual nodes). Moreover, these findings may provide a recipe for engineers when constructing a network where oscillation suppression in a specific configuration (chimera death with a certain number of clusters in the coherent domain) is desired.

2.4 The Role of Time Delay

In the present section, we study the role of time delay for coherence-incoherence patterns occurring in a deterministic network of nonlocally coupled Stuart-Landau oscillators.[2] In particular, we aim to find out under what conditions the lifetime of amplitude chimeras be enhanced or reduced. It has been shown theoretically for phase oscillators in nonlocally coupled ring networks that chimera states are long-living transients, and their lifetime increases exponentially with the system size as the number of oscillators increases [129, 167, 195], similar to transient chaos in spatially extended systems [281–283]. Controlling the lifetime and basin of attraction of chimera states is therefore crucial in practical application, and some progress in this direction has been made only recently [102, 144, 146]. It is known that generally time-delayed feedback or coupling is a powerful method to control the stability in nonlinear systems [284] and networks [29]. Here, we investigate the influence of time delay on chimera states in delay-coupled oscillator networks, specifically, the dynamical regimes and the lifetime of amplitude chimeras in a ring network of Stuart-Landau oscillators with nonlocal time-delayed coupling. We already know that amplitude chimeras represent saddle states in the underlying phase space of the network [103]. In contrast to classical chimeras [129] for amplitude chimeras, the transient time decreases and saturates for large system size [102].

This section is organized as follows. First, we introduce the model equations and describe the characteristics of the occurring amplitude chimera states and the related chimera death steady states (Sect. 2.4.1). Then we generalize the global Kuramoto order parameter as a measure to characterize the transient time of chimera states

[2]Portions of the following text have been re-printed from [104] with the permission of American Physical Society.

towards different types of completely coherent states, in particular, phase-lag synchronization (traveling wave), and other coherent states with more complicated waveform structures (Sect. 2.4.2). In Sect. 2.4.3, a detailed analysis of the dynamical regimes and the transient time of amplitude chimeras is provided in the case of (i) no delay in the coupling, (ii) constant delay, (iii) deterministic time-varying delay, and (iv) distributed delay in the coupling. The conclusions are given in Sect. 2.4.4.

2.4.1 Time-Delay Model

We investigate nonlocally coupled ring networks of N oscillators with different types of delay in the coupling introduced via the delay operator $\mathscr{D}(\cdot)$. As already discussed in Sect. 2.2, the local dynamics of the nodes is described by the Stuart-Landau oscillator, i.e., the normal form of a supercritical Hopf bifurcation. The dynamical equations are given by

$$\begin{aligned}\dot{z}_j =& \left[\lambda + i\omega - |z_j|^2\right] z_j \\ &+ \frac{\sigma}{2P} \sum_{k=j-P}^{j+P} \left[\mathrm{Re}(\mathscr{D}[z_k(t)]) - \mathrm{Re}(z_j(t))\right],\end{aligned} \tag{2.18}$$

where $z_j = x_j + iy_j \in \mathbb{C}, \lambda, \omega \in \mathbb{R}$. As discussed previously (Sect. 2.2), the variable z_j describes the j-th node ($j = 1, 2 \ldots N$, all indices mod N), σ is the coupling strength, and P is the number of coupled neighbors in each direction on a ring. In polar coordinates, $z_j = r_j e^{i\theta_j}$, where $r_j = |z_j|$ and $\theta_j = \arg(z_j)$, the dynamics of the uncoupled system is given by $\dot{r}_j = (\lambda - r_j^2)r_j$ and $\dot{\theta}_j = \omega$. For $\lambda > 0$, the uncoupled delay-free system exhibits self-sustained limit cycle oscillations with radius $r_0 = \sqrt{\lambda}$ and frequency ω. Since in our simulations we fix $\omega = 2$, the period of oscillations is $T = \pi$, which we will also refer to as intrinsic period. Depending on the type of the delay in the coupling, the delay operator $\mathscr{D}$ could act upon the state function $z(t)$ in different forms, e.g., $\mathscr{D}_1[z(t)] = z(t - \tau)$ in the case of *constant* delay, $\mathscr{D}_2[z(t)] = z(t - \tau(t))$ for *time-varying* delay, with time dependence given by the function $\tau(t)$, or $\mathscr{D}_3[z(t)] = \int_0^\infty G(t')z(t - t')\mathrm{d}t'$ for *distributed* delay, where $G(t)$ is a kernel characterizing the delay distribution. As discussed in Sect. 2.2, we consider here the coupling only in the real parts, since this breaks the rotational S^1 symmetry of the system which is a necessary condition for the existence of nontrivial steady states $z_j \neq 0$ and thus for oscillation death [43, 46, 285]. Therefore, the symmetry-breaking form of the interaction between the oscillators induces a set of inhomogeneous fixed points in addition to the homogeneous fixed point at the origin $r_j = 0$.

We recall that in the instantaneous coupling case $\mathscr{D}_0[z(t)] = z(t)$, the system (2.18) demonstrates chimera behavior with respect to amplitude dynamics, i.e., amplitude chimeras [75, 76, 101–104], where one part of the network is oscillating with spatially coherent amplitude, while the other displays oscillations with spatially incoherent amplitudes and centers of mass (Fig. 2.23a–d). In the present section, for all parameter values, we choose the same initial condition (Fig. 2.23a, b) which is an amplitude chimera for $P = 1, \sigma = 14$, and is constructed by simulating the system starting from a fully antisymmetric state: the first half of the nodes $j \in (1, ..., N/2)$ is set to $(x_j, y_j) = (-0.9, -0.9)$, whereas the second half $j \in (N/2 + 1, ..., N)$ is set to$(x_j, y_j) = (0.9, 0.9)$.

We recall the main characteristic features of amplitude chimera pattern. First, within its incoherent domain, the center of mass of each oscillator is shifted away from the origin, while the nodes belonging to the coherent population oscillate around the origin and with larger amplitude. This can be clearly seen from the phase portrait of amplitude chimera state (Fig. 2.23d). It is important to note that in the amplitude chimera regime, the mean phase velocities are the same for all elements of the network. This is due to the fact that the phases are correlated not only within the coherent domain, but also within the incoherent part. Therefore, amplitude chimera is characterized by chimera behavior only with respect to amplitude dynamics, in contrast to amplitude-mediated chimeras, where both amplitude and phase are in a chimera state [99, 110]. Amplitude chimeras (Fig. 2.23c) represent long-living transients and can be observed for hundreds or even thousands of oscillation periods before a synchronized regime is reached.

Further, we recall the definition and the main properties of another chimera pattern detected in system (2.18). We already know that in the deterministic regime, it shows chimera behavior of steady states that is called *chimera death* [75, 76, 101] (Fig. 2.23e, f). In the chimera death regime, the oscillations die out in a peculiar way. In the network of identical elements, two groups are formed: (i) spatially coherent oscillation death, where the neighboring elements are correlated since they populate the same branch of the inhomogeneous steady state, and (ii) spatially incoherent oscillation death, where the sequence of populated branches of the inhomogeneous steady state of neighboring elements is completely random (Fig. 2.23e). The term "coherent/incoherent" here refers to the coherence/incoherence in space. It turns out that the variation of the coupling range for fixed coupling strength in system (2.18) results in the formation of clusters within the coherent part of chimera death pattern: with decreasing coupling range, the number of clusters in the coherent spatial domain is increased (a 3-cluster chimera death is illustrated in Fig. 2.23e). This feature is common for chimera patterns. Next, we provide a possible explanation. For the phenomenon of oscillation death, it has been demonstrated that the oscillations are suppressed due to a Turing bifurcation [43, 285]. A similar mechanism might apply to chimera death state. Further, it is known that reaction-diffusion spatial patterns arising from an instability in a uniform medium (Turing patterns) are characterized by the square root dependence of the pattern wavelength on the diffusion coefficient in the medium [286]. For the ring networks considered here, the coupling range can be treated as a measure analogous to the diffusion in the medium. It has been found that it rescales the diffusion coefficient in the continuum limit ([287], see Sect. 2.2.1, Equation (5)). Consequently, it determines the wavelength of the pattern and, therefore, the number of clusters of the chimera pattern. An explanation along the same lines has been previously reported in [73, 78, 79].

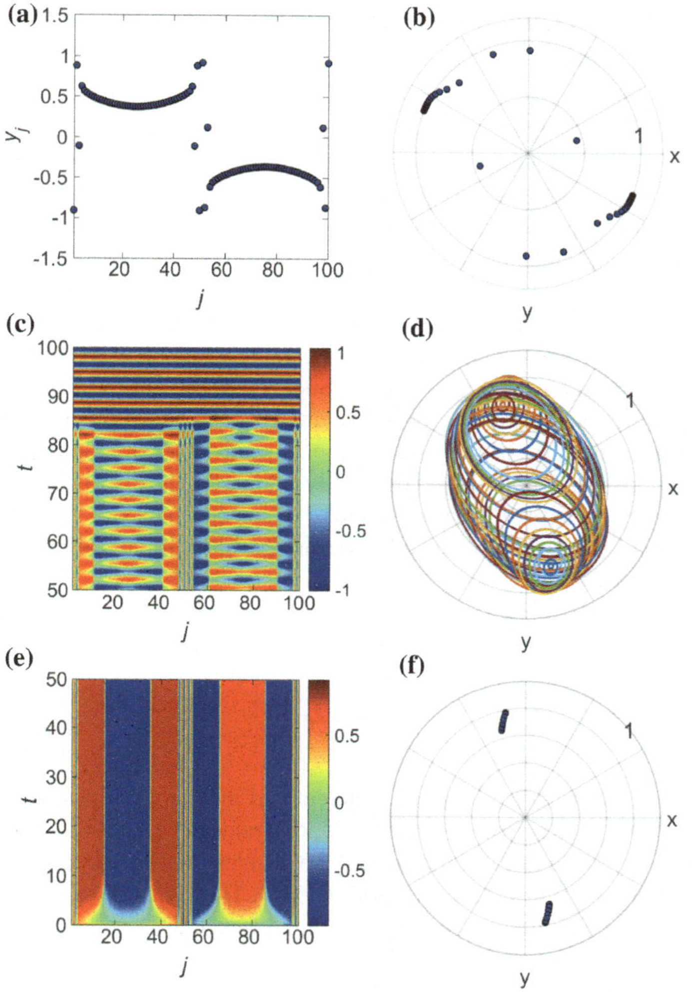

Fig. 2.23 **a, b** Initial condition ($t \leq 0$) used in the simulations: Snapshot of amplitude chimera state $y_j = \mathrm{Im}(z_j)$ (panel a) and phase portrait in the complex z-plane (panel b) for $P = 1, \sigma = 14, \tau = 0$. **c, d** Space-time plot of $y_j = \mathrm{Im}(z_j)$ showing the collapse of initial amplitude chimera towards in-phase synchronized regime at $t \approx 84$ (panel c) and phase portrait of all oscillators for the time window $t \in [60, 65]$ (panel d) for $P = 16$, $\sigma = 7$, $\tau = 0$. **e, f** Space-time plot of $y_j = \mathrm{Im}(z_j)$ showing the collapse of initial amplitude chimera towards 3-cluster chimera death state (panel e) and phase portrait of all oscillators for the time window $t \in [60, 90]$ (panel f) for $P = 31, \sigma = 9$, $\tau = 0$. Other parameters: $\lambda = 1, \omega = 2, N = 100$. In our simulations, and in the rest of the section, we use MATLAB `ddesd` numerical routine for integrating delay-differential equations with general (time-dependent and state-dependent) delays [104]

2.4.2 Characterizing the Transition from Incoherence to Coherence

As discussed in Sect. 2.3.3, the state of the network at a given time can be quantitatively described by introducing a global order parameter [129], but also other measures have been suggested [168]. This is relevant for defining the transient times of various states. In particular, we investigate the transition from partially incoherent states, such as amplitude, phase, or amplitude-mediated chimeras towards different coherent states, such as in-phase synchronization, phase-lag synchronization (traveling waves), and more complicated waveforms with different degree of coherence. In Sect. 2.3.3, we have used the modified order parameter based on the centers of mass to detect the transition from amplitude chimeras to coherent oscillating state [see (2.8)]. In the present section, we use phase difference to adjust the order parameter.

We define a snapshot of the network dynamics as a set of values $\{z_1(t), z_2(t), \dots z_N(t)\}$ of the variables at a fixed time t. By examining the distribution of the state vectors $z_j(t) = |z_j| \exp(i\theta_j)$ in the complex z plane, one can extract the information on the dynamics at a given instant and, in particular, trace the appearance of spatially coherent structures.

First, we use the global Kuramoto order parameter

$$R^0 = \left| \frac{1}{N} \sum_{j=1}^{N} e^{i\theta_j} \right|, \tag{2.19}$$

as a measure to distinguish between coherent and incoherent states of the system, which we refer to as a *zeroth-order* mean-field parameter in what follows. For complete in-phase synchronization, the phases θ_j of all elements are constant, resulting in $R^0 = 1$. For complete desynchronization, $R^0 = 0$. There are, however, other coherent states such as antiphase clustering (standing waves), where two groups of nodes oscillating in antiphase are separated by elements which are almost stationary. For the detection of such coherent states or different types of traveling waves, R_0 cannot be used (see Fig. 2.24).

For coherent traveling waves (constant phase-lag synchronization or splay state), the phases θ_j are *not* constant, but the *phase differences* between adjacent oscillators are constant. For the characterization of such states, we introduce here the *first-order* mean-field parameter

$$R^1 = \left| \frac{1}{N} \sum_{j=1}^{N} e^{i\Delta\theta_j} \right|, \tag{2.20}$$

as an extension of the Kuramoto order parameter R^0, where instead of the phases θ_j, the phase differences between adjacent oscillators are used:

$$\Delta\theta_j = \theta_{j+1} - \theta_j. \tag{2.21}$$

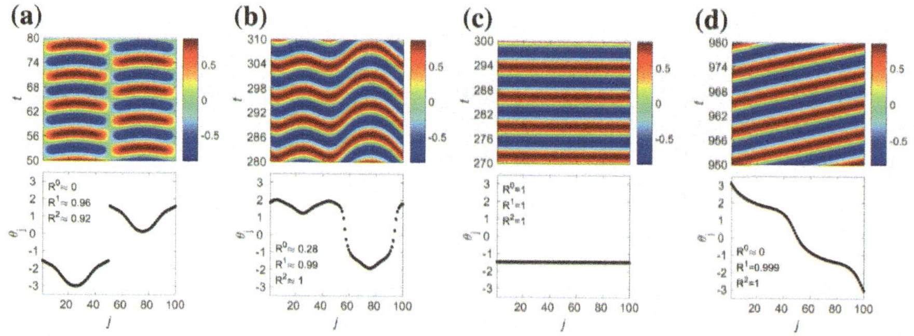

Fig. 2.24 Coherent structures in a ring network of $N = 100$ Stuart-Landau oscillators for $\sigma = 3$ and constant delay $\tau = \pi/2$. Upper panels: space-time plots of the variable $y_j = \mathrm{Im}(z_j)$. Lower panels: snapshots of the phase θ_j at time $t = t_s$. **a** antiphase clustering, $P = 7$, $t_s = 66$. **b** distorted sine-like wave, $P = 2$, $t_s = 296.5$. **c** zero-lag synchronization, $P = 23$, $t_s = 275$. **d** imperfect phase-lag synchronization (traveling wave), $P = 7$, $t_s = 959.8$. Corresponding mean-field parameters at $t = t_s$: **a** $R^0 \approx 0$, $R^1 = 0.96$, $R^2 = 0.92$; **b** $R^0 = 0.28$, $R^1 = 0.99$, $R^2 = 1.00$; **c** $R^0 = R^1 = R^2 = 1.00$; **d** $R^0 \approx 0$, $R^1 = 0.999$, $R^2 = 1.00$. Other parameters: $\lambda = 1$, $\omega = 2$ [104]

Since for the traveling waves $\Delta\theta_j \simeq const$, this regime is characterized by $R^1 \approx 1$, while R^0 in this case cannot be used as an indicator of coherence ($R^0 \approx 0$) (see Fig. 2.24d).

By examining the phase snapshots in the $(\theta_j(t), j)$ plane at a given time t, we observe that an in-phase synchronized regime has a constant θ profile, i.e., a horizontal line (see Fig. 2.24c). On the other hand, in the ideal case of no fluctuations of the phase differences, the phase-lag synchronization (traveling wave) is characterized by an inclined line in the (θ_j, j) plane, such that the *first derivative* $d\theta_j/dj$ is constant (see Fig. 2.24d for an example of an imperfect traveling wave). Strictly speaking, this derivative is taken in the continuum limit $N \to \infty$. In our finite difference case, we have

$$\frac{\Delta\theta_j}{\Delta j} = \Delta\theta_j \tag{2.22}$$

since $\Delta j = (j + 1) - j = 1$, which motivates the definition of R^1 in (2.20).

For the description of more complicated waveforms (e.g., sine-shaped waves and other distorted waveforms) occurring after an amplitude chimera collapse, further generalization of these coherence measures to higher orders is possible (see Fig. 2.24b). The distorted waves also represent coherent states. However, their occurrence is difficult to detect by the previously introduced order parameter. Therefore, we introduce the *second-order* mean-field parameter:

$$R^2 = \left| \frac{1}{N} \sum_{j=1}^{N} e^{i\Delta^2\theta_j} \right|, \tag{2.23}$$

where now the difference of the phase differences is used,

$$\Delta^2\theta_j = \Delta\theta_{j+1} - \Delta\theta_j, \tag{2.24}$$

which, according to (2.21) can be written as

$$\Delta^2\theta_j = \theta_{j+2} - \theta_{j+1} - (\theta_{j+1} - \theta_j) = \theta_{j+2} - 2\theta_{j+1} + \theta_j. \tag{2.25}$$

The *second-order* mean-field parameter $R^2 = 1$ characterizes second-order wave-like coherent structures with constant curvature.

For higher order coherent structures, one may define the *n-th order* mean-field parameter:

$$R^n = \left| \frac{1}{N} \sum_{j=1}^{N} e^{i\Delta^n\theta_j} \right|, \tag{2.26}$$

with the n-th order forward difference:

$$\Delta^n\theta_j = \sum_{k=0}^{n} (-1)^k \binom{n}{k} \theta_{j+n-k}. \tag{2.27}$$

One can use the mean-field parameters R^n defined above for the description of various coherent regimes of the network independently of its topology and the fact whether the nodes are identical or not [104]. Representing a natural extension of the global Kuramoto order parameter, they are complementary measures which allow to discriminate between different transient patterns and to define their lifetime. Their application is not confined to amplitude dynamics, they can be used to describe transient phenomena in phase oscillator networks.

Next, we illustrate how the mean-field parameters can be used to detect the transition from an amplitude chimera to coherent state. In Fig. 2.25, we show the space-time plots of $\mathrm{Im}(z_j)$, $|z_j|$, $\arg(z_j)$, and the time evolution of R^0, R^1, R^2, and R^3 for the Stuart-Landau ring network of $N = 100$ elements and constant time-delayed coupling. As initial conditions, we take amplitude chimera pattern as in Fig. 2.23a, b. We observe a transient amplitude chimera state which transforms into a coherent traveling wave at $t \approx 1280$ (Fig. 2.25) that develops into a constant phase-lag synchronization pattern at $t > 1500$. As discussed above, the zeroth-order parameter R^0 cannot be used to detect the collapse of the chimera state, since it shows irregular oscillatory behavior in time. The higher order mean-field parameters $R^1, R^2, \ldots,$ on the contrary, are highly sensitive to transitions from partial incoherence to coherent wave-like patterns. Therefore, they serve as an appropriate measure for the indication of the chimera collapse.

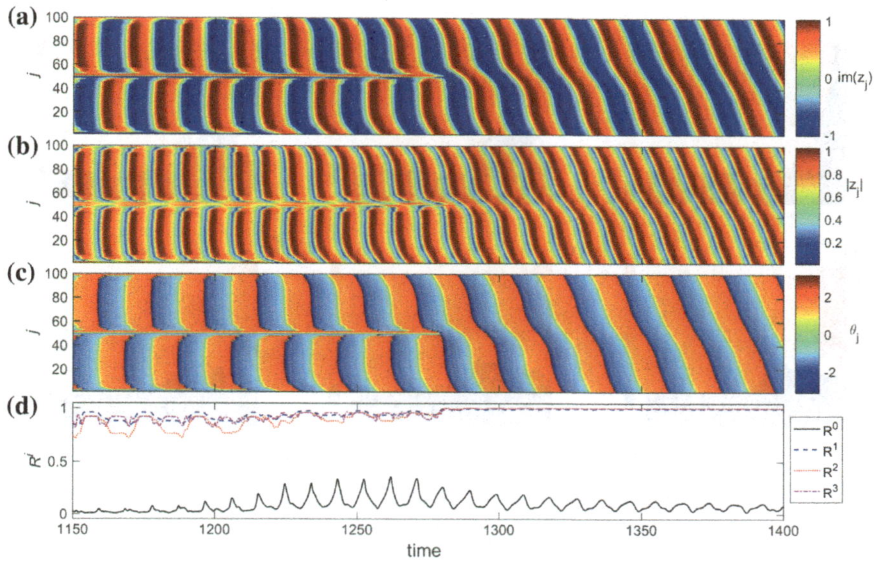

Fig. 2.25 Space-time plots showing gradual collapse of amplitude chimera towards a phase-lag synchronized regime: **a** $y_j = \text{Im}(z_j)$; **b** $|z_j|$; **c** $\theta_j = \arg(z_j)$; **d** Time series of mean-field parameters R^0 (solid, black), R^1 (dashed, blue), R^2 (dotted, red), R^3 (dash-dotted, magenta). Parameters: $P = 5, \sigma = 8, \tau = \pi, \lambda = 1, \omega = 2, N = 100$ [104]

2.4.3 *The Impact of Various Time Delay Types*

Further, we investigate the lifetime of amplitude chimeras and the influence of time-delayed coupling on the system dynamics. To define the transient time, we apply the first-order mean-field parameter R^1 and use the criterion $R^1 > 0.98$ to identify the transition to a coherent pattern. For different local dynamics and oscillator network, this criterion should be appropriately adjusted by preparatory inspection of a few samples of dynamic states.

Instantaneous coupling ($\tau = 0$)

As a reference case, we use the map of regimes in (2.18) for instantaneous coupling ($\tau = 0$) in the plane of coupling range P and coupling strength σ in Fig. 2.26a. The two dominating regions are coherent states (SYNC) represented by in-phase and phase-lag synchronized oscillations (traveling waves), and chimera death states (CD) with different number of clusters. In our numerical simulations, we use a ring of $N = 100$ Stuart-Landau elements with $\lambda = 1$, $\omega = 2$, and integrate the network until $t = 5000$ time units. As initial condition, we take a snapshot of an amplitude chimera pattern for $P = 1$ (Fig. 2.23a, b).

In Fig. 2.27, the lifetimes of amplitude chimeras are indicated by the color code. Panels (a)–(f) correspond to the respective delay times of Fig. 2.26. The black region shows chimera death regime. From panel (a) ($\tau = 0$), it can be seen that the lifetime

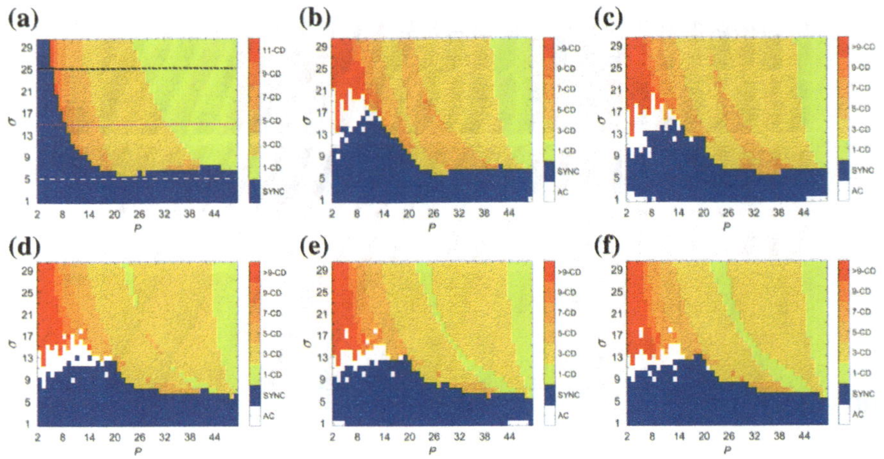

Fig. 2.26 Map of dynamic regimes in a ring network of $N = 100$ delay-coupled Stuart-Landau oscillators in the plane of coupling range P and coupling strength σ for constant values of the time delay: **a** $\tau = 0$; **b** $\tau = \pi/4$; **c** $\tau = \pi/2$; **d** $\tau = \pi$; **e** $\tau = 3\pi/2$; **f** $\tau = 2\pi$. Color code: 1-CD: 1-cluster chimera death; 3-CD: 3-cluster chimera death; n-CD: n-cluster chimera death; SYNC: coherent states (synchronized oscillations, traveling waves, etc.); AC: amplitude chimera and related incoherent states. Other parameters: $\lambda = 1$, $\omega = 2$. Simulation time $t = 5000$ [104]

of amplitude chimera state is significantly larger for low values of P and strong coupling strength σ. The lifetime of amplitude chimera regime for the instantaneous coupling does not exceed 800 time units ($\approx 255T$). Therefore, the amplitude chimera state is not visible in Fig. 2.26a after a simulation time of $t = 5000$.

Consequently, without time delay in the coupling, the initial amplitude chimera is characterized by a relatively short lifetime (at most a few hundred oscillation periods) and transforms into one of two different possible asymptotic states: asymptotic coherent states, in particular, in-phase synchronized oscillations (Fig. 2.23c, d), or chimera death states with several clusters (Fig. 2.23e, f).

Constant time-delay coupling

Time-delayed coupling influences the dynamical states in the network and the lifetime of amplitude chimera pattens. We analyze the regimes of the system (2.18) for increasing time-delay values: $\tau = \pi/4; \pi/2; \pi; 3\pi/2; 2\pi$ (Fig. 2.26b–f, respectively). The delay times we use are integer or non-integer multiples of the intrinsic period of the system $T = \pi$. The corresponding lifetime diagrams are depicted in Fig. 2.27b–f, respectively, where the color code shows the lifetime of the amplitude chimera (AC). Initial conditions are snapshots of amplitude chimera states. The history function for $t < 0$ is taken to be constant and equal to the initial condition at $t = 0$. The lifetime of amplitude chimera states is significantly enlarged in the presence of time delay: they are still present for $t = 5000$ at certain coupling parameter values σ and P (see white regions in panels (b)–(f) of Figs. 2.26 and 2.27). Moreover, time delay induces various long-living partially incoherent states related to ampli-

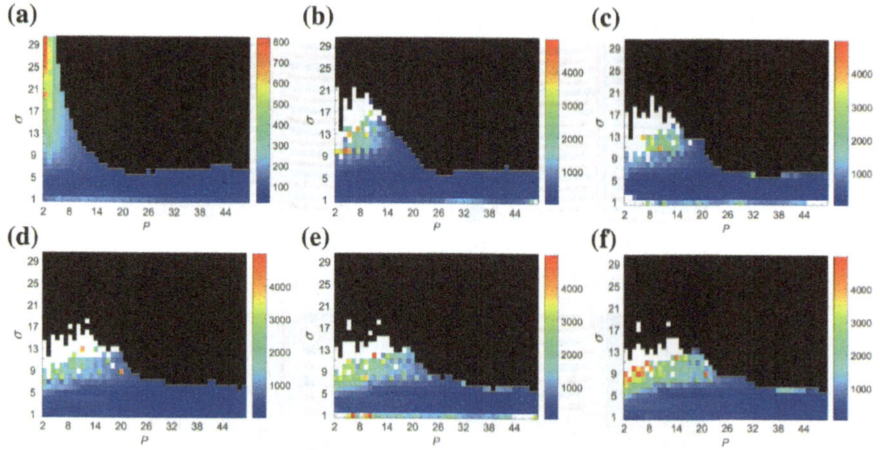

Fig. 2.27 Lifetime of amplitude chimeras in the plane of coupling range P and coupling strength σ, corresponding to Fig. 2.26. Color code indicates the time of transition from partially incoherent states (amplitude chimera) to coherent states; the white region denotes amplitude chimeras and related incoherent states living longer than $t = 5000$, and the black region shows stable steady states (chimera death states). Other parameters as in Fig. 2.26 [104]

tude chimeras (Fig. 2.28). These spatiotemporal patterns include a typical symmetric amplitude chimera with two coherent domains of the same spatial width oscillating in antiphase, separated by spatially incoherent regions consisting typically of few oscillators (Fig. 2.28a). Depending on parameter values, the amplitude chimera pattern can be asymmetric, consisting of two phase-antiphase coherent domains with different spatial widths (Fig. 2.28b). Further, one of the coherent domains can stop oscillating and transform into an oscillation death state (Fig. 2.28c). Finally, the width of coherent oscillating region can be significantly smaller than the oscillation death domain (Fig. 2.28d). Similar partially incoherent patterns have been previously shown for other models [84, 105, 178].

The presence of small time delay in the coupling already results in the appearance of amplitude chimera states that are observed for networks with small number of nearest neighbors P and strong coupling strength σ (Figs. 2.26b and 2.27b). As time delay is increased, amplitude chimeras shift towards the intermediate range of σ. In more detail, amplitude chimera region evolves into an irregular region located around the middle of the σ interval and the first third of the P interval ($P < 20$). Moreover, for increasing τ, the in-phase synchronized regime detected for small number of nearest neighbors P and strong coupling strength σ disappears. Instead, chimera death state with large number of clusters is observed (red region in Fig. 2.26). Amplitude chimera patterns are still present as the delay time is further increased. Their position in the (P,σ) parameter plane does not change significantly and they exhibit an occasional subsidiary appearance for weak coupling strength σ and different number of nearest neighbors within the interval $P \in [2, 20]$.

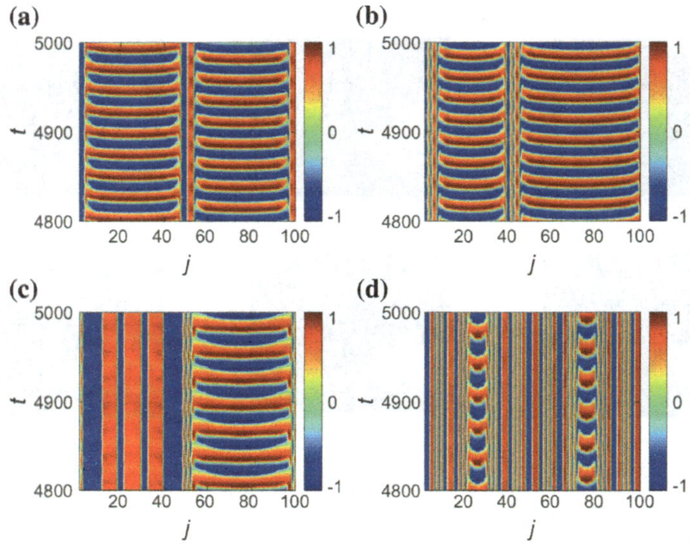

Fig. 2.28 Examples of partially incoherent space-time patterns related to amplitude chimera states for constant time delay $\tau = \pi$ in the coupling and integration time $t = 5000$: **a** Symmetric amplitude chimera, $\sigma = 11$, $P = 5$; **b** Asymmetric amplitude chimera, $\sigma = 11$, $P = 4$; **c** Partial amplitude chimera, $\sigma = 13$, $P = 19$; **d** Partial oscillation death, $\sigma = 15$, $P = 4$. Other parameters: $\lambda = 1$, $\omega = 2$, $N = 100$ [104]

To gain more insight into the regimes observed in the network for varying time delay, we investigate dynamical states in the (P, τ)-plane for different values of the interaction strength σ. In more detail, we select three representative σ values, indicated by the horizontal lines in Fig. 2.26a (dashed, white line for $\sigma = 5$; dotted, red line for $\sigma = 15$; dash-dotted, black line for $\sigma = 25$). Figure 2.29 shows the (P, τ) diagram for $\sigma = 25$. We can clearly see that the region of synchronized states for small P is rapidly replaced by amplitude chimera patterns for increasing time delay. Interestingly, amplitude chimera regime disappears at $\tau \approx 0.4$ and we observe chimera death states with the number of clusters in the coherent domain exceeding 9. At the same time, the region of 1-cluster chimera death (1-CD) is strongly reduced and partially replaced by 3-cluster chimera death pattern.

As the strength of the interaction between network elements becomes smaller ($\sigma = 15$), the synchronization region for $P < 14$ starts to transform into the amplitude chimera regime already for $\tau \approx 0.2$ (Fig. 2.30). For increasing τ, amplitude chimera states are still observed in this region, being further partially replaced by chimera death patterns with large number of clusters in the coherent domain.

Further, we calculate the lifetime of amplitude chimera patterns for $\sigma = 5$ in the (P, τ)-plane to understand how time-delayed coupling influences the synchronized dynamics (Fig. 2.31). The lifetime of amplitude chimeras becomes larger for increasing delay time for each P value. However, this enlargement is more rapid for smaller number of nearest neighbors P.

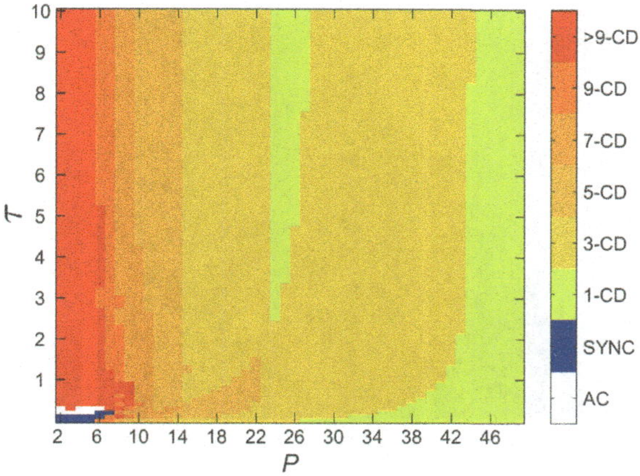

Fig. 2.29 Map of dynamic regimes in the plane of coupling range P and time delay τ for coupling strength $\sigma = 25$. Color code: 1-CD: 1-cluster chimera death; 3-CD: 3-cluster chimera death; n-CD: n-cluster chimera death. SYNC: coherent states (synchronized oscillations, traveling waves, etc.). AC: amplitude chimeras and related partially incoherent states. Other parameters: $\lambda = 1$, $\omega = 2$, $N = 100$ [104]

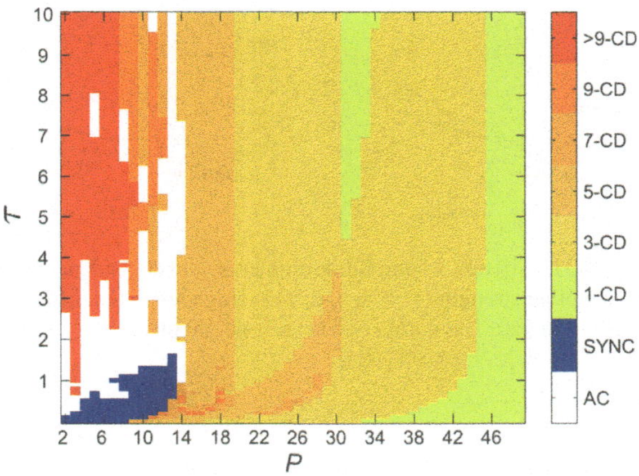

Fig. 2.30 Same as Fig. 2.29 for coupling strength $\sigma = 15$ [104]

Other values of coupling strength σ support the trend of the lifetime prolongation for the amplitude chimera patterns by time delay. Next, we calculate the lifetime of amplitude chimera states in dependence on time delay for different values of coupling strength $\sigma = 4, 5, 6, 7$ (see panels (a)–(c) in Fig. 2.32). We compare the results obtained for three values of the coupling range: $P = 2$, $P = 3$, $P = 5$ (Fig. 2.32a, b, c, respectively). For the small number of nearest neighbors P in the network, the lifetime of amplitude chimera states grows faster (Fig. 2.32a). Therefore, an appropriate choice of the time delay allows to achieve the desired value of the amplitude chimera lifetime.

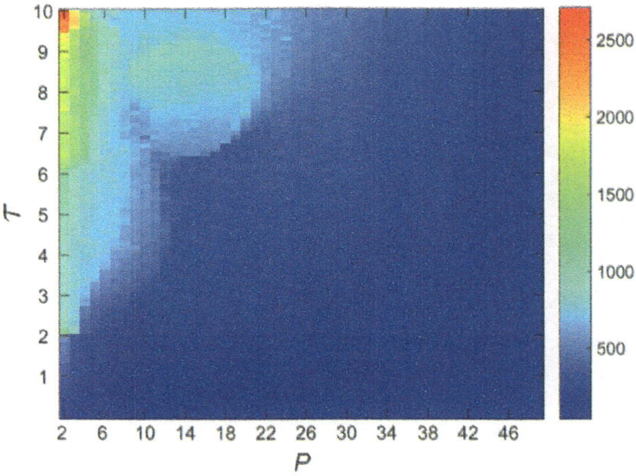

Fig. 2.31 Lifetime of amplitude chimeras in the plane of coupling range P and time delay τ for coupling strength $\sigma = 5$. Color code indicates the time of transition from partially incoherent states (amplitude chimera) to coherent states (in-phase synchronization). Other parameters: $\lambda = 1, \omega = 2$, $N = 100$ [104]

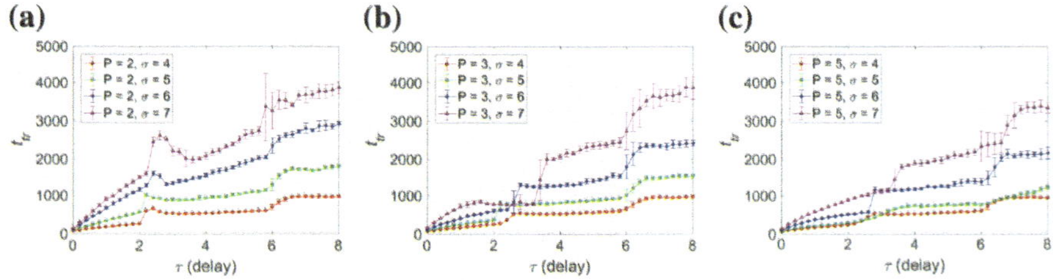

Fig. 2.32 Lifetime of amplitude chimeras t_{tr} in dependence on time delay τ for four selected values of coupling strength ($\sigma = 4, 5, 6, 7$) averaged over ten different initial conditions that favor amplitude chimeras. The number of nearest neighbors: **a** $P = 2$; **b** $P = 3$; **c** $P = 5$. Other parameters: $\lambda = 1, \omega = 2, N = 100$ [104]

Interestingly, in the parameter region where amplitude chimera-related patterns are observed, we detect a peculiar delay-induced state that we call a "breathing" amplitude chimera. It appears at around $t \approx 4500$ and exists much longer than the simulation time displayed in Fig. 2.33. We examine the space-time plot of $\mathrm{Im}(z_j)$, $|z_j|$, $\arg(z_j)$, and the temporal evolution of the global order parameters $R^0, \ldots R^3$ for the "breathing" amplitude chimera with two coherent and two incoherent parts. The size (spatial width) of the two coherent domains of the amplitude chimera changes in time in a periodic way (Fig. 2.33a–c). Furthermore, these oscillations occur in antiphase for the two coherent domains, i.e., when one coherent cluster attains a maximum width, the other has a minimum, and vice versa. The periodicity is inherited also in the time evolution of the order parameters (Fig. 2.33d). Specifically, we have integrated the system until $t = 20000$ time units ($\approx 6370T$) and observed a sustained

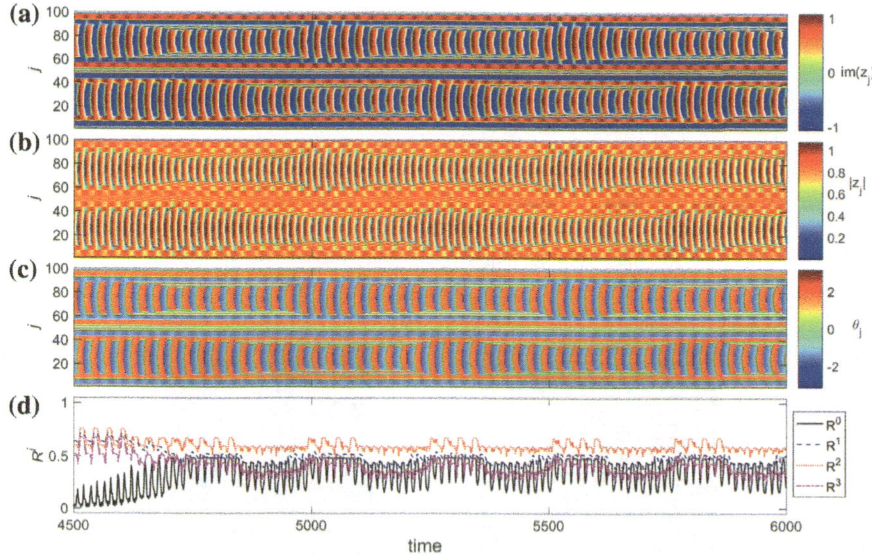

Fig. 2.33 Space-time plots for a long-living "breathing" amplitude chimera. **a** $y_j = \mathrm{im}(z_j)$; **b** $|z_j|$; **c** $\theta_j = \arg(z_j)$; **d** Time series of mean-field parameters R^0 (solid, black), R^1 (dashed, blue), R^2 (dotted, red), R^3 (dash-dotted, magenta). Parameters: $P = 12$, $\sigma = 13$, $\tau = \pi$, $\lambda = 1$, $\omega = 2$, $N = 100$ [104]

"breathing" amplitude chimera, with a breathing period of each coherent domain approximately equal to 500 time units ($\approx 159T$).

The results presented above are obtained for specially prepared initial conditions leading to the amplitude chimera pattern with two equally sized spatially coherent domains (symmetric amplitude chimera state). It is important to note that increasing the lifetime of amplitude chimeras by implementing time delay in the coupling is not restricted to these types of initial conditions. The essential enhancement of the chimera lifetime by means of delay provides a powerful tool for designing a desired multicluster as well as asymmetric amplitude chimera states by appropriately choosing initial conditions and making these pattern long-living by adding delay into the coupling. To illustrate this, we take spatially asymmetric initial conditions (Fig. 2.34a, b). Without coupling delay, the asymmetric multicluster amplitude chimera pattern that evolves from these initial conditions disappears very fast, typically living only for a few time units (Fig. 2.34c). Depending on coupling parameters P and σ, the final state is either in-phase synchronized regime (Fig. 2.34c), or multicluster oscillation death (Fig. 2.34d). The transition from amplitude chimera to zero-lag synchronization for undelayed coupling ($\tau = 0$), $P = 2$ and $\sigma = 14$ occurs gradually, starting from $t \approx 20$. The amplitude chimera breaks down, forming a coherent pattern that subsequently evolves into a complete zero-lag synchronization around $t \approx 50$ (Fig. 2.35).

To understand how time delay influences the lifetime of multicluster amplitude chimeras, we examine the map of regimes (Fig. 2.36a–c) and the corresponding life-

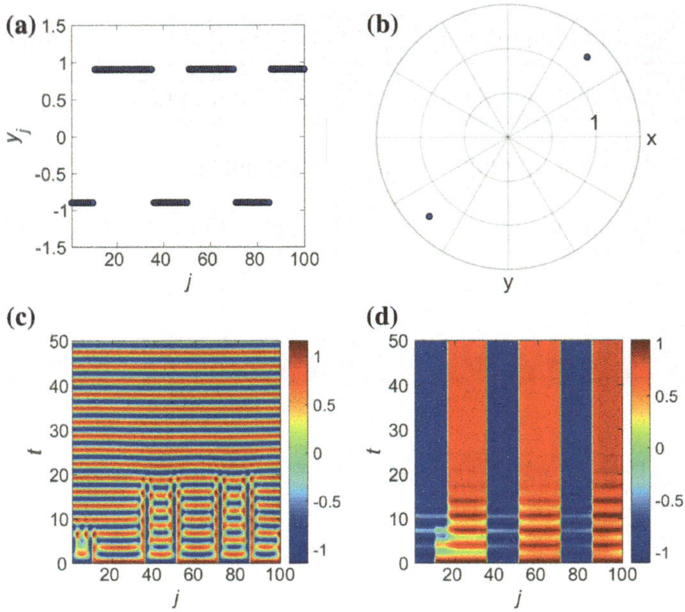

Fig. 2.34 Asymmetric initial condition ($t \leq 0$): **a** Snapshot of the variable $y_j = \mathrm{Im}(z_j)$, **b** phase portrait in the complex z-plane. **c** Space-time plot of $y_j = \mathrm{Im}(z_j)$ showing the collapse of initial asymmetric multicluster amplitude chimera towards in-phase synchronized regime at $t \approx 20$ for $P = 2$ and $\sigma = 14$. **d** Collapse into an asymmetric 3-cluster oscillation death state for $P = 20$, $\sigma = 12$. Other parameters: $\tau = 0$, $\lambda = 1$, $\omega = 2$, $N = 100$ [104]

time diagrams (Fig. 2.36d–f) of multicluster amplitude chimera patterns in the (P, σ) parameter plane for different values of delay time: $\tau = 0$ (instantaneous coupling); $\tau = \pi/4$; $\tau = \pi$. The presence of time delay in the coupling leads to the occurrence of stable multicluster amplitude chimeras and related partially incoherent patterns (white region) that live longer than the simulation time. The spatiotemporal dynamics of these partially incoherent patterns becomes much richer than that of the simple amplitude chimera states. Here we provide four examples (Fig. 2.37). The oscillation death region changes non-monotonically with increasing delay time: first, it expands, but then significantly shrinks in favor of the synchronized regime.

Time-varying delayed coupling

Furthermore, we investigate the influence of time-varying delay on the network dynamics. In particular, we choose a periodic deterministic modulation of the time delay around a nominal (average) delay value τ_0 in the form of a sawtooth-wave modulation [44]:

$$\tau(t) = \tau_0 + \varepsilon \left[2 \left(\frac{\varpi t}{2\pi} \bmod 1 \right) - 1 \right], \tag{2.28}$$

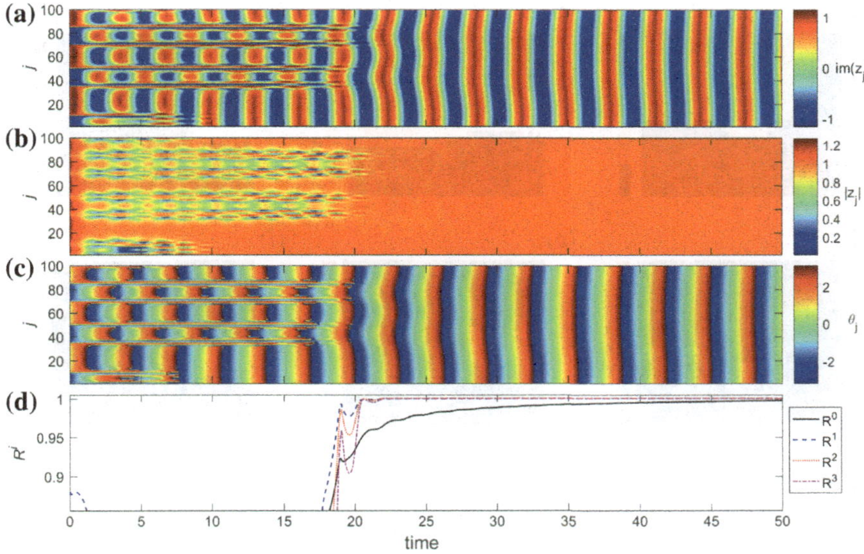

Fig. 2.35 Space-time plots showing gradual collapse of an asymmetric multicluster amplitude chimera towards a zero-lag synchronized state: **a** $y_j = \text{Im}(z_j)$; **b** $|z_j|$; **c** $\theta_j = \arg(z_j)$; **d** Time series of mean-field parameters R^0 (solid, black), R^1 (dashed, blue), R^2 (dotted, red), R^3 (dash-dotted, magenta). Parameters: $P = 2$, $\sigma = 14$, $\tau = 0$, $\lambda = 1$, $\omega = 2$, $N = 100$ [104]

and also in the form of a square-wave modulation:

$$\tau(t) = \tau_0 + \varepsilon \,\text{sgn}[\sin(\varpi t)], \tag{2.29}$$

where ε and ϖ are the amplitude and the angular frequency of the corresponding delay modulations, respectively. In particular, for a sawtooth-wave modulation of the delay given by (2.28), we analyze the dynamical states of the network (Fig. 2.38a–c) and the corresponding lifetimes of partially incoherent states (Fig. 2.38d–f). We compare the results for different modulation amplitudes: $\varepsilon = \pi/2$ (Fig. 2.38a, d); $\varepsilon = 3\pi/4$ (Fig. 2.38b, e); $\varepsilon = \pi$ (Fig. 2.38c, f). The nominal delay is set to $\tau_0 = \pi$, and the angular frequency of the modulation is fixed to $\varpi = 10$. The initial conditions and the history function are chosen the same as in the constant delay case (Fig. 2.23a, b). The related results for a square-wave modulation of the delay (2.29) are shown in Fig. 2.39. Note that the sawtooth-wave modulation of the delay does not have a significant influence on the various regimes if compared to the constant delay case (Fig. 2.27). There is, however, an occasional appearance of partially incoherent states at small values of coupling strength σ and different values of the number of nearest neighbors P. They survive the simulation time, but the main region of amplitude chimeras around $\sigma = 13$ and small P is mostly unchanged with increasing modulation amplitude (Fig. 2.38). The square-wave delay modulation is rather interesting, since in this case by increasing the modulation amplitude, the domains correspond-

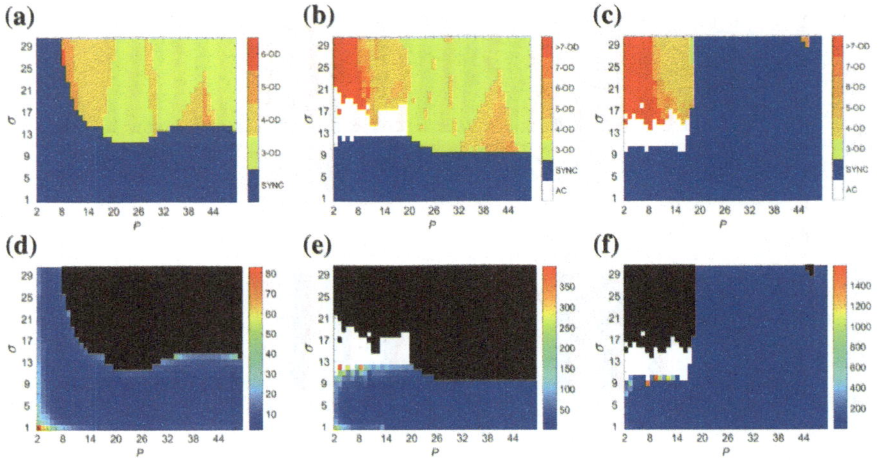

Fig. 2.36 Map of dynamic regimes **a–c** and corresponding chimera lifetimes **d–f** for multicluster amplitude chimeras in the plane of coupling range P and coupling strength σ for constant delay coupling. Time delay: **a**, **d** $\tau = 0$; **b**, **e** $\tau = \pi/4$; **c**, **f** $\tau = \pi$. Color scale in **a–c**: 3-OD: 3-cluster oscillation death; 4-OD: 4-cluster oscillation death; n-OD: n-cluster oscillation death. SYNC: coherent states (in-phase synchronized oscillations, traveling waves, etc.). AC: multicluster amplitude chimera and related partially incoherent states. Color code in lifetime diagrams **d–f** indicates the time of transition from partially incoherent states (multicluster amplitude chimera) to coherent states. The white region denotes amplitude chimeras and related partially incoherent states; the black region denotes stable steady states (oscillation death). Other parameters: $\lambda = 1$, $\omega = 2$, $N = 100$. Simulation time $t = 5000$ [104]

ing to partially incoherent states become drastically reduced, almost disappearing for larger values of modulation amplitude. The impact of the modulation of the coupling delay on the network dynamics in the square-wave case becomes more visible in the parameter plane of the coupling range P and the amplitude of delay modulation ε for constant coupling strength $\sigma = 15$ (Fig. 2.40). One can clearly see that increasing the modulation amplitude ε results in a sequence of appearance and disappearance of the amplitude chimera regions. Such behavior is a characteristic feature of systems under square-wave delay modulation, and it has already been reported, for instance, in variable-delay feedback control with respect to the sequence of stability islands for successful fixed-point control [288].

Distributed-delay coupling

Next, we study distributed-delay coupling. It has been previously reported that a time-varying delay system with a high-frequency modulation of the time delay is effectively equivalent to a distributed-delay system with a related delay distribution in the interval of delay variation. Moreover, this holds both analytically and numerically with respect to the *steady-state solutions* of the dynamical equations of the delayed system [44, 289]. Taking this into account, to check if this still holds for other dynamical regimes of the network, we analyze distributed-delay coupling kernels corresponding to the ones for the time-varying delayed coupling considered above in

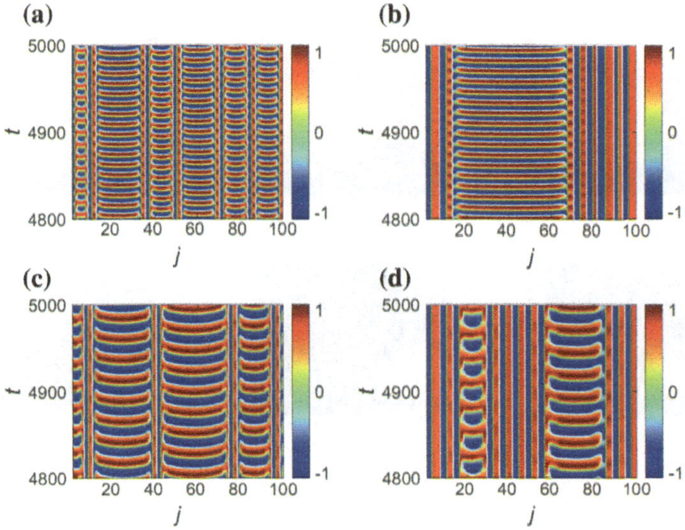

Fig. 2.37 Examples of partially incoherent dynamical patterns related to multicluster amplitude chimeras that survive the simulation time for constant time delay τ in the coupling. Parameters: **a** $\sigma = 13$, $P = 2$, $\tau = \pi/4$; **b** $\sigma = 14$, $P = 5$, $\tau = \pi/4$; **c** $\sigma = 11$, $P = 2$, $\tau = \pi$; **d** $\sigma = 12$, $P = 5$, $\tau = \pi$. Other parameters as in Fig. 2.36 [104]

the high-frequency limit of the delay modulation. Thus, we take a uniform distribution kernel:

$$G(t') = \begin{cases} \frac{1}{2\varepsilon}, & t' \in [\tau_0 - \varepsilon, \tau_0 + \varepsilon] \\ 0, & \text{elsewhere} \end{cases} \tag{2.30}$$

and a two-peak distribution kernel:

$$G(t') = \frac{\delta(t' - \tau_0 + \varepsilon) + \delta(t' - \tau_0 - \varepsilon)}{2}, \tag{2.31}$$

where $\delta(\cdot)$ denotes the Dirac delta function. In this case, τ_0 is the mean time delay of each distribution, and ε is the distribution width. They correspond to the average (nominal) delay value and the modulation amplitude, respectively, in the time-varying delay coupling case. The uniform distribution kernel corresponds to the high-frequency limit of a sawtooth-wave modulation of the coupling delay, and a two-peak distribution kernel represents the high-frequency limit of a square-wave modulation of the coupling delay. In the previous case of time-varying delayed coupling, the delay modulation frequency was chosen as $\varpi = 10 = 5\varpi_0$, where $\varpi_0 = 2\pi/T = 2$ is the intrinsic angular frequency of the uncoupled system. For these parameter values, the time-varying delay systems can be considered in the high-frequency limit, which is

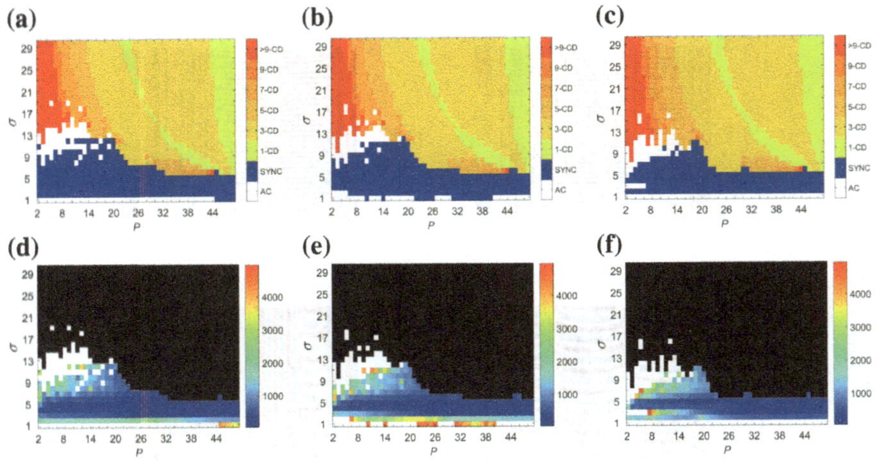

Fig. 2.38 Map of dynamic regimes **a–c** and corresponding chimera lifetimes **d–f** in the plane of coupling range P and coupling strength σ, for time-varying delay coupling with a sawtooth-wave modulation. Nominal time delay $\tau_0 = \pi$, modulation frequency $\varpi = 10$. Modulation amplitude: **a**, **d** $\varepsilon = \pi/2$; **b**, **e** $\varepsilon = 3\pi/4$; **c**, **f** $\varepsilon = \pi$. Color scale in **a–c**: 1-CD: 1-cluster chimera death; 3-CD: 3-cluster chimera death; n-CD: n-cluster chimera death. SYNC: coherent states (in-phase synchronized oscillations, traveling waves, etc.). AC: amplitude chimera and related partially incoherent states. Color code in transient time diagrams **d–f** indicates the time of transition from partially incoherent states (amplitude chimera) to coherent states, the white region denotes amplitude chimeras and related partially incoherent states, and the black region denotes stable steady states (death states). Other parameters: $\lambda = 1$, $\omega = 2$, $N = 100$ [104]

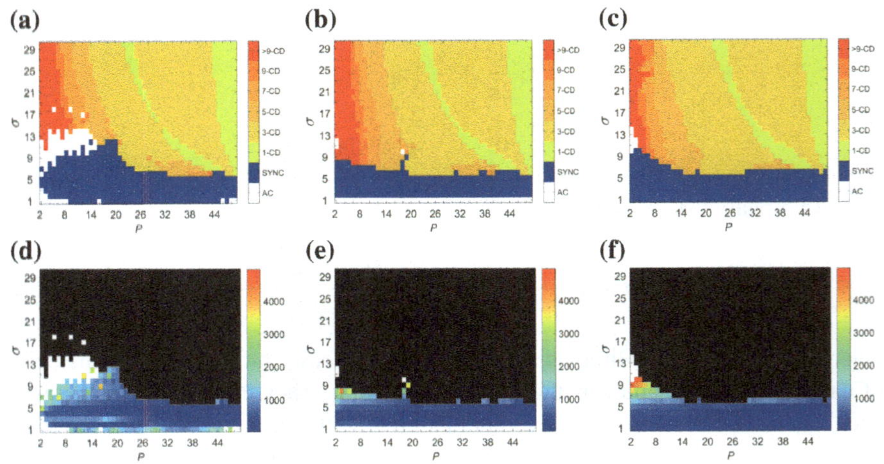

Fig. 2.39 Same as Fig. 2.38 for time-varying delay coupling with a square-wave modulation [104]

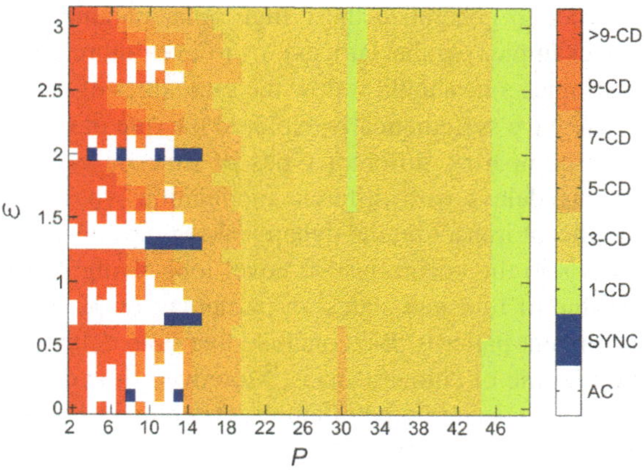

Fig. 2.40 Map of dynamic regimes in the plane of coupling range P and modulation amplitude ε, for time-varying delay coupling with a square-wave modulation, and constant value of the coupling strength $\sigma = 15$. Color scale and other parameters as in Fig. 2.38 [104]

confirmed by our numerical simulations. The obtained diagrams demonstrate that the system dynamics with distributed-delay coupling indeed corresponds to the dynamics with time-varying delay coupling with a high-frequency delay modulation. The simulations provide excellent matching of the maps of dynamic regimes and chimera lifetimes between the systems with distributed-delay coupling and time-varying delay coupling. Since the dynamic regimes include various synchronous and asynchronous solutions, and combinations of both, we arrive at a conclusion that approximating the high-frequency time-varying delay system by a distributed-delay system with a corresponding distribution kernel is quite general, extending well beyond steady-state solutions of the complex network dynamics.

2.4.4 Summary

In this section, we have investigated the impact of time delay in a ring network of Stuart-Landau oscillators coupled nonlocally and through the real part of the complex variable. In more detail, we have analyzed various dynamical regimes of the network, focusing on the transition from amplitude chimera states to phase-lag synchronization (traveling waves) and higher order coherent patterns. Moreover, we have developed a measure which generalizes global Kuramoto mean-field order parameter and can be used for detection of the transition from partially incoherent states (e.g., amplitude chimera pattern) to any type of coherent regime. In the derivation of these higher order mean-field parameters, we have used the notion that the continuity of the phase snapshot indicates the occurrence of spatially coherent states. In this respect, different degrees of coherence are related to different degrees of smoothness of the phase angle snapshot. For a ring network, i.e., in the one-dimensional case we have considered here, different degrees of smoothness of the snapshot curve are represented by higher order spatial derivatives of the phase angle in the continuum limit, which upon discretization leads to the defining equations for higher order mean-field parameters.

In order to generalize these high-order parameters to two-dimensional networks (for example, regular lattices), or to even more general high-dimensional network topologies, one should follow the same lines of reasoning.

We have systematically explored the role of time delay implemented in the coupling comparing different types of time delays: constant, distributed, and time-varying delays with different modulation types. We found that time delay has a significant impact on the dynamical regimes observed in the network. In particular, it leads to the occurrence of novel long-living patterns and allows to enhance the lifetime of transient states, in particular, amplitude chimeras and related partially incoherent patterns. We conclude that time delay is a powerful tool for controlling the lifetime of chimera states. Moreover, time delay provides a possibility of constructing a desired type of amplitude chimera pattern. For instance, one can induce an amplitude chimera state with a certain number of clusters or asymmetric cluster configuration. By appropriately adjusting the modulation of the coupling delay (e.g., square-wave modulation), or equivalently, changing the type and the parameters of the distributed-delay kernel (e.g., two-peak distribution kernel), it is possible to reduce the lifetime of amplitude chimeras. Additionally, we have shown numerically that at high-frequency delay modulation, the system with time-varying delay coupling is equivalent to distributed-delay coupling with related delay-distribution kernels. Consequently, time delay in the coupling allows to control chimera states and their lifetimes in networks of coupled oscillatory units.

2.5 The Role of Noise

An important and challenging question related to the dynamics of chimera patterns is whether they persist under perturbations.[3] The robustness of classical phase chimeras with respect to heterogeneous natural frequencies of the oscillators [171, 172, 179] or heterogeneous connections [115] has been previously investigated. It has been found that the specific way in which links are removed from the network defines how strongly the chimera lifetimes are affected. Also in the FitzHugh-Nagumo model, the robustness of chimeras with respect to heterogeneous frequencies and topologies has been studied [80]. The robustness of chimera patterns with respect to external *noise* has been for the first time considered in [102]. This problem is especially relevant in the light of experimental realizations where noise is inevitable. It is known that even at a relatively low intensity, noise can significantly influence the behavior of a nonlinear dynamical system. Under certain conditions noise can cause the increase of coherence, e.g., in coherence resonance [135, 141, 142], and on the contrary, it can also induce irregular, chaotic dynamics. Stochasticity appears due to the intrinsic fluctuations in a system, or alternatively can be implemented as an external random control force. The influence of noise strongly depends on its characteristics such as

[3]Portions of the following text have been re-printed from [102] with the permission of American Physical Society.

intensity, for example, which can be treated as bifurcation parameters of a stochastic system. Therefore, the investigation of noisy dynamics is on the one hand significant for the understanding of the processes occurring in nature and on the other hand relevant from the point of view of control.

In the present section, we investigate how external noise influences the occurrence of amplitude chimeras and chimera death in networks of nonlocally coupled Stuart-Landau oscillators. In particular, we study how stochasticity affects the lifetime of amplitude chimeras. For this purpose, we use modified order parameters introduced in Sect. 2.3.3 to distinguish between chimera patterns and completely coherent dynamics like in-phase synchronization or traveling waves.

The section is organized as follows. First, we introduce the model (Sect. 2.5.1). Then, we analyze amplitude chimeras and chimera death without noise (Sect. 2.5.2). In Sect. 2.5.3, the influence of noise upon the transient times of chimeras is investigated. Section 2.5.4 discusses maps of regimes summarizing the results, and Sect. 2.5.5 draws the conclusions.

2.5.1 *Stochastic Model*

We consider a network of N Stuart-Landau oscillators under the impact of external Gaussian white noise $\xi_j(t)$ with intensity $D > 0$:

$$\dot{z_j} = f(z_j) + \frac{\sigma}{2P} \sum_{k=j-P}^{j+P} (Rez_k - Rez_j) + \sqrt{2D}\xi_j(t), \tag{2.32}$$

where $j = 1, 2, \ldots, N$ and all indices are modulo N (see Sect. 2.2). The coupling and the noise are only applied to the real parts. In many real-world oscillator systems, only one variable is accessible for coupling. For instance, the diffusive coupling through only one variable naturally arises in biological and biochemical systems. The simplest coupling scheme is hence to choose the real parts of the complex variable z. Here $\xi_j(t) \in \mathbb{R}$ is additive Gaussian white noise [265], i.e., $\langle \xi_j(t) \rangle = 0,\ \forall j$, and $\langle \xi_i(t)\xi_j(t') \rangle = \delta_{ij}\delta(t - t'),\ \forall i, j$, where δ_{ij} denotes the Kronecker-Delta and $\delta(t - t')$ denotes the Delta-distribution. Hence the noise is spatially uncorrelated. In this section, we use fully symmetric initial conditions with $Q = \frac{N}{4}$ and $V = 0.1$ to generate amplitude chimeras (see Sect. 2.3.4).

2.5.2 *Deterministic Chimera Patterns*

As shown in Sect. 2.3 in the absence of noise and time delay, various different states can be found in the network given by (2.2). Which particular state actually arises depends on the specific values of the coupling parameters and the initial conditions.

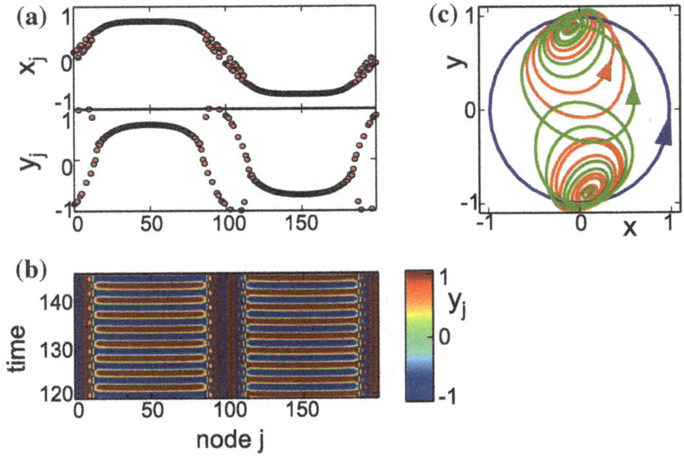

Fig. 2.41 Amplitude chimera state in the deterministic system with $N = 200$ nodes, for coupling range $P/N = 0.04$ and coupling strength $\sigma = 18$: **a** snapshot (top: x_j, bottom: y_j), **b** space-time plot, **c** phase plot in the complex plane: trajectories of 12 nodes of the coherent domains (unit circle, blue) and 12 nodes of the incoherent domains (red and green), the arrows indicate the direction of the motion. Initial condition: see Sect. 2.3.4. Other parameters: $D = 0$, $\lambda = 1$, $\omega = 2$ [102]

Among the possible states, two different types of asymptotically stable states can be found, on the one hand oscillatory states, and on the other hand steady-state patterns which are related to oscillation death. The latter are represented by completely coherent or completely incoherent oscillation death patterns, as well as by chimera death patterns. The asymptotically stable oscillatory states appear in two different spatiotemporal patterns: in-phase synchronized oscillations and coherent traveling waves. Besides these, long lasting oscillatory transients with interesting features occur, i.e., amplitude chimera states. We demonstrate that all these states can also be observed under the influence of noise. Before an asymptotic oscillatory state (a completely in-phase synchronized oscillation or a coherent traveling wave) is approached, *amplitude chimera* states can appear as long transients, potentially lasting for hundreds or even thousands of oscillation periods. We recall the main features of amplitude chimeras. In contrast to classical chimeras, in the amplitude chimera state, all nodes (including the ones within the incoherent domains) oscillate with the same approximate period, $T = \frac{2\pi}{\omega}$, and a spatially correlated phase, but they show spatially incoherent behavior with respect to the *amplitudes* in part of the system. Figure 2.41 shows an exemplary amplitude chimera configuration. The nodes within the two coherent domains (here $13 \leq j \leq 85$ and $113 \leq j \leq 185$) perform synchronized oscillations, all with the same amplitudes. The coherent domains always appear pairwise, such that for every time t, all nodes within one coherent domain have opposite phases than all nodes of the other, antipodal domain. Hence they always fulfill the "antiphase partner" condition $z_j = -z_{j+N/2}$, $j \mod N$, assuming even N. As visible in Fig. 2.41c, the trajectories in the complex plane of all nodes are cycles,

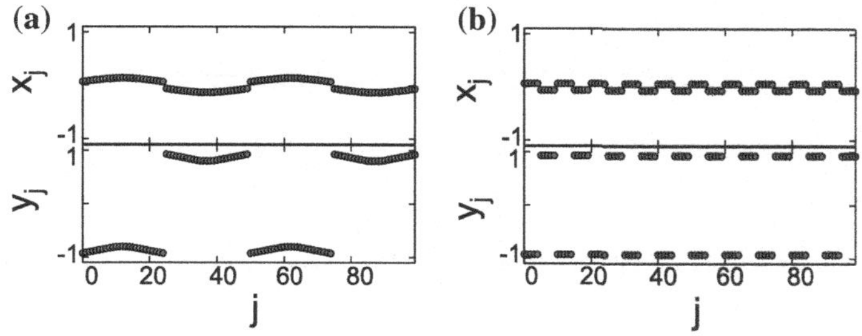

Fig. 2.42 Snapshots of coherent oscillation death states: **a** coupling strength $\sigma = 18$ and coupling range $P/N = 0.14$ (2-cluster), **b** $\sigma = 8$, $P/N = 0.04$ (10-cluster). Initial condition: nodes $0 \le j \le 24$ and $50 \le j \le 74$ are set to $(x_j, y_j) = (0.1, -1)$, all other nodes are set to $(x_j, y_j) = (-0.1, +1)$. Other parameters: $N = 100$, $D = 0$, $\lambda = 1$, $\omega = 2$ [102]

illustrating that all nodes have periodic dynamics in time. This is a fundamental difference between the classical chimera states where a part of the network demonstrates chaotic temporal behavior. The nodes of the coherent domains all oscillate on a perfect circle around the origin. Both coherent domains are represented by one single blue line in Fig. 2.41c, which is at the same time the trajectory of all nodes when the completely in-phase synchronized oscillatory solution is approached. The two antipodal coherent domains are separated by incoherent domains. There, neighboring nodes can be in completely different states at a given time t. Their trajectories are deformed circles, whose centers are shifted from the origin. In the incoherent domain, the sequence of nodes that oscillate around centers in the upper and lower half-plane, respectively, is completely random. The ratio of the domain sizes can vary strongly depending upon the parameters, but does not depend upon the initial conditions. Transient amplitude chimeras with very narrow incoherent domains can be observed, as well as with broad ones (Sect. 2.3.5).

If the coupling strength and coupling range exceeds certain values, the oscillations of the Stuart-Landau nodes can be suppressed due to the stabilization of a new inhomogeneous steady state created by the coupling. Instead of performing oscillations, each node approaches a fixed point close to one of the following two branches: $(x^{*1}, y^{*1}) \approx (-0.1, +0.85)$ or $(x^{*2}, y^{*2}) \approx (+0.1, -0.85)$ (for $\lambda = 1$), and remains there for all times. The oscillation death states exhibit a huge variety of spatial patterns, including multiple coherent and multiple incoherent oscillation death states [75, 76]. Two exemplary configurations of completely coherent oscillation death patterns are shown in Fig. 2.42 (2-cluster and 10-cluster oscillation death), see also [45]. The oscillation death regime is characterized by very high multistability. Among the oscillation death states, *chimera death* patterns can be found, which combine the characteristics of both phenomena: chimera state and oscillation death. We recall that these patterns consist of coexisting domains of coherent and incoherent populations of the inhomogeneous steady-state branches. Within the incoherent domains, the population of the two branches (upper and lower) follows a random sequence, as, for example, visible in Fig. 2.43. Within the coherent domains,

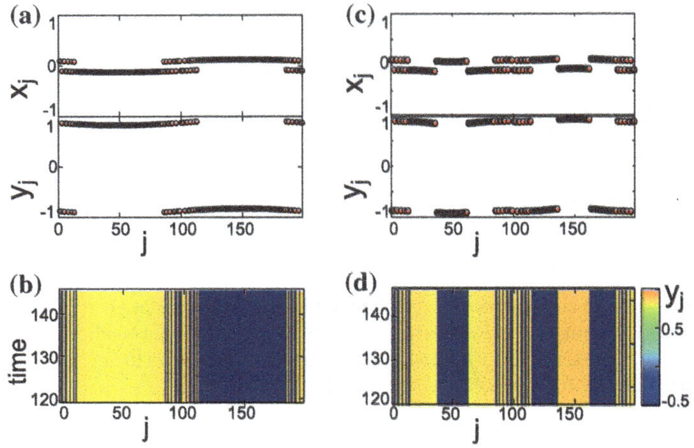

Fig. 2.43 **a, b** 1-cluster chimera death (1-CD) for coupling range $P/N = 0.4$, **c, d** 3-cluster chimera death (3-CD) for $P/N = 0.2$. Snapshots x_j, y_j are shown in panels **a**, **c**, space-time plots in panels **b**, **d**. Initial condition: see Sect. 2.3.4. Parameters: $N = 200$, $\sigma = 18$, $D = 0$, $\lambda = 1$, $\omega = 2$ [102]

the number of clusters of neighboring nodes that populate the same branch of the inhomogeneous steady state can vary. An m-cluster chimera death state (m-CD), with $m \in \{1, 3, 5, 7, 9, ...\}$, is characterized by the occurrence of m clusters within each coherent domain. The coherent domains always appear pairwise with antiphase symmetry $z_j = -z_{j+N/2}$, similarly to the coherent domains of the amplitude chimera configurations. Our numerical results confirm that the stable oscillation death patterns fulfill the "antiphase partner" condition.

2.5.3 *Influence of Noise on Transient Times*

Next, we study the robustness of chimera states with respect to external noise. By using the same initial conditions which lead to amplitude chimera states and chimera death in the deterministic case, we also observe these states in (2.2) in the presence of noise in a wide range of the coupling parameters. The stochastic equations were integrated with the well-established Euler-Maruyama scheme with step size 10^{-4} (see p. 340 in [290].). Figure 2.44 shows one exemplary configuration for an amplitude chimera which occurs in a system under the impact of noise of intensity $D = 5 \cdot 10^{-3}$.

In general, the transient times of amplitude chimeras decrease with increasing noise intensity. Figure 2.45 shows the average transient times and the corresponding standard deviations in dependence of the noise intensity D, for three choices of the coupling strength σ, in a semilogarithmic plot. The average is over 50 different fully symmetric initial conditions (with $Q = \frac{N}{4}$, see Sect. 2.3.4) drawn from different realizations of the associated random distribution. For each one of those realizations

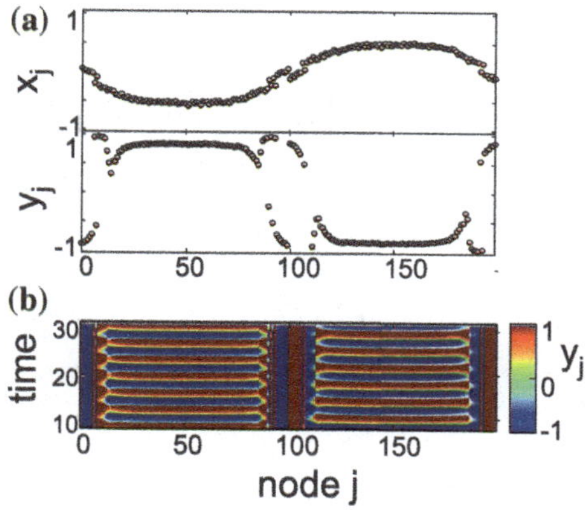

Fig. 2.44 Amplitude chimera for noise intensity $D = 5 \cdot 10^{-3}$: **a** snapshot (top: x_j, bottom: y_j), **b** space-time plot. Parameters: $N = 200$, $P/N = 0.04$, $\sigma = 19$, $\lambda = 1$, $\omega = 2$. Initial condition: see Sect. 2.3.4 [102]

of the initial conditions, a different realization of the Gaussian white noise $\xi_j(t)$ is considered. The average transient times show a clear linear decrease as a function of the logarithmic noise intensity. This behavior is found throughout the range $6 \leq \sigma \leq 24$, i.e., $t_{\text{tr}} = -\frac{1}{\mu} \, ln(D) + \eta$ with slope $-\frac{1}{\mu}$ and axis intercept η. This gives the scaling law

$$D \sim e^{-\mu t_{\text{tr}}}. \tag{2.33}$$

The lines in Fig. 2.45 show the linear fits, and the inset depicts the slope in dependence on the coupling strength σ. For the same set of 50 initial conditions, Fig. 2.46a depicts the mean transient time in dependence of the coupling strength for four differ-

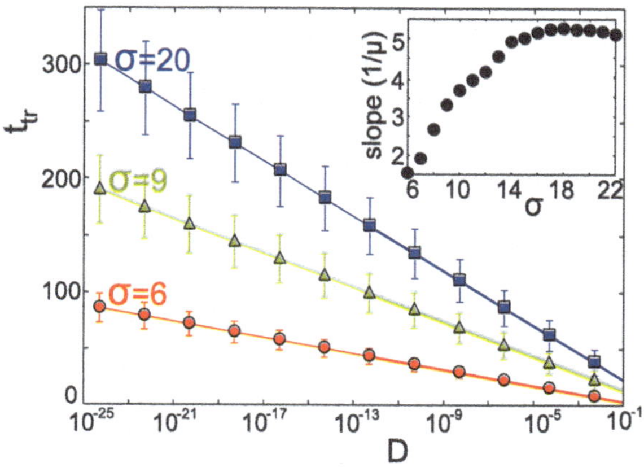

Fig. 2.45 Transient times of amplitude chimeras t_{tr} versus noise strength D (log-scaled) for different values of coupling strength σ. Symbols: average over 50 fully symmetric initial conditions (with $Q = \frac{N}{4}$, each associated with a different realization of the random force $\xi(t)$; error bars: standard deviations; lines: linear fits from (2.33). Inset: slope versus σ. Parameters: $N = 100$, $P/N = 0.04$, $\lambda = 1$, $\omega = 2$ [102]

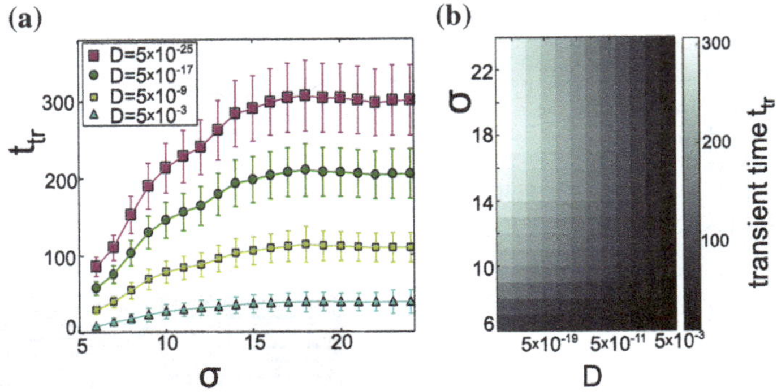

Fig. 2.46 Transient times of amplitude chimeras t_{tr} averaged over 50 initial conditions and noise realizations (the same set of initial conditions as used in Fig. 2.45): **a** t_{tr} versus coupling strength σ for different noise intensities. Symbols: mean transient times; error bars: standard deviations. The lines serve as a guide to the eye. **b** t_{tr} in the plane of coupling strength σ and noise intensity D. Other parameters: $N = 100$, $P/N = 0.04$, $\lambda = 1$, $\omega = 2$ [102]

ent noise intensities D, and Fig. 2.46b shows a color-coded density plot of the mean transient times of amplitude chimeras in the (σ, D)-plane. The transient times generally decrease with increasing noise, and increase with increasing coupling strength up to a saturation value at about $\sigma \approx 15$. The vertical error bars in panel (a) show that the transient times are less sensitive to the initial condition, the larger the noise is. We generally find that the spread of the amplitude chimera lifetimes for different initial conditions (and different noise realizations) is smaller with increasing noise strength.

We recall that transient amplitude chimeras can last for thousands of oscillation periods until they disappear. Even under disturbance by external noise they persist for a significant time. Noise does not essentially change their spatial configuration. If noise throws the system onto an adjacent trajectory in the underlying high-dimensional phase space of the network, this does not normally lead to a flow into a completely different direction in phase space. Geometrically speaking, this shows that there are some attracting directions in phase space along which the system dynamics is pushed towards the amplitude chimera. Furthermore, amplitude chimeras can evolve out of initial configurations that do not show the characteristic coexistence of coherent and incoherent domains (see Sect. 2.3.4). In fact, they can be observed when completely incoherent initial configurations are used, as well as when the initial condition consists of two completely coherent parts. These dynamical properties indicate that the flow within a certain volume of the phase space is directed towards the amplitude chimera state. From the perspective of the amplitude chimera, there exist some associated stable directions. However, even in the absence of any external perturbation, for all system sizes, the amplitude chimera states disappear after some time, and the system approaches a coherent oscillatory state. Accordingly, there are also unstable directions in phase space (see Sect. 2.3.7

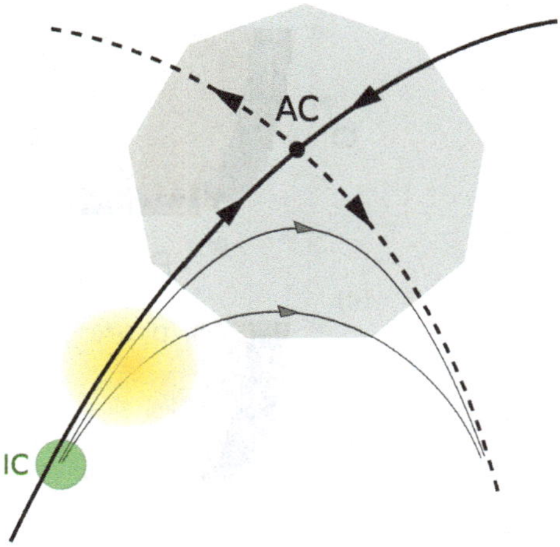

Fig. 2.47 Schematic phase-space structure of an amplitude chimera (AC) as a saddle-point. Thick solid lines: stable directions, thick dashed lines: unstable directions, thin solid lines: different trajectories, with arrows denoting the direction of time evolution. Gray shaded region: scheme of amplitude chimera configuration, green (dark gray) disk: set of initial conditions (IC), yellow (light gray) area: impact of Gaussian white noise [102]

for the detailed stability analysis). The structure of the phase space is schematically depicted in Fig. 2.47.

As discussed in Sect. 2.3.4 in the absence of noise, the amplitude chimera lifetime depends on the initial conditions. This can be explained by the fact that typically chimera states coexist with the completely synchronized regime that is characterized by a larger basin of attraction. The numerical experiments we have performed confirm the assumption that amplitude chimera is a saddle state consisting of a stable (solid lines in Fig. 2.47) and an unstable (dashed lines Fig. 2.47) manifold. The initial conditions resulting in amplitude chimera regime can then be seen as a volume confined in the phase space (green disk in Fig. 2.47). Amplitude chimeras correspond to trajectories that start from this initial conditions and pass the saddle-point from the stable direction towards the unstable manifold. Consequently, the lifetime of an amplitude chimera is determined by the chosen trajectory: the closer it gets to the saddle-point, the longer is the lifetime. It means that the lifetime is defined by the time the network stays in the vicinity of the saddle where coherent and incoherent parts coexist before it escapes to the in-phase synchronized state along the direction of the unstable manifold. Such a phase-space scenario explains the sensitivity of transient times to initial conditions since they define the particular path the system takes. For increasing network size N, the phase-space dimension of the system increases, and, therefore, it becomes more likely that the distance of the initial conditions to the stable manifold is larger, which leads to decreasing times the system spends in the vicinity of the saddle. This fact explains the decrease of the lifetimes of the amplitude chimera state for increasing N (Fig. 2.15).

Further, the results of our numerical simulations of the stochastic model (2.4) demonstrate that Gaussian white noise significantly reduces the sensitivity of the amplitude chimera lifetime to initial conditions. In the absence of noise, we have tested a set of realizations of initial conditions leading to different lifetimes of amplitude chimera. In the case when noise of relatively weak intensity $D = 5 \cdot 10^{-13}$ is

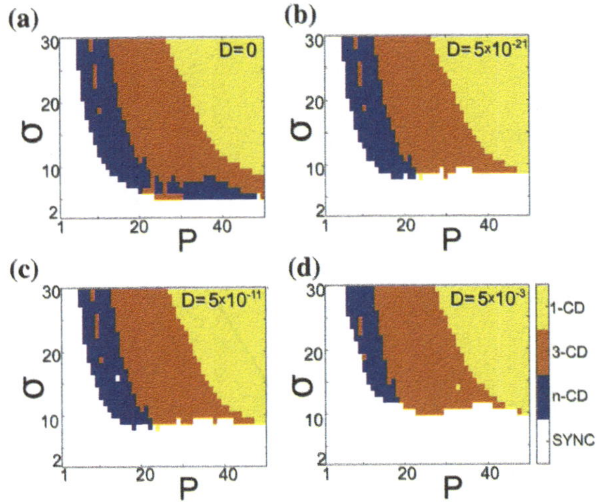

Fig. 2.48 Map of dynamic regimes in the plane of coupling strength σ and coupling range P for noise intensities: **a** $D = 0$, **b** $D = 5 \cdot 10^{-21}$, **c** $D = 5 \cdot 10^{-11}$, **d** $D = 5 \cdot 10^{-3}$. Color code: 1-cluster chimera death (1-CD), 3-cluster chimera death (3-CD), multicluster chimera death (n-CD, $n > 3$), in-phase synchronized oscillations and coherent traveling waves (SYNC). Initial condition: snapshot of an amplitude chimera calculated for $D = 0$, $P = 4$, $\sigma = 14$, $t = 150$. Maximum simulation time: $t = 5000$. Parameters: $N = 100$, $\lambda = 1$, $\omega = 2$ [102]

present, all realizations produce amplitude chimera states with similar lifetime. This supports the results of the stability analysis for the deterministic case (see Sect. 2.3.7) and our vision of the amplitude chimera pattern as a saddle. The stochastic force continuously perturbs the system, making it randomly switch between different trajectories close to the saddle. Consequently, the dynamics of the network is not defined by a single trajectory, but rather determined by a set of trajectories belonging to the N-dimensional hyper-sphere. This essentially reduces the impact of initial conditions on the lifetime of amplitude chimera. We illustrate the impact of noise by yellow shading, indicating the random forces applied to the network at one instant of time (Fig. 2.47).

2.5.4 Maps of Dynamic Regimes

For a network of $N = 100$ coupled elements and a large range of the coupling parameters σ and P, we calculate the asymptotically stable state and the transient time of amplitude chimera. For each set of parameters (σ, P), we use the same amplitude chimera configuration as initial condition. We compare the resulting maps of regimes for four different values of noise intensity (Fig. 2.48). For very small coupling range P or very weak interaction strength σ, the asymptotic states are coherent oscillatory states, either in-phase synchronized oscillations or traveling waves (white region, labeled SYNC). For very small number of nearest neighbors, we detect amplitude chimera states as transients. For larger σ and P values, we observe chimera death patterns (yellow, orange, and blue regions) with one coherent domain (1-CD), or for slightly smaller P, with three (3-CD), or more (n-CD with $n > 3$) coherent domains. For all considered noise intensities, we find a chimera death regime (1-CD, 3-CD, n-CD), as well as a coherent oscillatory state (SYNC).

The region of chimera death patterns is highly multistable. The border between the oscillatory region and the chimera death state is almost independent of initial conditions. On the contrary, for many values of (σ, P), the particular type of chimera death pattern is defined by initial conditions. It is important to note that still there is a clear trend that the m-CD state with ($m < k$) are more likely to appear for larger coupling ranges than the k-CD states ($k, m \in \{1, 3, n\}$). This trend is especially pronounced for large valued of interaction strength. The dynamic regimes are influenced by noise in different ways. The borders between chimera states with different number of clusters are almost independent of the stochastic force. No noise-induced switching between the different types of chimera death states is detected. The stochasticity has no impact on the asymptotic chimera death state. With increasing noise intensity, the border between the oscillatory region and the death state is shifted towards stronger interaction strength. Therefore, noise pushes the network out of the deterministic inhomogeneous steady state inducing oscillatory dynamics in a parameter range where without noise the steady state is asymptotically stable. Noise intensity D defines the size of this parameter range. To illustrate the comparison, we show the boundaries between the oscillatory region and the chimera death regime for different values of noise intensity (Fig. 2.49). In general, stronger noise leads to larger oscillatory regions.

Transient amplitude chimera states are detected in the oscillatory regime. Figure 2.50 illustrates the lifetimes of amplitude chimeras, resulting from the same simulations described above. In general, the lifetime decreases with decreasing strength of interaction, and with increasing intensity of noise, as demonstrated already in

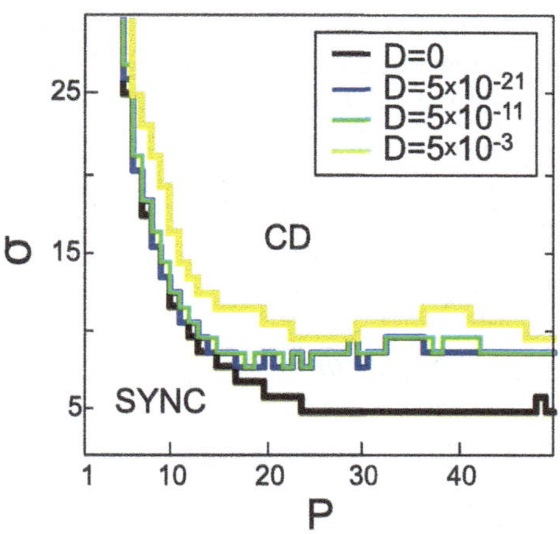

Fig. 2.49 Boundary between the oscillatory regime and the chimera death regime for different noise intensities D, extracted from the maps of dynamic regimes shown in Fig. 2.48 [102]

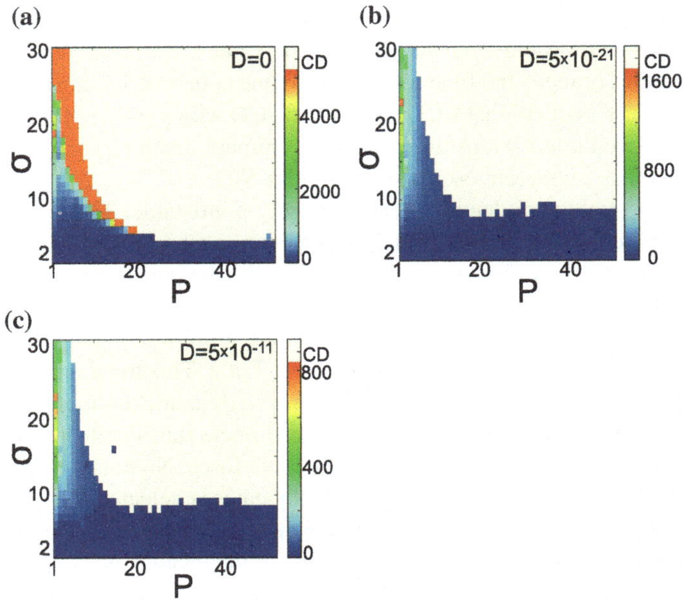

Fig. 2.50 Transient times of amplitude chimeras t_{tr} in the plane of coupling strength σ and coupling range P, for the noise intensities: **a** $D = 0$, **b** $D = 5 \cdot 10^{-21}$, **c** $D = 5 \cdot 10^{-11}$. System parameters, initial condition and simulation time as in Fig. 2.48 [102]

Figs. 2.45 and 2.46 for a restricted range of coupling parameters. Interestingly, Fig. 2.50b ($D = 5 \cdot 10^{-11}$) and c ($D = 5 \cdot 10^{-21}$) look very similar up to rescaling of the transient times. Therefore, the impact of the stochastic force on the network dynamics is rather independent of the strength and range of the coupling.

Without noise (Fig. 2.50a) there is a regime of high values of the coupling strength, at the border between the oscillatory region and the chimera death regime, where the transient amplitude chimeras last longer than the maximum simulation time of $t =$ 5000 (bright orange). For several values of (σ, P) in this region, we have simulated much longer time series until $t = 40{,}000$ (more than 12,700 oscillation periods T), and have found that the amplitude chimeras still exist. However, they disappear much earlier as soon as a tiny amount of external noise is applied. Therefore, the amplitude chimera states are unstable in this region. The extremely long transient times might simply be related to the choice of initial conditions in the deterministic network.

2.5.5 *Summary*

In the present section, the robustness of chimera states with respect to noise has been investigated for a paradigmatic network of oscillators. We have presented numerical results demonstrating that transient amplitude chimeras and chimera death states in a ring network of identical Stuart-Landau oscillators with symmetry-breaking coupling continue to exist in the presence of Gaussian white noise. Transient amplitude chimeras occur in the same range of coupling parameters as in the deterministic case. The key quantity we use to measure their robustness is the transient time. The latter decreases logarithmically with the applied noise intensity. For a constant noise intensity, the transient times increase with the coupling strength up to a saturation value. The width of the incoherent domains relative to the overall system size is not affected by external noise, and is independent of the used initial condition and the total system size.

The chimera death patterns also persist under the impact of stochastic forces. However, the coupling parameter regime where they occur is reduced with increasing noise intensity. The boundary between the coherent oscillatory regime and the chimera death regime is shifted towards higher values of the coupling strength. That means that the system favors oscillatory behavior for a larger coupling parameter regime. In contrast, this boundary appears to be independent on the particular realization of the initial condition. The number of clusters within the coherent domains appears to be unaffected by the external noise, but depends on the particular initial condition.

Our numerical findings for the stochastic case can be explained in terms of the structure of the phase space. They further confirm the results of the deterministic stability analysis and the representation of amplitude chimeras as saddle states in the phase space of the network. This elucidates the behavior of their lifetime, and explains that generally the initial conditions become less important under the influence of noise.

2.6 Conclusions

We have investigated two types of chimera states for a paradigmatic network of oscillators under the influence of noise and time delay. We have presented numerical results demonstrating that amplitude chimeras and chimera death states in a ring network of identical Stuart-Landau oscillators with symmetry-breaking coupling continue to exist in the presence of Gaussian white noise or if time delay is introduced to the coupling. In particular, in the presence of time delay, the lifetime of amplitude chimera patterns is essentially enlarged. Noise allows to decrease the lifetime of amplitude chimeras. Thus, the lifetime of amplitude chimeras can be controlled by tuning the noise intensity and the value of time delay, which, therefore, play the role of control parameters.

Chapter 3
Coherence-Resonance Chimeras in Ring Networks

Abstract In the present chapter, we consider noise-induced chimera patterns called coherence-resonance chimeras. These peculiar states combine features of coherence resonance and chimera states and are characterized by the coexistence of two different domains separated in space: one part of the network is spiking coherently in space, while the other exhibits incoherent spiking, i.e., the spiking of neighboring nodes is uncorrelated. We explain the formation mechanism of noise-induced chimeras and discuss time-delayed feedback control of these patterns. Specifically, we focus on the role of noise and time delay for the chimera states occurring in ring networks of FitzHugh-Nagumo neurons in the excitable regime.

3.1 Introduction

One of the challenging issues concerning chimera states is their behavior in the presence of random fluctuations, which are unavoidable in real-world systems. They appear due to intrinsic noise or alternatively can be implemented as an external random control force. On the one hand, random fluctuations are often seen as nuisance degrading the characteristics of a system. On the other hand, however, noise can play a constructive role and gives rise to new dynamic behavior, e.g., stochastic bifurcations, stochastic synchronization, or *coherence resonance* [134–137, 291–297]. The counterintuitive phenomenon of coherence resonance describes ac non-monotonic behavior of the regularity of noise-induced oscillations in the excitable regime, leading to an optimal response in terms of regularity of the excited oscillations for an intermediate noise intensity. The robustness of chimeras with respect to external noise has been studied only very recently (Sect. 2.5). An intriguing question is whether the constructive role of noise in nonlinear systems, manifested, for example, in the counterintuitive increase of temporal coherence due to noise in coherence resonance, can be combined with the chimera behavior in spatially extended systems and networks. Coherence resonance, originally discovered for excitable systems like the FitzHugh-Nagumo model, has been investigated theoretically and experimentally in various systems and network topologies [131, 133–142, 295, 298–307]. In particular, it has been shown that coherence resonance can be observed and even modulated

A. Zakharova, *Chimera Patterns in Networks*, Understanding Complex Systems,
https://doi.org/10.1007/978-3-030-21714-3_3

by applying time-delayed feedback in excitable [299, 303, 308, 309] as well as in non-excitable systems [138, 139].

In the present chapter, we investigate an effect which combines coherence resonance and chimera states in a network of nonlocally coupled excitable elements. We demonstrate that chimera behavior can be observed in excitable systems and not in oscillatory systems and show that the presence of noise is a crucial condition for this case. Moreover, we disclose the constructive role of noise for chimera states and describe a novel type of coherence resonance, which has been discovered recently and called *coherence-resonance chimeras* [131–133]. In these spatiotemporal patterns, coherence resonance is associated with spatially coherent and incoherent behavior, rather than purely temporal coherence or regularity measured by the correlation time. Additionally, the noise-based control mechanism that we describe in this chapter reveals an alternative direction for chimera control complementary to recent deterministic control schemes [126, 144–147]. Further, applying time-delayed feedback, we demonstrate how to control coherence-resonance chimeras by adjusting delay time and feedback strength.

The sections are organized as follows. First, we introduce the model (Sect. 3.2). Then, we investigate coherence-resonance chimeras for the case without time delay (Sect. 3.3). Further, we discuss time-delayed feedback control of the chimera patterns (Sect. 3.4). Specifically, we focus on the role of noise and time delay for the chimera states occurring in ring networks of FitzHugh-Nagumo systems in excitable regime. Section 3.5 summarizes the results.

3.2 FitzHugh-Nagumo Model

In the present chapter, we investigate FitzHugh-Nagumo (FHN) system, a paradigmatic model for excitable dynamics, originally suggested to characterize the spiking behavior of neurons [310–313]. Its fields of application range from neuroscience and biological processes [295, 314] to optoelectronic [315] and chemical [237] oscillators and nonlinear electronic circuits [48]. In Sect. 3.3, we consider a ring of N nonlocally coupled FHN systems in the presence of Gaussian white noise without time delay [131, 132]:

$$\begin{aligned} \varepsilon \frac{du_i}{dt} &= u_i - \frac{u_i^3}{3} - v_i + \frac{\sigma}{2R} \sum_{j=i-R}^{i+R} [b_{uu}(u_j - u_i) + b_{uv}(v_j - v_i)], \\ \frac{dv_i}{dt} &= u_i + a_i + \frac{\sigma}{2R} \sum_{j=i-R}^{i+R} [b_{vu}(u_j - u_i) + b_{vv}(v_j - v_i)] + \sqrt{2D}\xi_i(t), \end{aligned} \tag{3.1}$$

where u_i and v_i are the activator and inhibitor variables, respectively, $i = 1, ..., N$ and all indices are modulo N, $\varepsilon > 0$ is a small parameter responsible for the timescale separation of fast activator and slow inhibitor, a_i defines the excitability threshold. For an individual FHN element, it determines whether the system is excitable ($|a_i| > 1$), or oscillatory ($|a_i| < 1$). For $|a_i| < 1$, the system is in the oscillatory regime where the steady state is unstable and self-sustained oscillations are observed. For $|a_i| > 1$, the system is in the excitable regime and characterized by a locally stable steady state. It is important to note that the FHN system can also exhibit excitability in the oscillatory regime prior to canard explosion ($0.995 < a < 1$), where a sufficiently large perturbation can trigger a spike on top of small-amplitude subthreshold oscillations emerging from the supercritical Hopf limit cycle bifurcation at $a = 1$. In the present chapter, we are mainly interested in the excitable regime. Therefore, we assume that all elements are in the excitable regime close to the threshold ($a_i \equiv a = 1.001$). The coupling is characterized by the following parameters: σ is the coupling strength, R is the number of nearest neighbors, and $r = R/N$ is the coupling range. The form of the coupling of (3.1) is inspired from neuroscience [79, 316–318], where strong interconnections between neurons are found within a range R, but much fewer connections exist at longer distances. Further, $\xi_i(t) \in \mathbb{R}$ is Gaussian white noise, i.e., $\langle \xi_i(t) \rangle = 0$ and $\langle \xi_i(t)\xi_j(t') \rangle = \delta_{ij}\delta(t - t')$, $\forall i, j$, and D is the noise intensity. Equation (3.1) contains not only direct, but also cross-couplings between activator (u) and inhibitor (v) variables, which is modeled by a rotational coupling matrix [79]:

$$\boldsymbol{B} = \begin{pmatrix} b_{uu} & b_{uv} \\ b_{vu} & b_{vv} \end{pmatrix} = \begin{pmatrix} \cos\phi & \sin\phi \\ -\sin\phi & \cos\phi \end{pmatrix}, \tag{3.2}$$

where $\phi \in [-\pi; \pi)$. In the present chapter, we fix the parameter $\phi = \pi/2 - 0.1$ for which chimeras have been found in the deterministic oscillatory regime [79].

Next, we study nonlocally coupled ring networks of N FHN systems in the presence of noise and time-delayed feedback [133] (see Sect. 3.4):

$$\begin{aligned} \varepsilon \frac{du_i}{dt} &= u_i - \frac{u_i^3}{3} - v_i + \frac{\sigma}{2R} \sum_{j=i-R}^{i+R} [b_{uu}(u_j - u_i) + \\ &+ b_{uv}(v_j - v_i)] + \gamma\varepsilon(u_i(t) - u_i(t - \tau)), \\ \frac{dv_i}{dt} &= u_i + a_i + \frac{\sigma}{2R} \sum_{j=i-R}^{i+R} [b_{vu}(u_j - u_i) + \\ &+ b_{vv}(v_j - v_i)] + \sqrt{2D}\xi_i(t), \end{aligned} \tag{3.3}$$

where the feedback term is characterized by time delay τ and strength γ.

3.3 Coherence-Resonance Chimeras Without Time Delays

In real-world systems, chimera states might play a role, e.g., in power grids [12], in social systems [165], in modular neural networks [187], in the unihemispheric sleep of birds, dolphins [156] and humans [157], or in epileptic seizures [122, 160, 161]. In the context of the latter two applications, it is especially relevant to explore chimera states in neuronal networks under conditions of excitability. Chimera states have previously been reported for neuronal networks in the oscillatory regime, e.g., in the FitzHugh-Nagumo system [79], or a network of oscillatory elements containing a block of excitable units [248]. In the purely excitable regime, they have been detected for the first time in [131].

The excitable regime of the FitzHugh-Nagumo system which we consider in the present chapter is fundamentally different from the previously investigated oscillatory regime [79], and the chimera states presented here have very different features as compared to those found previously in the oscillatory regime. It is known, for instance, that synchronization mechanisms for noise-induced oscillations below the Hopf bifurcation and for deterministic limit cycle oscillations above the Hopf bifurcation are crucially different [295, 319, 320].

The phenomenon of coherence resonance, originally discovered for excitable systems like the FitzHugh-Nagumo model, implies that noise-induced oscillations demonstrate the best temporal regularity for an optimum intermediate value of noise intensity. A question naturally arising in this context is whether noise can also have a beneficial effect on chimera states. Here we show the constructive role of noise for chimeras. In particular, we establish a connection between two intriguing counterintuitive phenomena which have been previously studied independently in different scientific communities: coherence resonance and chimera states.

In this section, we first discuss the phenomenon of coherence resonance in a single FHN system (Sect. 3.3.1). Further, we compare chimera states occurring in oscillatory and excitable regimes and define coherence-resonance chimeras (Sect. 3.3.2). Next, we provide an explanation of the alternation mechanism of coherence-resonance chimeras by analyzing the time evolution of coupling terms and nullclines (Sect. 3.3.3). Then we investigate the patterns observed in the network in the presence of strong noise (Sect. 3.3.4). Section 3.3.5 discusses dynamic regimes for varying coupling parameters. Finally, we introduce measures to quantify coherence-resonance chimeras (Sect. 3.3.6) and summarize the results on the dynamics without time delay in Sect. 3.3.7.

3.3.1 Coherence Resonance in a Single FitzHugh-Nagumo System

In the excitable regime ($|a| > 1$), a single FHN system rests in a locally stable steady state (point A in Fig. 3.1a) and upon excitation by noise beyond a threshold emits a spike, i.e., performs a long excursion in phase space (line B in Fig. 3.1a), before returning to the rest state. With increasing noise, the excitation across threshold occurs more frequently, and thus the interspike intervals become more regular. On the other hand, with increasing noise the deterministic spiking dynamics becomes smeared out. The best temporal regularity is observed for an optimum intermediate value of noise intensity and the corresponding counterintuitive phenomenon is known as coherence

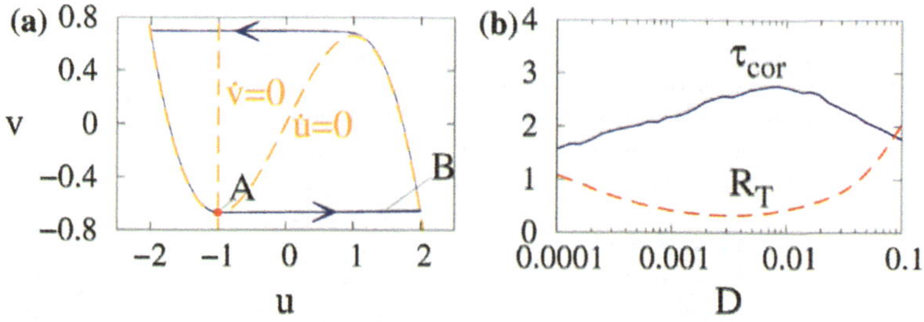

Fig. 3.1 Single FHN system: **a** Schematic phase portrait with activator and inhibitor nullclines $\dot{u} = 0$ and $\dot{v} = 0$ respectively (dashed lines). A is a stable steady state. Parameters: $\varepsilon = 0.01$, $a = 1.001$, $D = 0.0001$. **b** Coherence resonance: Dependences of τ_{cor} (solid line) and R_T (dashed line) on the noise intensity D. Parameters: $\varepsilon = 0.05$, $a = 1.001$ [132]

resonance [134, 135, 291]. There are different temporal correlation measures used to detect coherence resonance [135, 321]. For instance, the optimal value of noise intensity typically corresponds to the maximum of the correlation time $\tau_{\text{cor}}(D)$ or the minimum of the normalized standard deviation of interspike intervals $R_T(D)$, see Fig. 3.1b. Such behavior has been shown theoretically and experimentally in a variety of excitable systems, like lasers with saturable absorber [322], optical feedback [323, 324], and optical injection [304], semiconductor superlattices [308, 325], or neural systems [135, 295] and recently in non-excitable systems as well [136–139, 296].

To characterize spatial coherence and incoherence of chimera states, one can use a local order parameter [69, 167]:

$$Z_k = \left| \frac{1}{2\delta_Z} \sum_{|j-k|\leq\delta_Z} e^{i\Theta_j} \right|, \quad k = 1, \dots N \tag{3.4}$$

where the geometric phase of the jth element is defined by $\Theta_j = arctan(v_j/u_j)$ [79] and $Z_k = 1$ and $Z_k < 1$ indicate coherence and incoherence, respectively.

3.3.2 *Chimera States in Oscillatory and Excitable Regimes*

Chimera states have been previously reported for the deterministic oscillatory regime of the FitzHugh-Nagumo system [79]. In more detail, for the oscillatory regime far from the threshold ($a = 0.5$) one can find chimera states: domains of coherent and incoherent oscillations clearly separated in space. This pattern is shown as a space-time plot color-coded by the variable u_i and by the local order parameter Z_i in Fig. 3.2a. While approaching the oscillatory threshold with increasing threshold parameter a, we observe shrinking of the incoherent domains (Fig. 3.2b), which completely disappear for $a > 0.8$ indicating the collapse of the chimera state. On the other hand, in the excitable regime without noise ($D = 0$) the network rests

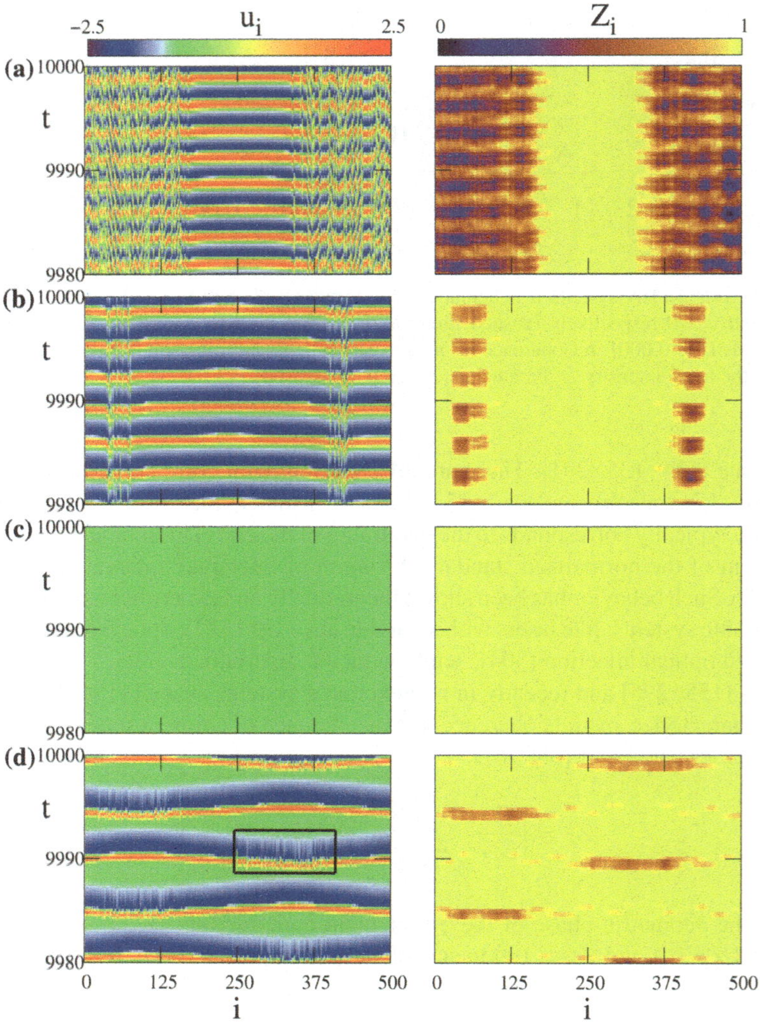

Fig. 3.2 Space-time plots (left column) and local order parameter (right column) for different values of excitability parameter and noise intensity **a** $a = 0.5$, $D = 0$, $r = 0.35$, $\sigma = 0.1$; **b** $a = 0.8$, $D = 0$, $r = 0.35$, $\sigma = 0.1$; **c** $a = 1.001$, $D = 0$, $r = 0.12$, $\sigma = 0.4$; **d** $a = 1.001$, $D = 0.0002$, $r = 0.12$, $\sigma = 0.4$. Initial conditions: randomly distributed on the circle $u^2 + v^2 = 4$. In all panels $\varepsilon = 0.05$ [132]

in a homogeneous steady state and, therefore, no chimera states occur (Fig. 3.2c). Once noise is introduced to the system and its intensity reaches a certain value ($0.000062 \leq D \leq 0.000325$), we detect the appearance of a spatiotemporal spiking pattern, which combines features of chimera states and coherence resonance and is essentially different from the one occurring in the deterministic oscillatory regime (Fig. 3.2d). This noise-induced state which we call *coherence-resonance chimera* has been discovered in [131] and is characterized by the coexistence of two different domains separated in space, where one part of the network is spiking coherently in space while the other exhibits incoherent spiking, i.e., the spiking of neighboring nodes is uncorrelated. In order to quantify coherence and incoherence for this pattern, we calculate the local order parameter Z_i (right panel in Fig. 3.2d). It can be clearly seen that the islands of desynchronization corresponding to the incoherent domains are characterized by values of the order parameter noticeably below unity (dark patches).

One important feature, which distinguishes coherence-resonance chimeras from the deterministic chimeras observed in the oscillatory network is that they are manifested in partially coherent and partially incoherent excitation waves. Therefore, the appearance of this pattern can be explained by analyzing the propagation and termination of excitation waves in a ring. From the incoherent domain marked with a black rectangle in the space-time plot (left panel in Fig. 3.2d), two very fast counter-propagating excitation waves emanate, and as they propagate they become coherent and as they meet again on the antipodal position on the ring they annihilate. Subsequently, at that position around $i = 50$, another incoherent domain is born, which again generates two fast counterpropagating coherent excitation waves, and so on.

3.3.3 *Alternating Behavior of Coherence-Resonance Chimeras*

Another characteristic feature of this stochastic chimera pattern is its alternating behavior which is absent in the oscillatory regime without noise. In more detail, the incoherent domain of the chimera pattern switches periodically its position on the ring, although its width remains fixed. Previously, alternating chimera behavior has been reported for a deterministic oscillatory medium with nonlinear global coupling [243]. The alternating behavior of coherence-resonance chimeras has been previously described in [131], and the mechanism of the alternation has been disclosed in [132].

In what follows, we explain why the coherent and incoherent spiking alternates between the two groups of the network elements. For this purpose, we analyze the time evolution of the coupling term for every node of the network. Taking into account that the system (3.1) involves both direct and cross-couplings between activator (u) and inhibitor (v) variables, in total we have four coupling terms which we consider separately. It turns out that the coupling terms form patterns shown as space-time plots in Fig. 3.3a–d. Therefore, the action of the coupling is not homogeneous: it

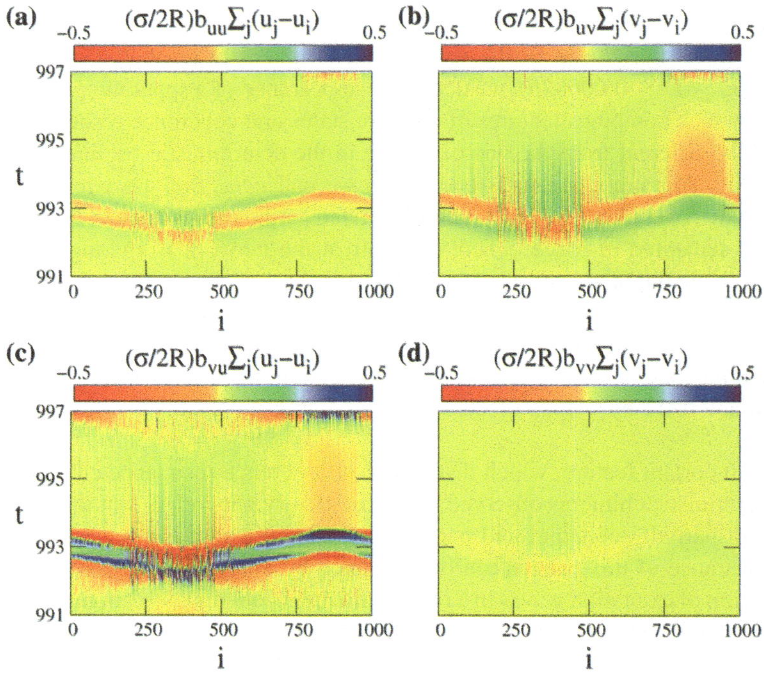

Fig. 3.3 Space-time plots of coupling terms for u and v variables in the coherence-resonance chimera regime: **a** direct coupling for the u variable, **b** cross-coupling for the u variable, **c** cross-coupling for the v variable, **d** direct coupling for the v variable. Parameters: $\varepsilon = 0.05$, $a = 1.001$, $\sigma = 0.4$, $r = 0.12$, $D = 0.0002$ [132]

is stronger for a certain group of nodes at a certain time (green and red regions in Fig. 3.3b, c), while the rest of the network is not influenced by the coupling (yellow regions in Fig. 3.3). Moreover, these patterns are more pronounced for cross-couplings (Fig. 3.3b, c) since the contribution of the off-diagonal elements of the coupling matrix (3.2) is much stronger than that of the diagonal elements. The coupling acts as an additional term which modifies and shifts the threshold parameter a which is responsible for the excitation. Consequently, the probability of being excited by noise is much higher for the nodes for which the excitation threshold becomes lower due to coupling. Note that the sign of the coupling term alternates between the two groups of oscillators; specifically, at the end of the quiescent period, just before spiking starts, it is positive (green-blue region) for the group which has previously spiked coherently, and negative (red region) elsewhere. Thus that group starts spiking first (in a random way). This explains the alternating behavior of coherence-resonance chimeras, since the group of nodes for which the threshold gets lower due to coupling changes its location on the ring network periodically.

Next, we investigate a temporal sequence of snapshots of the variable u_i and phase portraits in the (u_i, v_i)-plane (Fig. 3.4). The middle nodes from $i = 170$ to $i = 420$ are marked in orange (light), while the rest of the network elements is marked in green (dark). We start with the state where all the elements of the network are located close to the steady state $u_i \approx -1$ (panel a). A little bit later one node from the middle group $i = 269$ (red dot) gets excited by noise (panel b) and starts its excursion in phase space. Further, the whole middle group incoherently joins the excursion in phase space (phase portrait in panel c). As the excitation rapidly moves to the left and to the right from the middle group, it becomes more and more coherent (panel d). This phase in the time evolution corresponds to spiking. Note that the nodes from the incoherent domain start their journey in the phase space first (desynchronized spiking), while the nodes from the coherent domain catch up later but more synchronously. Next, all the FHN elements jump back to the left branch of the activator nullcline in phase space (Fig. 3.4e) and return along the nullcline slowly and rather synchronously to the steady state (Fig. 3.4f). Subsequently, the steps described above repeat, however, with the coherent and incoherent domains interchanged (Fig. 3.4g).

To further deepen our understanding of this alternation, we study the impact of the coupling on activator and inhibitor nullclines for selected nodes of the system (3.1). In particular, we investigate a temporal sequence of phase portraits for the nodes $i = 269$ (red dot) and $i = 1$ (blue dot) which belong to the incoherent and coherent domains, respectively, during the observation time (Fig. 3.5). We start with the state where all the elements of the network are located close to the steady state and the nullclines of the node $i = 269$ remain unchanged (Fig. 3.5a). A little bit later (panel b), the vertical inhibitor nullcline of this node is shifted to the left due to the positive coupling term and, therefore, the node can be excited more easily by noise. This is due to the fact that the network elements do not change their location in the vicinity of the steady state, while at the same time the excitation threshold for some particular nodes moves to the left together with the vertical inhibitor nullcline. Therefore, these nodes (in particular, $i = 269$ in Fig. 3.5c) are now located to the right of the excitation threshold, become more sensitive to noise and consequently get excited. The node $i = 269$ which is excited first separates from the rest of the network elements and starts its journey in phase space. Then some other nodes (belonging to the incoherent domain of coherence-resonance chimera) for which the threshold also becomes lower due to coupling get excited incoherently by noise and go on excursion in phase space. At the same time, the nullclines for the nodes from the coherent group remain unchanged (right column in Fig. 3.5a–c) and therefore, they start their journey later being pulled coherently by the incoherent group. It is important to note that the coupling also influences the activator nullcline and shifts it as shown in Fig. 3.5d, once the spiking is well under way.

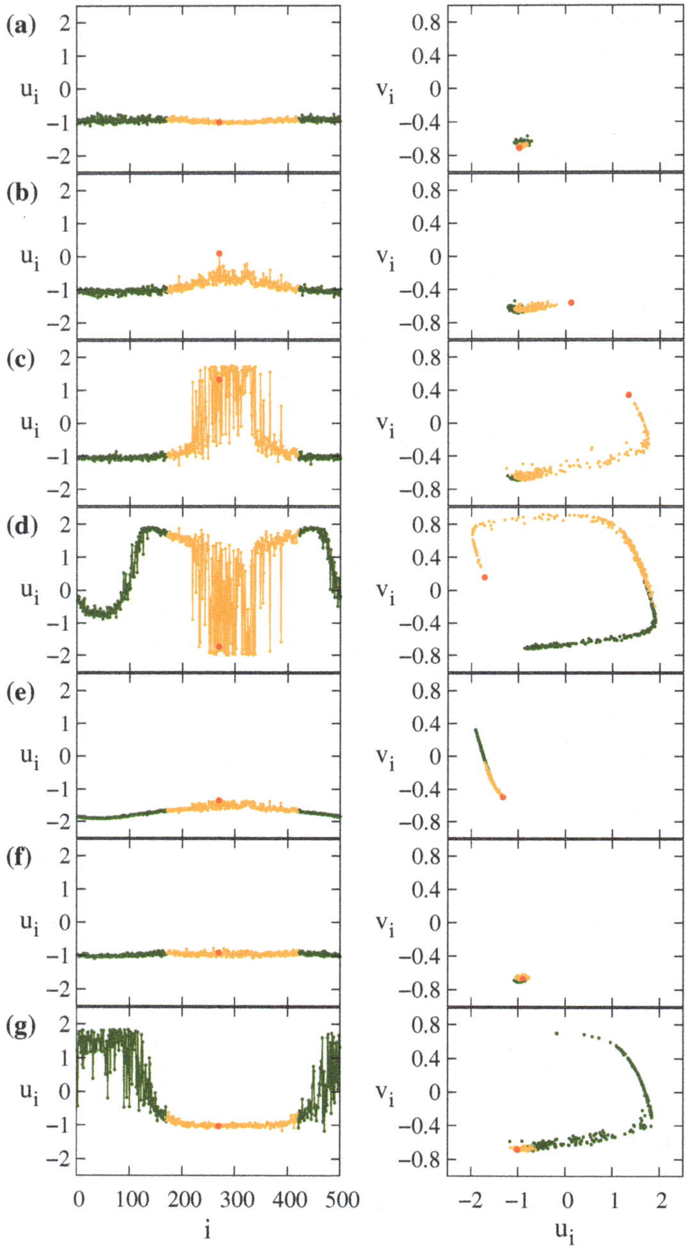

Fig. 3.4 Time evolution of coherence-resonance chimera: Snapshots (left column) and corresponding phase space (right column) for **a** $t = 995.7$, **b** $t = 996.1$, **c** $t = 996.4$, **d** $t = 996.9$, **e** $t = 998.5$, **f** $t = 1000.4$, **g** $t = 1001.3$. The node $i = 269$ is marked in red. Other parameters: $\varepsilon = 0.05$, $a = 1.001$, $D = 0.0002$, $\sigma = 0.4$, $r = 0.12$ [132]

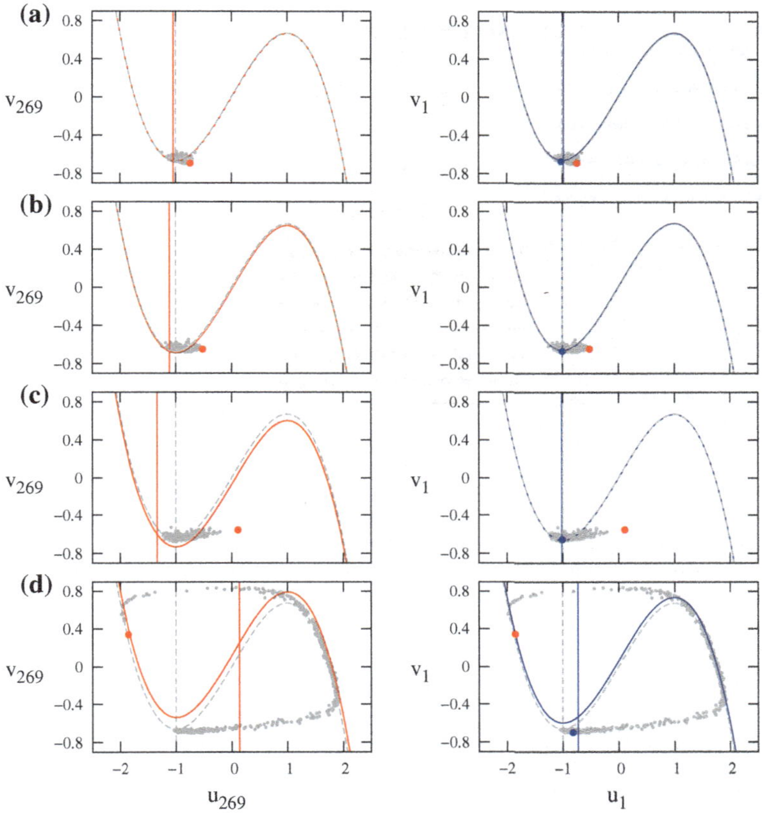

Fig. 3.5 Activator and inhibitor nullclines $\dot{u}$ and $\dot{v}$, respectively, for the selected nodes $i = 269$ (left column) and $i = 1$ (right column) of the system (3.1) in the coherence-resonance chimera regime for **a** $t = 995.90$, **b** $t = 996.00$, **c** $t = 996.10$, **d** $t = 996.8$. Parameters: $\varepsilon = 0.05$, $a = 1.001$, $\sigma = 0.4$, $r = 0.12$, $D = 0.0002$ [132]

3.3.4 Network Dynamics in the Presence of Strong Noise

Coherence-resonance chimeras appear to be a persistent phenomenon, which continues to exist for at least $T_{int} = 10^5$ dimensionless integration time units, which corresponds to ≈35,000 intrinsic periods. This discloses the constructive role of noise for the considered pattern in contrast to amplitude chimeras, which tend to have shorter lifetimes monotonically decreasing with increasing noise [102].

However, strong noise destroys coherence-resonance chimeras. For noise intensity $D > 0.000325$, the system (3.1) is incoherent in space but still very regular (approximately periodic) in time (Fig. 3.6a). In the case of even stronger noise, for instance, $D = 0.1$ (Fig. 3.6b), the behavior becomes incoherent in time and even more incoherent in space. Therefore, coherence-resonance chimeras appear for intermediate

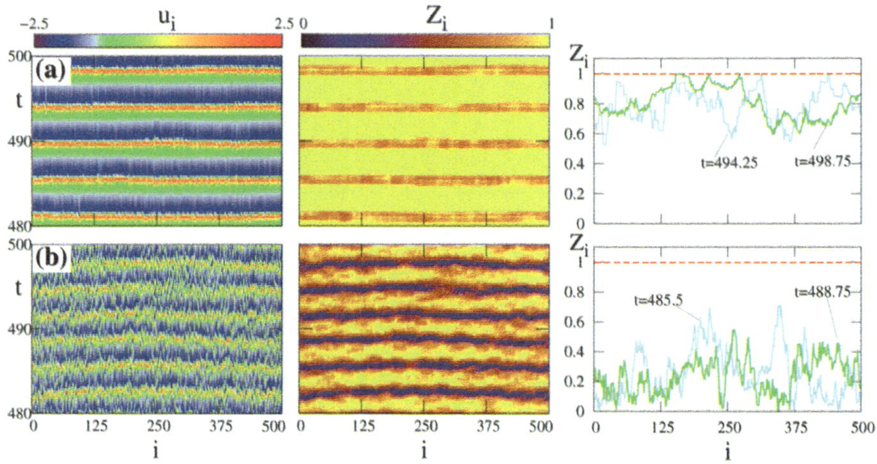

Fig. 3.6 Space-time plots of activator u_i (left column), local order parameter Z_i (middle column, $\delta_Z = 25$; coherence is identified by $Z_i > 1 - 10^{-6}$ for numerical purposes), and line scan of Z_i at fixed time (right column) for different values of noise intensity D: **a** $D = 0.0004$: incoherent in space but periodic in time, **b** $D = 0.1$: incoherent in space and time. Horizontal dashed line indicates the value $Z_k = 1$. Parameters: $\varepsilon = 0.05$, $a = 1.001$, $\sigma = 0.4$, $r = 0.12$ [132]

values of noise intensity, which is a characteristic signature of coherence resonance. Note that coherence-resonance chimeras occur in a network at much lower values of the noise intensity than coherence resonance in a single FHN system. This is due to the strong coupling of each element with its neighbors.

3.3.5 *Dynamic Regimes: The Impact of Coupling Parameters*

To gain an overview of the different regimes in the network, we fix the values of parameters ε, a, D, N, and tune r and σ (Fig. 3.7). As it has been shown in [131] strong coupling and a large number of nearest neighbors force the network to rest in the homogeneous steady state (region a). For weaker coupling and almost the whole range of r values, we detect spiking patterns which are approximately periodic in time and incoherent in space (region b). Coherence-resonance chimeras occur above a certain threshold $\sigma \approx 0.2$. Depending on the coupling range r, we find coherence-resonance chimeras with one, two, and three incoherent domains (regions c, d, and e, respectively). Therefore, the number of the incoherent domains can be increased by decreasing the coupling range r for fixed value of the coupling strength σ, which is a typical feature of "classical chimera states," cf. e.g., [69, 75, 78, 79, 151]. Coherence-resonance chimeras with two and three incoherent domains are shown in Fig. 3.8a and Fig. 3.8b, respectively.

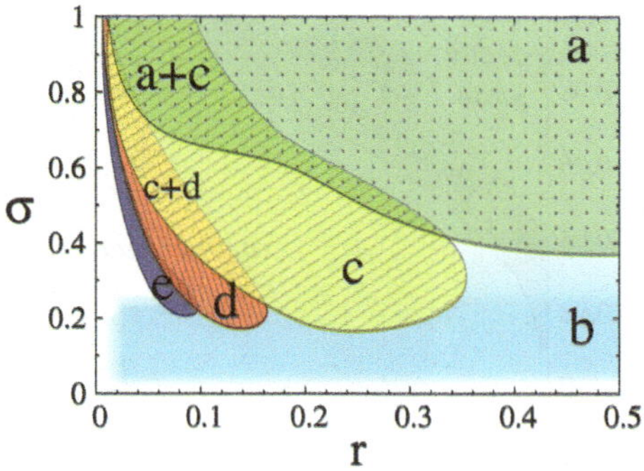

Fig. 3.7 Dynamic regimes in the (r, σ) parameter plane: **a** steady state (green dotted), **b** incoherent in space and periodic in time (blue plain), **c** coherence-resonance (CR) chimera with one incoherent domain (yellow cross-hatched) **d** CR chimera with two incoherent domains (orange cross-hatched) **e** CR chimera with three incoherent domains (purple plain). Multistability is also indicated. Other parameters: $\varepsilon = 0.05$, $a = 1.001$, $D = 0.0002$, $N = 500$ [132]

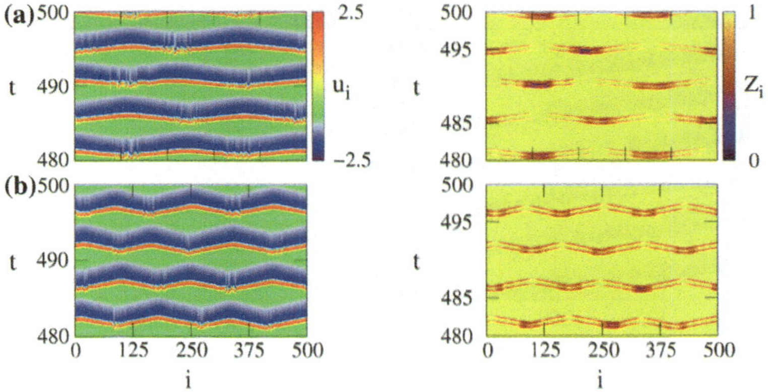

Fig. 3.8 Space-time plots (left column) and local order parameter (right column) for different values of coupling range **a** $r = 0.07$: coherence-resonance chimera with two incoherent domains, **b** $r = 0.04$: coherence-resonance chimera with three incoherent domains. Other parameters: $\varepsilon = 0.05$, $a = 1.001$, $D = 0.0002$, $\sigma = 0.4$ [132]

It is important to note that near the borders of the different regimes multistability is observed (regions a+c and c+d in Fig. 3.7), and the initial conditions determine the particular pattern.

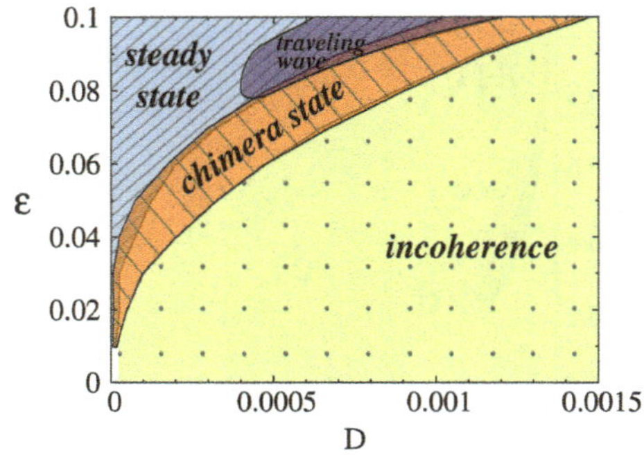

Fig. 3.9 Dynamic regimes in the (ε, D) parameter plane: incoherent in space and periodic in time (yellow dotted); coherence-resonance (CR) chimera with one incoherent domain (orange cross-hatched); steady state (blue cross-hatched); traveling waves (purple cross-hatched). Other parameters: $a = 1.001$, $N = 100$, $\sigma = 0.4$, $r = 0.2$ [132]

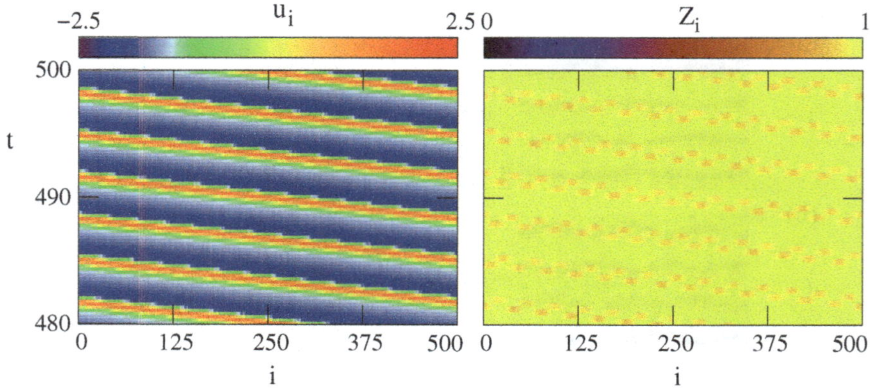

Fig. 3.10 Traveling wave patterns. Parameters: $\epsilon = 0.09$, $D = 0.0006$, $r = 0.2$, $\sigma = 0.4$, $a = 1.001$, $N = 500$ [132]

3.3.6 Characterization of Coherence-Resonance Chimera

To understand how the behavior of coherence-resonance chimeras depends on the parameters of the FitzHugh-Nagumo system, we investigate first the impact of the timescale separation parameter ε. For $\varepsilon = 0.05$, coherence-resonance chimeras are observed for intermediate values of noise intensity ($0.000062 \leq D \leq 0.000325$). To find out whether this also holds for other values of ε, we analyze the patterns occurring in the network in the (ε, D)-plane (Fig. 3.9). Indeed, we detect noise-induced chimera states for a wide range of the timescale separation parameter $0.01 \leq \varepsilon \leq 0.1$. For increasing ε, stronger noise is required to achieve coherence-resonance chimeras and at the same time the interval of noise values within which they occur is enlarged.

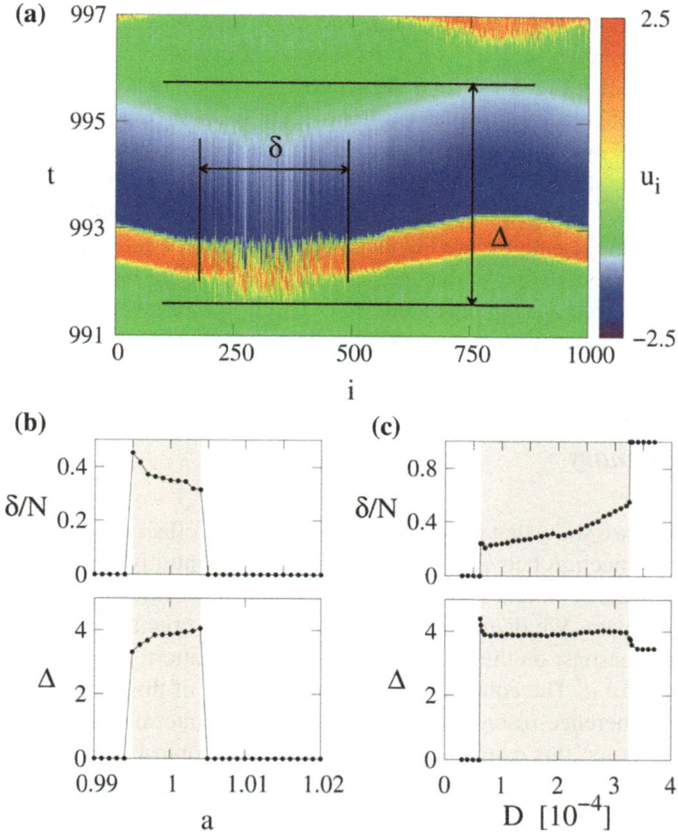

Fig. 3.11 Characterization of CR chimera: **a** Space-time plot defining active time Δ and size δ of the incoherent domain. **b**, **c** Dependence of δ/N and Δ upon excitation threshold a for $D = 0.0002$ (**b**) and upon noise intensity D for $a = 1.001$ (**c**). Gray region corresponds to the existence of CR chimeras. Other parameters: $\varepsilon = 0.05$, $N = 1000$, $\sigma = 0.4$, $r = 0.2$ [132]

Additionally, for large values of the timescale separation parameter $\varepsilon > 0.075$, there occurs a regime of coherent traveling waves (Fig. 3.10).

To further deepen our understanding of coherence-resonance chimeras, we analyze the impact of the excitation threshold a. Since chimera states in the deterministic FHN model have been previously observed only in the oscillatory regime for $|a| < 1$ [79], we investigate if coherence-resonance chimeras are sensitive to the choice of a. For that purpose, we consider two characteristic quantities: (i) the normalized size of the incoherent domain δ/N, where δ is the number of elements in the incoherent domain (Fig. 3.11a); (ii) the active time span of the chimera Δ, which measures the time from the excitation of the first node belonging to the incoherent domain till the return of the last node to the rest state (Fig. 3.11a). This is analogous to the pulse duration for the single FHN model [135] but takes into account that different nodes spike at distinct moments of time depending on the domain. As it has been shown in [131] for increasing a, the incoherent domain size δ/N shrinks (top panel in Fig. 3.11b) and the active time span Δ increases (bottom panel). Interestingly, coherence-resonance chimeras occur for both oscillatory and excitable regimes of FHN systems, but they exist only for a restricted interval of the threshold parameter a (shaded region $0.995 \leq a \leq 1.004$ in Fig. 3.11b). To the left of this interval, the dynamics is completely synchronized in space and periodic in time, while to the right the patterns are incoherent in space and periodic in time (similar to Fig. 3.6a). Figure 3.11c shows that δ/N increases with noise intensity D (top panel of Fig. 3.11c), while Δ is independent of D within the interval of existence of the coherence-resonance chimeras $0.000062 \leq D \leq 0.000325$ (bottom panel).

3.3.7 *Summary*

In conclusion, we show that noise can have a beneficial effect on chimera states and establish a connection between two intriguing counterintuitive phenomena: coherence resonance and chimera states. Therefore, we call the resulting pattern coherence-resonance chimera. We demonstrate that noise plays a crucial role for this pattern for two main reasons: on the one hand, it induces the pattern and on the other hand allows to control it. The coherence-resonance attribute of this pattern discloses the first aspect: coherence-resonance chimeras appear for intermediate values of noise intensity. However, this can also be viewed from the control perspective: by properly choosing the noise intensity, we achieve the desired regime of the network: steady state, coherence-resonance chimera, or other patterns. Indeed, by fine-tuning the noise intensity, we can adjust the size δ of the incoherent domain of the chimera pattern. While the active time span remains fixed for all noise intensities within the interval of existence, the size of the incoherent domain δ essentially grows with increasing noise intensity. An important aspect is also that these novel coherence-resonance chimeras in a neural network under the influence of noise exhibit alternating chimera behavior, i.e., the coherent and incoherent domains switch position periodically. We explain the mechanism of this alternating behavior by examining the coupling terms and the resulting shift of the nullclines.

3.4 Time-Delayed Feedback Control of Chimera States

It is well known that in the presence of time delay simple dynamical systems can exhibit complex behavior, such as delay-induced bifurcations [326], delay-induced multistability [327], stabilization of unstable periodic orbits [328], or stationary states [329], to name just a few examples. Time delays can be implemented into the feedback or into the coupling in the case of interacting units. It has been previously shown that time-delayed feedback allows to modulate coherence resonance in excitable [299,

303] as well as in non-excitable systems [138, 139]. In particular, for appropriate choices of feedback delay, either suppression or enhancement of coherence resonance can be achieved.

Chimera states have been investigated for noisy systems [102] and delayed systems as well. In general, chimera patterns tend to form clusters in the presence of time delay [208, 330]. The role of time-delayed coupling has been previously investigated in two-population networks of oscillators [207]. In particular, it has been reported that coupling delay induces globally clustered chimera states in which the coherent and incoherent regions span both populations [331, 332]. Further, chimera states have been detected in multilayer networks with time-delayed coupling [150, 251, 333]. Experimental evidence for chimera states in systems with time delay has been provided for chemical oscillators [88] and electronic or optoelectronic systems [92, 153]. Internal delayed feedback has been shown to induce chimeras in systems of globally coupled phase oscillators [108] and laser networks [93]. Chimera states in the presence of both delayed feedback and noise have been investigated in [143].

In this section, we investigate the interplay of noise and time-delayed feedback in a network of nonlocally coupled excitable elements and mainly focus on the role of feedback for coherence-resonance chimeras.[1] A distinctive feature of this type of chimera is that it is induced by noise and occurs in a certain restricted interval of noise intensity and system parameters. The question we address in the present section is whether these intervals can be increased by introducing time-delayed feedback. By exploring the impact of time delay, we uncover the mechanisms to control coherence-resonance chimeras by time-delayed feedback. Our results show that applying time-delayed feedback promotes the occurrence of coherence-resonance chimeras and induces new regimes. In particular, we show that feedback increases the parameter intervals of existence of chimera states and has a significant impact on their alternating dynamics leading to the appearance of novel patterns, which we call *period-two coherence-resonance chimera.*

The section is organized as follows. First, we analyze coupling terms, delay term, and the nullclines for the regime of coherence-resonance chimeras in the presence of time-delayed feedback (Sect. 3.4.1). Next, we explore dynamic regimes for different values of feedback strength and feedback delay time (Sect. 3.4.2). Further, we study the impact of the feedback on the noise intensity range (Sect. 3.4.3) and the threshold parameter range (Sect. 3.4.4) where coherence-resonance chimeras exist. Moreover, we investigate period-two coherence-resonance chimeras that are induced by delay (Sect. 3.4.4). Section 3.4.5 summarizes the results on the dynamics with delay.

[1]Portions of the following text have been re-printed from [133] with the permission of AIP Publishing.

3.4.1 Coherence-Resonance Chimeras in the Presence of Time-Delayed Feedback

We consider a ring of N identical nonlocally coupled FitzHugh-Nagumo (FHN) systems with time-delayed feedback in the presence of Gaussian white noise (see (3.3) in Sect. 3.2). As in Sect. 3.3 here we also fix the coupling matrix parameter $\phi = \pi/2 - 0.1$. In the absence of time delay $\tau = 0$ chimera states have been found for this value of ϕ in both the deterministic oscillatory [79] and the noisy excitable regime [131, 132]. Moreover, it has been shown that chimera states occurring in the excitable regime [131, 132] are different from those detected in the oscillatory regime [79] (see Sect. 3.3.2). In the presence of Gaussian white noise, a special type of chimera state called *coherence-resonance chimera* appears in a ring of N nonlocally coupled excitable FHN systems (Fig. 3.12).

In the present section, to control these patterns we introduce time-delayed feedback to the activator variable in (3.3). For that purpose, we fix all the parameters of the system in the regime of coherence-resonance chimera and vary those characterizing the feedback term: γ and τ. For $\gamma = 0$ (3.3) demonstrate coherence-resonance chimeras with the period $T \approx 4.76$. This regime can also be observed in the presence of time-delayed feedback for $\gamma = 0.2$, $\tau = 1.0$ and is shown as a space-time plot color-coded by the variable u_i in Fig. 3.12a, b. One can clearly distinguish the regions of coherent and incoherent spiking.

To characterize spatial coherence and incoherence of chimera states, one can use the local order parameter Z_i (see Sect. 3.3.1). Figure 3.12c represents a space-time plot color-coded by Z_i and illustrates coexistence of coherent and incoherent domains with the latter characterized by values of Z_i noticeably below unity (dark regions).

As discussed in Sect. 3.3, one of the main features of these noise-induced chimera states is their alternating behavior which is absent in the oscillatory regime without noise. In more detail, the incoherent domain of the chimera pattern switches periodically its position on the ring, although its width remains fixed (Fig. 3.12b, c). This property has been described in Sect. 3.3.3 where the explanation based on the time evolution of the coupling term has been provided [132]. Taking into account that the system (3.3) involves both direct and cross-couplings between activator u and inhibitor v variables, in total we have four coupling terms. It turns out that coupling terms form patterns shown as space-time plots in Fig. 3.13a–d.

The crucial point is that the coupling acts as an additional term and shifts the nullclines of every individual element of the network. The coupling term with the strongest impact corresponds to cross-coupling for the variable v (Fig. 3.13c). It means that the coupling significantly influences the $\dot{u} = 0$ nullcline and shifts the threshold parameter a which is responsible for the excitation. As a result for a certain group of nodes, the threshold becomes lower due to coupling, and the probability of being excited by noise increases. Therefore, the elements of this group are the first to start the large excursion in the phase space and experience random spiking. The elements constituting the rest of the network spike coherently since they are pulled by already excited nodes and are, therefore, excited by coupling and not by

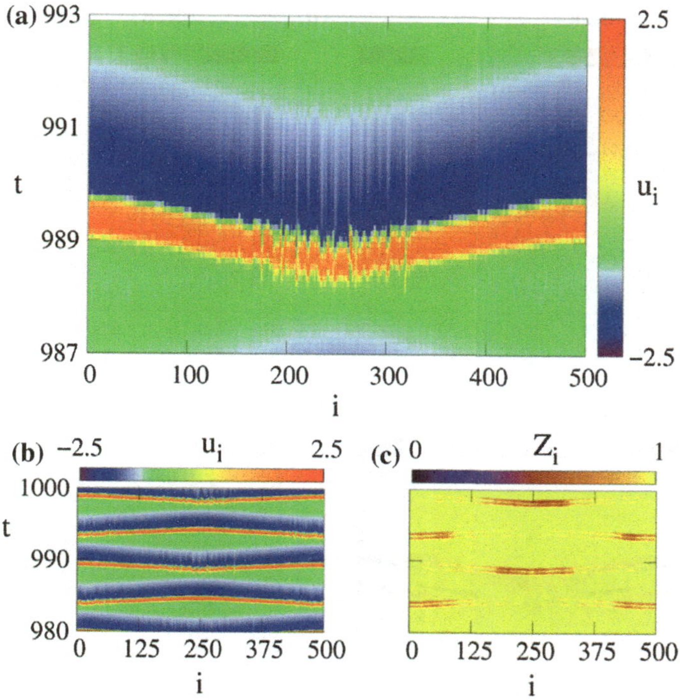

Fig. 3.12 **a**, **b** Space-time plots and **c** local order parameter for the coherence-resonance chimera. Initial conditions: randomly distributed on the circle $u^2 + v^2 = 4$. Parameters: $N = 500$, $\varepsilon = 0.05$, $\phi = \pi/2 - 0.1$, $a = 1.001$, $\sigma = 0.4$, $r = 0.2$, $D = 0.0002$, $\gamma = 0.2$, $\tau = 1.0$ [133]

noise. This scenario can also be obtained for the system (3.3) in the presence of time-delayed feedback (Fig. 3.13). Due to the feedback, an additional term appears in (3.3) and should be taken into account. Its evolution in time for all nodes of the network is shown in Fig. 3.13e. The color-code bar clearly indicates that the values of the feedback term are larger than those of the coupling terms. However, for the chosen value of delay time $\tau = 1.0$, the feedback does not have any essential impact on the behavior of coherence-resonance chimeras since it is less than the intrinsic period of oscillations $T = 4.76$ (Figs. 3.12 and 3.13).

For the better understanding of this alternating dynamics in the presence of time-delayed feedback, we study the impact of the coupling on activator and inhibitor nullclines for selected nodes of the system (3.3). In particular, we consider a sequence of phase portraits for the nodes $i = 241$ (red dot) and $i = 1$ (blue dot) which belong to the incoherent and coherent domains, respectively (Fig. 3.14). First, all the elements are located near the steady state (Fig. 3.14a). After a while, the vertical nullcline of the node $i = 241$ is shifted to the left of the value $u = -a = -1.001$ due to positive

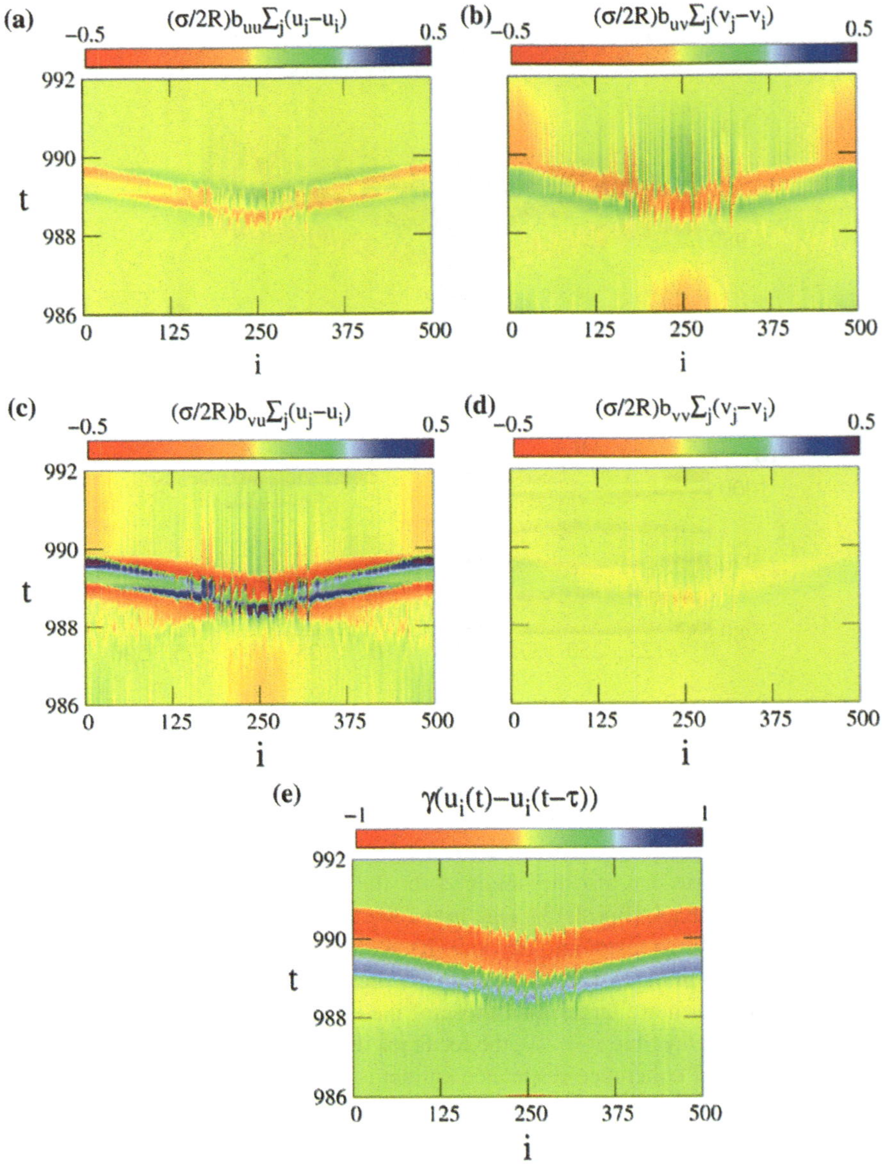

Fig. 3.13 Space-time plots of coupling terms for u_i and v_i variables in the coherence-resonance chimera regime: **a** direct coupling for the u_i variable, **b** cross-coupling for the u_i variable, **c** cross-coupling for the v_i variable, **d** direct coupling for the v_i variable. **e** Space-time plot of the delay term. Parameters: $N = 500$, $\varepsilon = 0.05$, $\phi = \pi/2 - 0.1$, $a = 1.001$, $\sigma = 0.4$, $r = 0.2$, $D = 0.0002$, $\gamma = 0.2$, $\tau = 1.0$ [133]

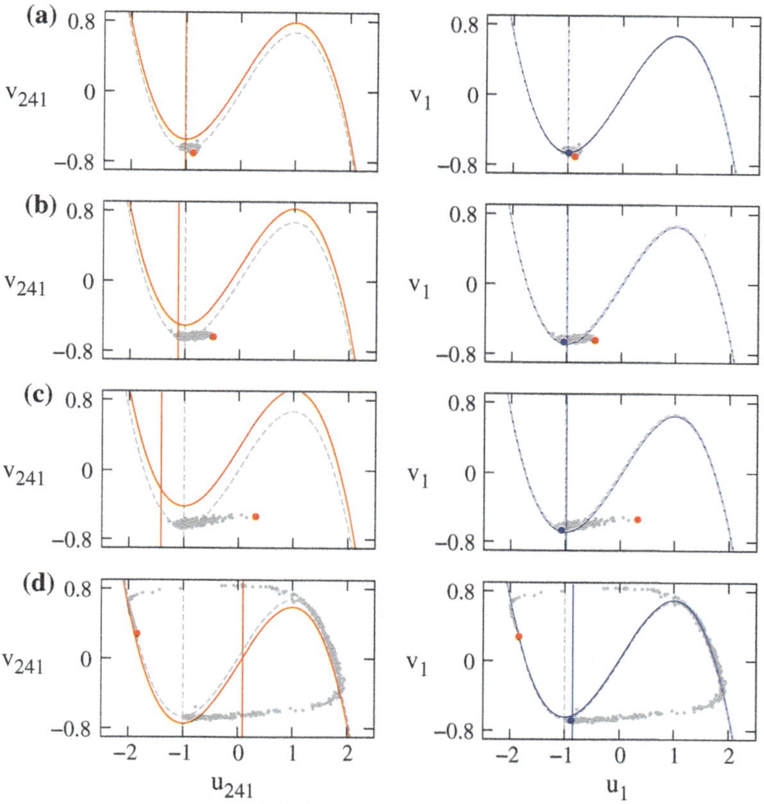

Fig. 3.14 Activator and inhibitor nullclines $\dot{u}_i$ and $\dot{v}_i$, respectively, for the selected nodes $i = 241$ (left column) and $i = 1$ (right column) of the system (3.3) in the coherence-resonance chimera regime for **a** $t = 997.4$, **b** $t = 997.6$, **c** $t = 997.7$, **d** $t = 998.4$. Parameters: $N = 500$, $\varepsilon = 0.05$, $a = 1.001$, $\sigma = 0.4$, $r = 0.2$, $D = 0.0002$, $\gamma = 0.2$, $\tau = 1.0$ [133]

coupling term (panel b). Consequently, this node can be more easily excited by noise (panel c). Due to nonlocal coupling, the excited node pulls its neighbors and they also start spiking. The coupling can also shift the vertical nullcline to the right of the value $u = -a = -1.001$ (panel d).

Similar results have been discussed in Sect. 3.3 for the case without time-delayed feedback [132]. Therefore, for the strength $\gamma = 0.2$ and time delay $\tau = 1.0$ the feedback does not have an impact on the nullclines. Consequently, coherence-resonance chimeras observed for small time delay of the feedback are the same as in the case without feedback.

3.4.2 Dynamic Regimes in the Presence of Time-Delayed Feedback

Since our main goal is to study the impact of time-delayed feedback, we now choose the parameters of the system in the regime of coherence-resonance chimera and vary only the feedback parameters γ and τ. For fixed feedback strength $\gamma = 0.4$ we observe the change of dynamic regimes by tuning the delay time τ. For $\tau = 3.6$ all the nodes of the network spike coherently, i.e., in-phase synchronization occurs (Fig. 3.15a). The feedback with $\tau = 2.2$ shifts the system into the regime which is incoherent in space and periodic in time: all the nodes demonstrate spiking behavior, but the spiking events of the neighboring nodes are not correlated (Fig. 3.15b).

To gain a general view of the dynamics in the network of nonlocally coupled noisy excitable elements in the presence of time-delayed feedback, we construct the map of regimes of the system (3.3) in the (γ, τ) parameter plane (Fig. 3.16). For visualization reasons, we have divided the map into two panels: panel (a) corresponds to the τ interval from 0 to 7 and includes the values $\tau \leq T$, where T is the period of the dynamics without delay ($T \approx 4.76$); panel (b) corresponds to larger values of τ including $\tau \approx 2T \approx 9.52$.

Note that the other parameters of the network are chosen in the coherence-resonance chimera state which now occurs only for certain intervals of delay time τ. We detect three main regions (yellow (light gray) in Fig. 3.16) separated by in-phase

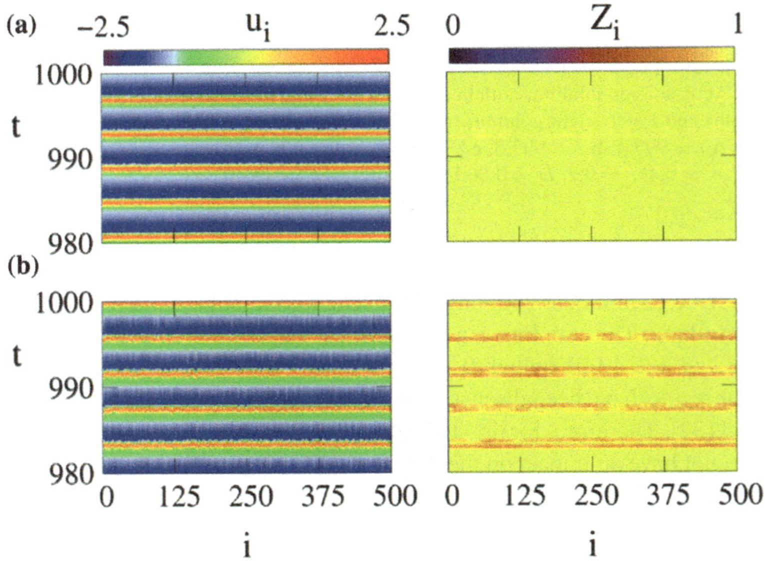

Fig. 3.15 Space-time plots for the variable u_i (left panels) and local order parameter Z_i in the regime of **a** complete in-phase synchronization for $\gamma = 0.4$, $\tau = 3.6$ and **b** spatial incoherence for $\gamma = 0.4$, $\tau = 2.2$. Other parameters: $N = 500$, $\varepsilon = 0.05$, $a = 1.001$, $\phi = \pi/2 - 0.1$, $D = 0.0002$, $r = 0.2$, $\sigma = 0.4$ [133]

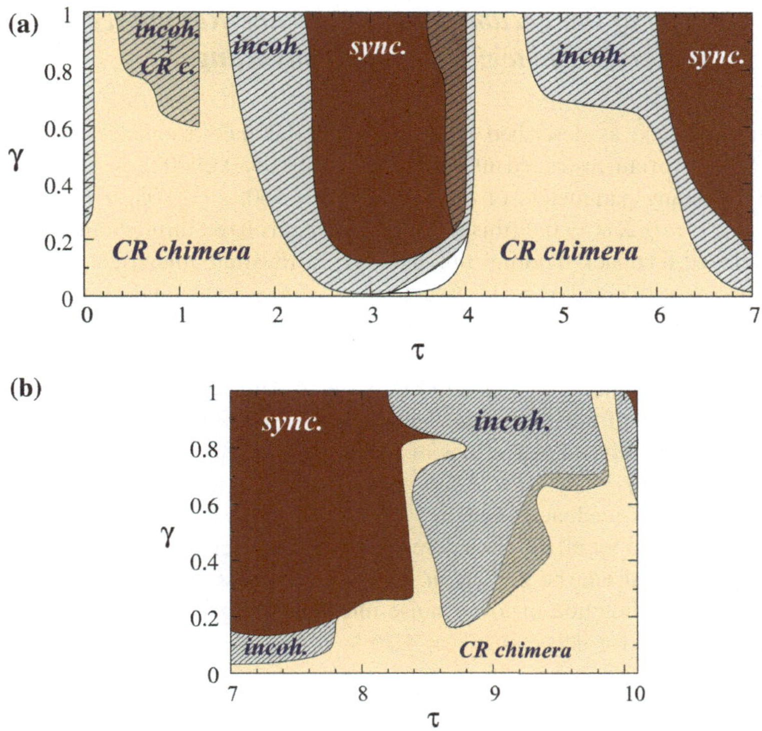

Fig. 3.16 Dynamic regimes in the (τ, γ) parameter plane. Red (dark gray) regions: in-phase synchronization (see space-time plot in Fig. 3.15a); hatched regions: spatial incoherence (see space-time plot in Fig. 3.15b); white region: steady state; yellow (light gray) regions: coherence-resonance (CR) chimeras. Parameters: $N = 500$, $\varepsilon = 0.05$, $a = 1.001$, $\phi = \pi/2 - 0.1$, $D = 0.0002$, $r = 0.2$, $\sigma = 0.4$ [133]

synchronization domains (red (dark gray) regions in Fig. 3.16) and regimes of spatially incoherent spiking (hatched regions Fig. 3.16). Although the map of regimes is dominated by various oscillatory patterns, for relatively small feedback strength $\gamma < 0.2$ and time delay $3.2 < \tau < 4$ we also observe a small regime of steady state (white region in Fig. 3.16a).

Moreover, the diagram is characterized by multistability since spatially incoherent spiking can coexist with chimera states or in-phase synchronization. The overall structure of the map resembles a sequence of synchronization tongues although there are no clear resonances for delay times equal to the multiples of the intrinsic period $T \approx 4.76$. Nevertheless, applying the feedback with delay time $\tau \approx 2T \approx 9.52$ does not change the dynamics dramatically, and the regime of coherence-resonance chimera is still observed for a wide range of feedback strength (Fig. 3.16).

3.4.3 *Impact of the Feedback on Coherence-Resonance Chimera Existence: Noise Intensity Range*

Without feedback, as described in Sect. 3.3.6, coherence-resonance chimeras are observed for a certain restricted interval of noise intensity $0.000062 \leq D \leq 0.000325$ for the following parameters of the system: $N = 500$, $\varepsilon = 0.05$, $a = 1.001$, $\phi = \pi/2 - 0.1$, $r = 0.2$, $\sigma = 0.4$ (this set of parameters is fixed throughout Sect. 3.4.3). Time-delayed feedback modifies this interval. To illustrate this effect, we consider two cases: $\gamma < 0.5$ and $\gamma > 0.5$ which allows for a better understanding of the impact of feedback strength on this interval. Also for the two values of parameter γ, we choose different delay times τ from all three regions of coherence-resonance chimeras shown in the (τ, γ) plane in Fig. 3.16a, b. Time-delayed feedback slightly changes the range of noise intensity values where chimera states occur in the system (3.3) for both considered values of feedback strength: $\gamma = 0.2$ (Fig. 3.17) and $\gamma = 0.6$ (Fig. 3.18).

For rather weak feedback strength $\gamma = 0.2$, the interval of existence of chimera patterns is enlarged for all the considered delay times. Interestingly, the right boundary of this interval can be shifted in the direction of stronger noise (Fig. 3.17b) as well as in the direction of lower noise intensities (Fig. 3.17a, d, e) and remains almost unchanged for delay time $\tau = 4.76 \approx T$ (Fig. 3.17c). Therefore, by appro-

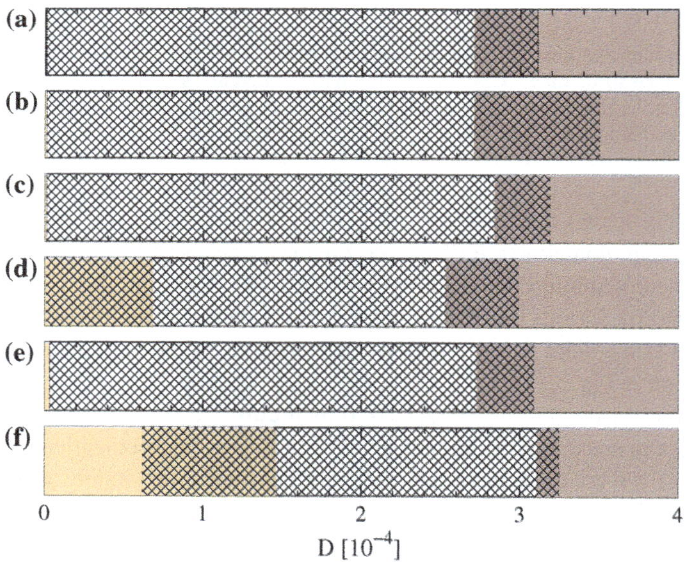

Fig. 3.17 Dynamic regimes depending on the noise intensity D for feedback strength $\gamma = 0.2$ and different values of delay time: **a** $\tau = 9.52$, **b** $\tau = 6.0$, **c** $\tau = 4.76$, **d** $\tau = 1.8$, **e** $\tau = 0.8$, **f** $\tau = 0$. Dynamic regimes: steady state (yellow/light gray); spatially incoherent spiking (pink/dark gray); coherence-resonance chimeras (hatching). Other parameters: $N = 500$, $\varepsilon = 0.05$, $a = 1.001$, $\phi = \pi/2 - 0.1$, $r = 0.2$, $\sigma = 0.4$ [133]

priately choosing the feedback delay time, one can adjust the value of noise intensity for which spatially incoherent spiking replaces coherence-resonance chimeras within the interval $0.00030 \le D \le 0.00035$ (Fig. 3.17). The transition from the steady state to coherence-resonance chimeras for increasing noise occurs at the left boundary (Fig. 3.17f) which is shifted by the feedback to smaller noise intensities (Fig. 3.17e). Furthermore, due to feedback, chimera states appear even at zero noise intensity (Fig. 3.17a–d). Therefore, time-delayed feedback promotes coherence-resonance chimeras not only by increasing the noise range where they exist, but also by inducing these patterns in the absence of noise. The largest range of D corresponds to $\tau = 6.0$ (Fig. 3.17b). It is important to note that on the borders of the intervals the multistability is observed. Chimera states can coexist with the steady state on the left border and with the regime of spatially incoherent spiking on the right border.

Large feedback strength $\gamma = 0.6$ can also shift the left boundary of the chimera interval to lower (Fig. 3.18a, b) and even zero (Fig. 3.18c) noise values. The multistability on the borders also occurs. Interestingly, for $\gamma = 0.6$ the chimera state overlaps with the complete synchronization regime on the left boundary and not with the steady state as in the case of weak feedback strength. The right boundary strongly depends on τ and shifts into the direction of lower noise intensities (Fig. 3.18a–c). The largest detected interval for $\gamma = 0.6$ corresponds to $\tau = 4.76 \approx T$ (Fig. 3.17b) and for $\tau = 9.52 \approx 2T$ we even observe shrinking of the interval (Fig. 3.18a).

If we compare the interval of chimera existence without time-delayed feedback $0.000062 \le D \le 0.000325$ (Figs. 3.17f and 3.18d) with the interval the most enlarged by the feedback $0.000001 \le D \le 0.00035$, it turns out that we achieve 33% improvement rate.

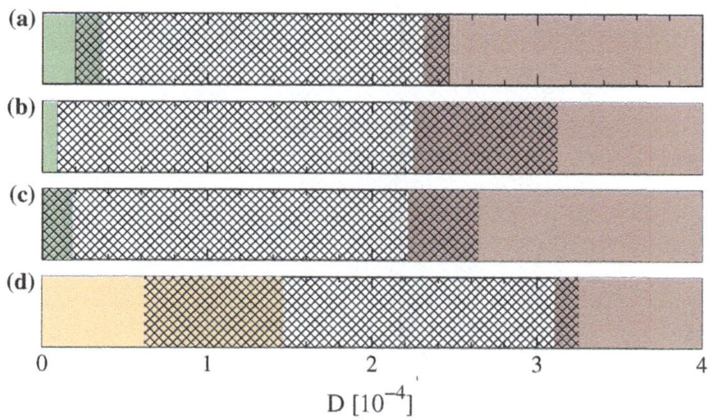

Fig. 3.18 Dynamic regimes depending on the noise intensity D for feedback strength $\gamma = 0.6$ and different values of delay time: **a** $\tau = 9.52$, **b** $\tau = 4.76$, **c** $\tau = 0.8$, (d) $\tau = 0$. Dynamic regimes: steady state (yellow/light gray); spatially incoherent spiking (pink/dark gray); synchronization (green/gray); coherence-resonance chimeras (hatching). Other parameters: $N = 500$, $\varepsilon = 0.05$, $a = 1.001$, $\phi = \pi/2 - 0.1$, $r = 0.2$, $\sigma = 0.4$ [133]

3.4.4 Impact of the Feedback on Coherence-Resonance Chimera Existence: Threshold Parameter Range

In Sect. 3.3.6, we have shown that coherence-resonance chimera can be obtained only in a small interval of a ($0.995 \le a \le 1.004$). To analyze the impact of time-delayed feedback, we again consider two cases: $\gamma = 0.2$ and $\gamma = 0.6$ and different values of delay time. Figure 3.19 corresponds to the case of small feedback strength $\gamma = 0.2$, and Fig. 3.20 illustrates the results for the case of larger feedback strength $\gamma = 0.6$.

For the two considered values of γ, time-delayed feedback significantly changes the range of the threshold parameter a where coherence-resonance chimeras exist. Moreover, in both cases this interval is increased the most when the delay time is equal to the intrinsic period of the system $\tau = 4.76 \approx T$ (Figs. 3.19c and 3.20b). However, smaller feedback strength allows for stronger enlargement of the interval: for $\gamma = 0.2$ and $\tau = 4.76$, it is $0.993 \le a \le 1.017$ and is more than doubled compared to the case without feedback $0.995 \le a \le 1.004$ (Fig. 3.19c).

While tuning the threshold parameter a, we observe multistability on the boundaries of the coherence-resonance chimera regime where this pattern coexists with spatially incoherent spiking (Figs. 3.19a–d and 3.20a–c). For increasing parameter a, the coherence-resonance chimeras disappear in the absence of feedback, and a steady state is observed (Figs. 3.19f and 3.20d). However, for $\gamma = 0.2$, $\tau \ge T$ (Fig. 3.19a–c)

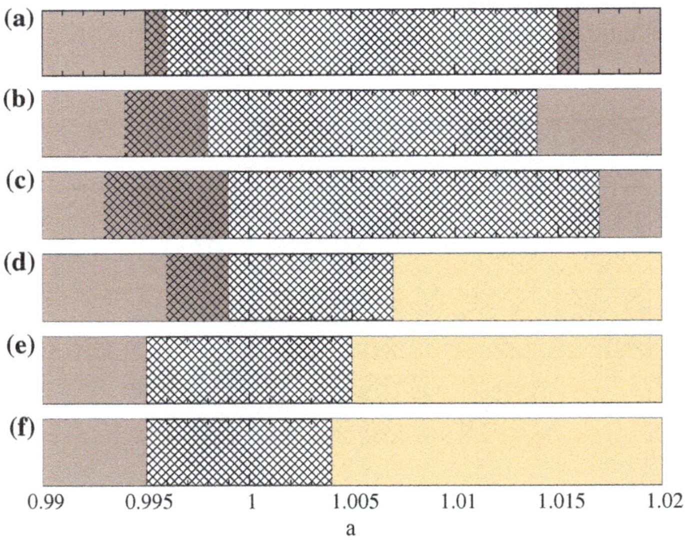

Fig. 3.19 Dynamic regimes depending on the threshold parameter a for feedback strength $\gamma = 0.2$ and different values of delay time: **a** $\tau = 9.52$, **b** $\tau = 6.0$, **c** $\tau = 4.76$, **d** $\tau = 1.8$, **e** $\tau = 0.8$, **f** $\tau = 0$. Dynamic regimes: steady state (yellow/light gray); spatially incoherent spiking (pink/dark gray); coherence-resonance chimeras (hatching). Other parameters: $N = 500$, $\varepsilon = 0.05$, $D = 0.0002$, $\phi = \pi/2 - 0.1$, $r = 0.2$, $\sigma = 0.4$ [133]

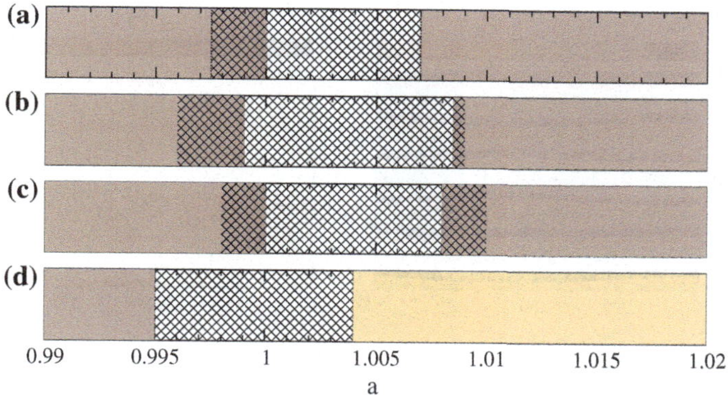

Fig. 3.20 Dynamic regimes depending on the threshold parameter a for feedback strength $\gamma = 0.6$ and different values of delay time: **a** $\tau = 9.52$, **b** $\tau = 4.76$, **c** $\tau = 0.8$, **d** $\tau = 0$. Dynamic regimes: steady state (yellow/light gray); spatially incoherent spiking (pink/dark gray); coherence-resonance chimeras (hatching). Other parameters: $N = 500$, $\varepsilon = 0.05$, $D = 0.0002$, $\phi = \pi/2 - 0.1$, $r = 0.2$, $\sigma = 0.4$ [133]

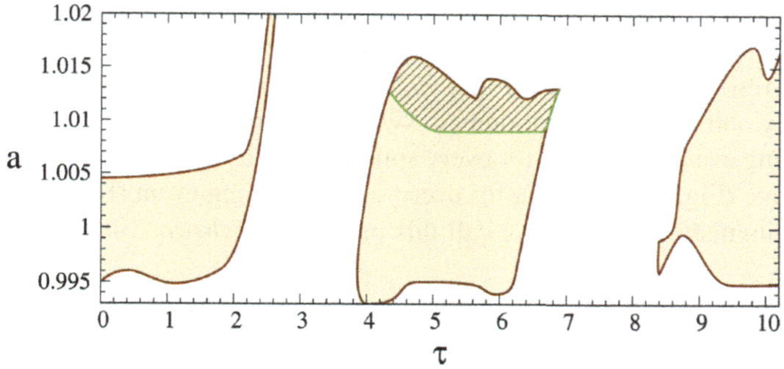

Fig. 3.21 Coherence-resonance chimera states in (a, τ)-plane (yellow/gray regions). The hatched region corresponds to the period-two coherence-resonance chimera. Parameters: $N = 500$, $\varepsilon = 0.05$, $D = 0.0002$, $\phi = \pi/2 - 0.1$, $r = 0.2$, $\sigma = 0.4$, $\gamma = 0.2$ [133]

and for all considered values of time delay in the case of strong feedback $\gamma = 0.6$ (Fig. 3.20a–c) the steady state is replaced by spatially incoherent spiking, i.e., the feedback induces oscillatory behavior of the network.

As it can be seen from Fig. 3.19, for decreasing delay time τ from 9.52 to 0, we observe a nonlinear modulation of the size of the a-interval of existence of chimera states. To gain more insight into this effect, we define the parameter range for which this pattern exists in the (a, τ) plane (Fig. 3.21). We detect isolated regions occurring for certain disconnected intervals of τ. The region centered at the time delay value close to the intrinsic period of the system $\tau = 4.76 \approx T$ clearly indicates the enlargement of the a-interval.

Interestingly, at the top of this region for $\tau \approx T$ and $a > 1.01$, we find a novel chimera regime (hatching in Fig. 3.21) which is induced by time-delayed feedback

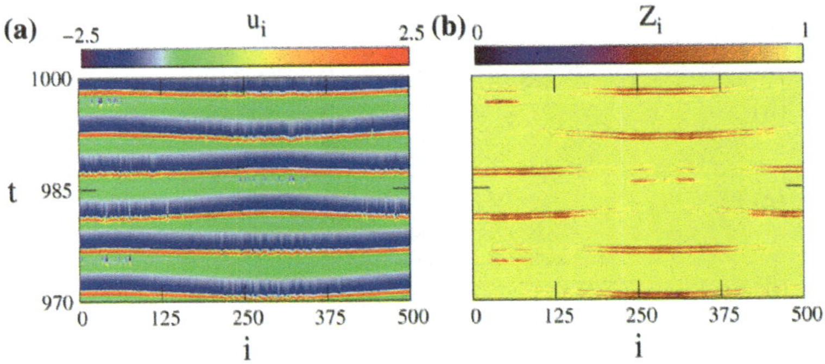

Fig. 3.22 Space-time plot for the variable u_i (**a**) and local order parameter Z_i (**b**) in the regime of period-two coherence-resonance chimera. Initial conditions: randomly distributed on the circle $u^2 + v^2 = 4$. Incoherent domains are marked by rectangles in panel (a). Parameters: $N = 500$, $\varepsilon = 0.05$, $a = 1.012$, $\sigma = 0.4$, $r = 0.2$, $D = 0.0002$, $\gamma = 0.2$, $\tau = 4.76$ [133]

and has not been observed for the system (3.3) without delay. The space-time plot for the variable u_i and the local order parameter indicate the coexistence in space of coherent and incoherent spiking as well as alternating behavior, typical features of coherence-resonance chimeras (Fig. 3.22). Furthermore, the alternation takes place periodically and the incoherent domain switches its position on the ring. However, the switching events occur not for every spiking cycle as in the coherence-resonance chimera state (Fig. 3.12b, c), but for every second spiking event (Fig. 3.22a, b). Due to this distinguishing feature, we call this pattern *period-two coherence-resonance chimera*.

To understand the mechanism of this alternation we consider a temporal sequence of the phase portraits of the system (3.3) with the nullclines indicated for four selected nodes of the network. As it can be seen from Fig. 3.22, the incoherent domain alternates between two regions: the first region corresponds to the nodes $i \in [0, 125], [450, 500]$ and the second region is $i \in [200, 375]$. For this reason, we choose nodes $i = 69$ and $i = 60$ from the first region and nodes $i = 231$ and $i = 284$ from the second region. Next, we analyze their dynamics during one period (Fig. 3.23).

We begin our observation when the incoherent spiking occurs in the second region $i \in [200, 375]$. The node $i = 231$ (red in Fig. 3.23 right column) starts the excursion in the phase space first since its nullclines are shifted and it can, therefore, be excited more easily by noise (right panel in Fig. 3.23a). At the same time, the elements $i = 60$ and $i = 69$ from the coherent domain (first region) rest in the steady state since their nullclines are unchanged (left panel in Fig. 3.23a). Next, the nullclines of the other nodes from the second region are modified (see node $i = 284$ (blue) in the right panel of Fig. 3.23b, c). Consequently, they are now also excited by noise and, therefore, incoherently (right panel in Fig. 3.23c), while the elements from the coherent domain still stay in the vicinity of the steady state with the nullclines unchanged (left panel in Fig. 3.23c). Further, when the nodes from the incoherent domain are well on the way (right panel in Fig. 3.23c), they pull the nodes from the first region that, therefore, also start spiking (left panel in Fig. 3.23c). Since they are excited not by noise but due to the pulling of the neighbors, their spiking is coherent. After performing a spike, all the nodes return to the steady state (Fig. 3.23d). This scenario is typical for coherence-resonance chimeras (see Fig. 3.14) for which the interchange of coherent and incoherent domains takes place during each subsequent excitation. However, this is not the case for period-two coherence-resonance chimera as we see below.

Next, we consider the second excitation for the elements of the network (3.3). At the next moment in time, a small group of nodes including $i = 60$ and $i = 69$ from the first region starts an excursion in the phase space due to noise (left panel in Fig. 3.23e), while the elements from the second region remain in the steady state (right panel in Fig. 3.23e). However, the spiking of this small group is weak since time-delayed feedback significantly shifts the nullclines (the cubic ones down and the vertical ones to the right) and does not allow the elements to make the full cycle in the phase space before going back to the steady state (left panel in Fig. 3.23f). At the same time for the nodes from the second region, the feedback shifts the cubic nullclines up and the vertical nullclines to the left making them more easily excitable by noise (right panel in Fig. 3.23f). Therefore, the incoherent spiking is again induced in the second region while the nodes from the first region are pulled coherently due to coupling (Fig. 3.23g). After that, all nodes return again to the steady state. Next during the third excitation, the nullclines for the elements from the first region are shifted in a way making the excitation threshold lower and, therefore, the spiking starts from the first region due to noise: the node $i = 60$ is the first to spike (left panel in Fig. 3.23h). Hence, finally the coherent and incoherent domains are interchanged and further the steps described above repeat with the only difference that the first region is now incoherent while the second corresponds to coherent spiking.

Thus, it is the time-delayed feedback that prevents the alternation for every spiking cycle. As it can be seen from Fig. 3.13 the largest coupling term corresponds to cross-coupling in the v-equation of the system (3.3). However, the contribution of the feedback term is significantly stronger than that of the coupling terms. For this reason, alternating behavior can only occur when the time-delayed feedback term is close to zero. To illustrate that we consider the impact of coupling terms and feedback upon the first and the second equation in system (3.3) in the regime of two-period coherence-resonance chimera (Fig. 3.24). This figure clearly indicates that the interchange of coherent and incoherent domains in the chimera pattern occurs when the feedback term is close to zero (line A in Fig. 3.24). On the other hand, the alternation fails when the feedback term is nonzero and the coupling term almost vanishes (line B in Fig. 3.24).

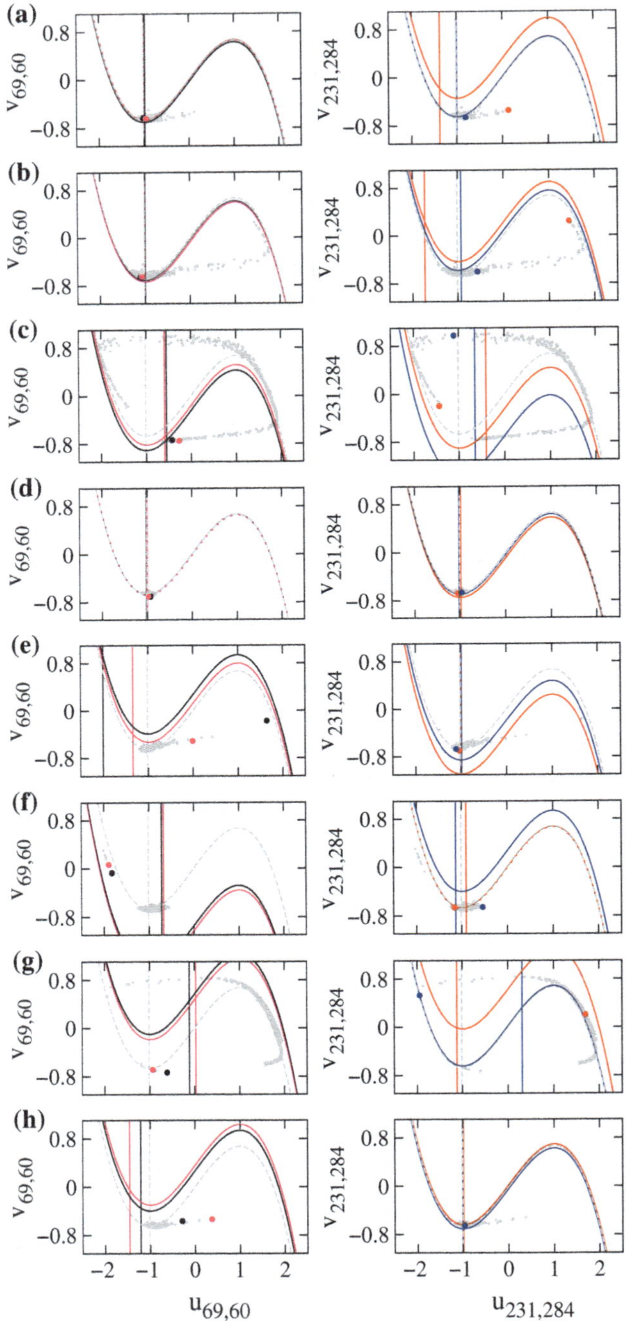
(a)
(b)
(c)
(d)
(e)
(f)
(g)
(h)
v69,60
v231,284
0.8
0
−0.8
−2 −1 0 1 2
u69,60
u231,284

◀**Fig. 3.23** Activator and inhibitor nullclines $\dot{u}$ and $\dot{v}$, respectively, for the selected nodes $i = 69$—black, $i = 60$—pink (left column) and $i = 231$—red, $i = 284$—blue (right column) of the system (3.3) in period-two coherence-resonance chimera regime for **a** $t = 969.9$, **b** $t = 970.15$, **c** $t = 970.9$, **d** $t = 974.5$, **e** $t = 975.0$, **f** $t = 975.9$, **g** $t = 976.65$, **h** $t = 980.55$. Parameters: $N = 500$, $\varepsilon = 0.05$, $a = 1.012$, $\sigma = 0.4$, $r = 0.2$, $D = 0.0002$, $\gamma = 0.2$, $\tau = 4.76$ [133]

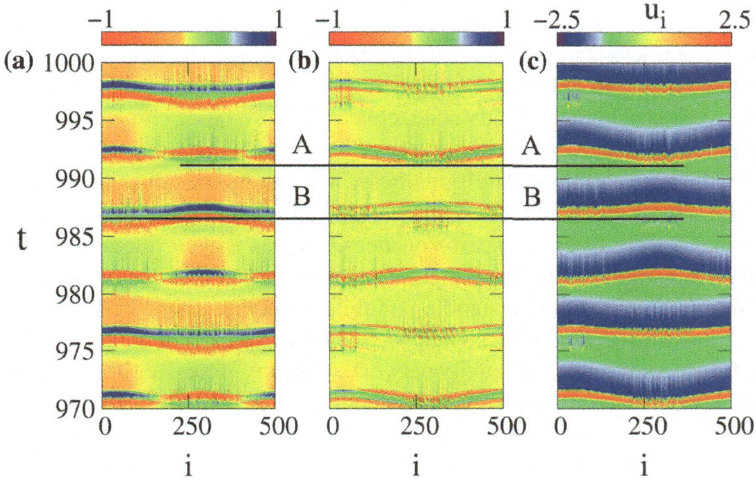

Fig. 3.24 Space-time plots of coupling terms for u and v variables in the period-two coherence-resonance chimera regime: **a** time-delayed feedback for the u variable, **b** coupling for the v variable. **c** Space-time plot of u_i variable. Parameters: $N = 500$, $\varepsilon = 0.05$, $a = 1.012$, $\sigma = 0.4$, $r = 0.2$, $D = 0.0002$, $\gamma = 0.2$, $\tau = 4.76$ [133]

3.4.5 Summary

In conclusion, we have investigated the impact of time-delayed feedback on the dynamics of a network of nonlocally coupled FitzHugh-Nagumo elements in the excitable regime in the presence of noise. Focusing on coherence-resonance chimera, we demonstrate that time-delayed feedback promotes this pattern: it allows for control of the range of parameter values where noise-induced chimera exists and in most cases increases this range. Moreover, the feedback induces coherence-resonance chimeras for vanishing noise intensities. Additionally, we show that the threshold parameter interval of coherence-resonance chimeras can be more than doubled by applying feedback with delay time close to the intrinsic period of the system. Furthermore, when the feedback delay coincides with the intrinsic period of the network, we find a novel feedback-induced regime which we have called period-two coherence-resonance chimera. We explain the alternating behavior of this novel pattern by analyzing the evolution of the nullclines due to the coupling and feedback terms of the network.

3.5 Conclusions

In the present chapter, we have investigated coherence-resonance chimeras, an intriguing spatiotemporal pattern that combines features of coherence resonance and chimera state. It occurs in networks of nonlocally coupled excitable elements and is induced by noise of intermediate intensity. Therefore, we disclose constructive role of noise for chimera states. Moreover, we show that noise can also be used for controlling the dynamical network. By adjusting the noise intensity, the desired regime of the network is achieved: steady state, coherence-resonance chimera, patterns that are incoherent in space but periodic in time and patterns characterized by incoherence in both space and time. Additionally, by fine-tuning the noise intensity, we can change the size of the incoherent domain of the chimera pattern.

A crucial feature of coherence-resonance chimeras is their alternating behavior. Such an interchange between the coherent and the incoherent domains of the chimera state is significant for the understanding of unihemispheric sleep, where the synchronization of neurons is known to switch between hemispheres of the brain, which are known to have a strong 2-community network structure. Here, we show that the alternating behavior can be caused in excitable media by stochasticity, which is inherent to real-world systems. Therefore, we propose that coherence-resonance chimeras which we uncover for a network of neuronal systems in stochastic environment might offer a natural possible explanation of the phenomenon of unihemispheric sleep.

Further, applying time-delayed feedback, we demonstrate how to control coherence-resonance chimeras by adjusting delay time and feedback strength. We show that time-delayed feedback is a powerful tool allowing to modify and in most cases increase the range of parameters where this noise-induced chimera pattern is observed. In particular, the threshold parameter interval of coherence-resonance chimeras can be more than doubled by applying feedback with delay time close to the intrinsic period of the system. Compared to the case without feedback, this provides an essential improvement which could be relevant for the experimental realization of coherence-resonance chimeras. Interestingly, we find that time-delayed feedback leads to the formation of period-two coherence-resonance chimera, a novel feedback-induced regime.

Since we consider a paradigmatic model for neural excitability in a noisy environment, which is inherent in real-world systems, we expect wide-range applications of these results to neuronal networks in general.

Chapter 4
Towards Realistic Topologies: Coherence, Incoherence, and Partial Synchronization Patterns

Abstract In the present chapter, we investigate network topologies beyond nonlocally coupled rings. In particular, we focus on multilayer networks, networks with power-law coupling kernel, and networks with fractal connectivities. First, we consider coherence resonance in multiplex networks. Then, we explore chimera states and solitary states in multiplex networks. Further, we study coherence-incoherence patterns in networks with power-law coupling kernel. Finally, we discuss the formation of chimera patterns in two types of networks with fractal connectivities: ring networks with fractal connectivities and networks with 2D modular fractal connections. We use FitzHugh-Nagumo model and Van der Pol oscillator to describe the individual node dynamics.

4.1 Introduction

The mutual interdependence of the nonlinear dynamics with the structure of the network is an important issue. In Chaps. 2 and 3, we have mainly focused on nonlocally coupled ring networks. However, the topologies of the real-world systems, both biological and man-made, are much more complex. Therefore, in the present chapter, we discuss more realistic topologies such as multilayer networks, networks with power-law coupling kernel, and networks with fractal connectivities. Besides chimera states, we investigate another mechanism of transition from completely synchronized to completely irregular behavior, solitary states. In the intriguing scenario of chimera states, the system spontaneously splits into coexisting domains of coherent (e.g., synchronized) and incoherent behavior, which are localized in space [61, 62]. In the regime of solitary states, which represent an alternative scenario, individual "solitary" oscillators leave the synchronous cluster at random positions in space [127, 128]. Moreover, in the present chapter, we study the regime of coherence resonance. While chimera states and solitary states are referred to as partial synchronization patterns, coherence resonance corresponds to the state of the network characterized by the best temporal regularity (coherence with respect to time) of noise-induced oscillations achieved for an intermediate optimal noise intensity.

A. Zakharova, *Chimera Patterns in Networks*, Understanding Complex Systems,
https://doi.org/10.1007/978-3-030-21714-3_4

What do we know about the behavior of chimera states, solitary states, and coherence resonance in the context of complex topologies? For the studies on chimera states, the topology of the underlying network plays a crucial role. Initially, it was believed that to induce chimeras a nonlocal coupling is essential. Later, it has been found that global coupling [93, 99, 100, 108, 110] and even local coupling [105, 111, 112] may give rise to chimera states. Moreover, chimeras have been found for irregular topologies [113–117] and hierarchical, quasi-fractal connectivities [80, 83, 118–123]. In the nonlocal coupling, one has two control parameters: coupling range and coupling strength. The former is controlled by the kernel of the nonlocal coupling function: in previous studies generally, a trigonometrically or exponentially decaying function or a rectangular kernel have been used. Here, we focus on the power-law coupling kernel that represents a more general and universal coupling scheme motivated from many real-world systems. Moreover, we discuss chimera states in multilayer networks [126] and networks with fractal connectivities [118, 122, 123]. For all three topologies discussed in this chapter, we investigate FitzHugh-Nagumo model. Additionally, for the fractal connectivity, we also use Van der Pol oscillator to describe the individual node dynamics. Previously, the impact of power-law coupling on chimera states has been investigated in ecological oscillators [84]. Solitary states have been reported for local, nonlocal, and global types of coupling in one-layer networks of Kuramoto oscillators with inertia [128, 334] and for nonlocal coupling in one-layer networks of FitzHugh-Nagumo oscillators [236]. Moreover, it has been shown that time delay in the coupling allows to induce and design solitary states in one-layer networks [335]. Here, we study solitary states for multilayer networks [126] and networks with power-law coupling kernel using the model of oscillatory FitzHugh-Nagumo neurons. Coherence resonance has been investigated theoretically and experimentally in various systems and network topologies [131, 133–141, 295, 298–307]. In the present chapter, we consider coherence resonance in multilayer networks of coupled FitzHugh-Nagumo systems in the excitable regime [142].

For networks of FitzHugh-Nagumo oscillators, it has been shown that fractal type of coupling results in chimera patterns with nested incoherent parts characterized by complex fragmented structure. Their specific structure depends on the complex interplay between the link geometry and the initial conditions [33, 80]. In networks of logistic maps with hierarchical connectivities, it has been shown that the symmetry of the base pattern promotes chimera states, and asymmetric connectivities result in complex nested chimera patterns [121]. Further, the existence of chimera states in a population dynamics model has been demonstrated and it turned out that quasi-fractal coupling topologies induce traveling multichimera states for this model [83]. Multichimera states have also been detected in networks of leaky integrate-and-fire model with excitatory coupling and hierarchical connectivity [119]. Here, we consider networks of Van der Pol oscillators and investigate the stepwise transition from a nonlocal to a quasi-fractal connectivities in ring networks [118]. Using the clustering coefficient and the symmetry properties to classify different topologies with respect to the occurrence of chimera states, we show that symmetric connectivities with large clustering coefficients promote the emergence of chimera states, while they are suppressed by slight topological asymmetries or small clustering coefficient.

Moreover, we explore the interplay of fractal topology and time-delayed coupling for FitzHugh-Nagumo model [123]. In the parameter plane of coupling strength and delay time, tongue-like regions of existence of chimera states alternating with regions of coherent dynamics are detected. By varying the time delay, one can deliberately stabilize desired spatiotemporal patterns in the system. The interplay of fractal topology and time-delayed coupling has been also considered for Van der Pol oscillator [120]. Finally, in the present chapter, we compare mathematically constructed network with modular fractal connectivity to an empirical structural neural connectivity derived from diffusion-weighted magnetic resonance imaging [122]. In more detail, for networks of FitzHugh-Nagumo neurons, we analyze the properties of chimeras and partially synchronized states and investigate the regions of their stability.

Why is it important to consider multilayer networks, networks with power-law coupling kernel and networks with hierarchical connectivity? *Multilayer networks* have been suggested to offer better representation of the topology and dynamics of real-world systems in comparison with isolated one-layer structures [213–215]. The nodes in such networks are distributed in different layers according to the type of the relation they share. For example, in the case of a neuronal network, the neurons can form different layers depending on their connectivity through a chemical link or by an ionic channel. In brain networks, different regions can be seen connected by functional and structural neural networks. Multilayer networks open up new possibilities of control allowing to regulate nonlinear systems by means of the interplay between dynamics and multiplexing. The alternative control mechanisms based on multiplexing have many advantages. For example, they allow controlling the dynamics of one layer by adjusting the parameters of the other layer. This is important from the point of view of applications since it is not always possible to directly access the desired layer while the network with which this layer is multiplexed may be adaptable.

The long-range *power-law interaction* is an ubiquitous form through which natural systems interact in physical and biological sciences. The examples are provided by electromagnetic and gravitational interactions [336], long-distance movements of butterflies [337], and the connectivity among the neurons in the animal brain that is controlled by long-range interaction with a specific algebraic scaling [338]. In spatial ecology, dispersal of species between different habitat patches is common. The spatial movements of most organisms are restricted and even for long-distance migrants in large networks not all the patches are likely to be accessible from a particular patch due to dispersal mortality, e.g., mites greatly suffer from dispersal mortality with increasing distance between patches [339]. Moreover, to estimate the density of long-distance dispersing populations, it is useful to consider that a proportion of population is distributed to the other connecting patches via a continuous geometric function so that more distant populations receive less migrants (e.g., inverse power law) [340].

Finally, recent findings in the area of neuroscience resulted in the growing interest in irregular coupling topologies. Diffusion tensor magnetic resonance imaging (DT-MRI) studies uncovered the peculiarities of the architecture in the neuron interconnectivity of the human and mammalian brain. In more detail, the connectivity of the neuron axons network represents a hierarchical (quasi-fractal) geometry [341–345].

These works serve as a motivation for the study of networks with *fractal connectivity* and, in particular, coherence-incoherence patterns occurring in such networks. Here, we focus on ring networks with fractal connectivities and networks with 2D modular fractal connections (Sect. 4.5).

This chapter is organized as follows. First, we consider coherence resonance in multiplex networks (Sect. 4.2). Then we investigate chimera states and solitary states in multiplex networks (Sect. 4.3). Further, we study coherence-incoherence patterns in networks with power-law coupling kernel (Sect. 4.4). Finally, we discuss the formation of chimera patterns in networks with fractal connectivities (Sect. 4.5). Section 4.6 summarizes the results of the present chapter.

4.2 Coherence Resonance in Multiplex Networks

Noise can play a constructive role and gives rise to new dynamic behavior, e.g., stochastic bifurcations, stochastic synchronization, or *coherence resonance* [134–137, 291–297]. In Chap. 3, we have investigated the constructive role of noise for chimera states in ring networks. Here, we study the phenomenon of coherence resonance in a multiplex network.[1] The counterintuitive phenomenon of coherence resonance describes a non-monotonic behavior of the regularity of noise-induced oscillations in the excitable regime, leading to an optimal response in terms of regularity of the excited oscillations for an intermediate noise intensity. One of the most relevant questions is related to control of coherence resonance. A well-studied mechanism of coherence-resonance control is based on time delay. In particular, it has been shown that coherence resonance can be modulated by applying time-delayed feedback in excitable [299, 303, 308, 309] as well as in non-excitable systems [138, 139].

In the present section, we focus on a network of coupled neurons, where the FitzHugh-Nagumo model in the excitable regime represents the local dynamics of each neuron. Thus far, for this paradigmatic model, delay-based mechanisms of controlling coherence resonance have been studied in a single [299] and in two coupled elements with time-delayed feedback [302]. Recently, the control of coherence resonance in a one-layer network of delay-coupled FitzHugh-Nagumo oscillators has also been investigated [141]. Moreover, time-delayed feedback control has been demonstrated for a special type of coherence resonance called coherence-resonance chimera occurring in a ring of nonlocally coupled excitable FitzHugh-Nagumo systems [131–133] (Sect. exrefchap3spssec:models). As we show in this section, multilayer networks allow to develop new control strategies for coherence resonance. There is a large set of open problems related to the dynamics of multilayer networks. Although a number of works on synchronization [23, 346–349] and pattern formation [23, 124, 125, 218, 223, 250, 350, 351] in multilayer networks has appeared

[1] Portions of the following text have been re-printed from [142] with the permission of AIP Publishing.

recently, the phenomenon of coherence resonance and, in particular, its control in multilayer structures is much less studied [142].

One of the most promising applications of the multilayer approach is the study of the brain [352]. The multilayer representation allows to model its structural and functional connectivity from a new perspective, i.e., combined with each other as layers of a complex multilayer structure. While the adjustment of the physical connections in neural networks has nowadays become feasible due to advances in brain surgery, the manipulation of the functional connectivity seems to be much more complicated. Moreover, while modeling the brain networks such phenomena as coherence resonance should be taken into account due to inherent brain noise [353]. We believe that the results presented in this section will stimulate further research on the brain applying the multilayer approach to stochastic dynamics. For example, they could be useful for the investigation of coherence resonance in stimulated neural networks [354] represented as multilayer structures.

In this section, we investigate the interplay between multiplexing and noise for a network of coupled FitzHugh-Nagumo systems in the excitable regime. In particular, focusing on the phenomenon of coherence resonance, we aim to develop control strategies based on multiplexing. We show that in a two-layer noisy network multiplexing allows for control of coherence resonance even if the coupling between the layers is rather weak. In more detail, we demonstrate that multiplexing induces coherence resonance for the parameter regimes where it is absent in an isolated layer. We consider two different situations where coherence resonance is not observed in an isolated layer: (i) the coupling strength is not optimal or (ii) there is no noise exciting the elements. We find that for both cases multiplexing with another layer results in the occurrence of coherence resonance in the desired layer. Interestingly, we discover a counterintuitive impact of multiplexing on the dynamics of the two-layer network when the noise is switched off in one of the layers: coherence resonance induced by multiplexing in the deterministic layer appears to be more pronounced than that in the noisy layer.

This section is organized as follows. After having introduced the model (Sect. 4.2.1), we begin our study with the dynamics of disconnected layers and consider an isolated one-layer network (Sect. 4.2.2). For a locally coupled ring of identical FitzHugh-Nagumo systems in the excitable regime, we study the role of noise intensity and coupling strength in inducing coherence resonance. In particular, we define the values of inter-layer coupling strength corresponding to coherence resonance. Previously, it has been shown that coherence resonance can be enhanced by choosing appropriate coupling strength, i.e., for coherence resonance in a network there is an optimal strength of interactions between the elements [141, 298, 355]. Further, we consider a multiplex network of two coupled rings where independent uncorrelated noise sources are introduced to every element in both layers (Sect. 4.2.3). Moreover, the coupling strength inside the layers is chosen such that the upper(first) ring does not demonstrate coherence resonance if it is isolated and the lower (second) ring shows well-pronounced coherence resonance. Next, we eliminate the intra-layer coupling mismatch, turn off the noise in one of the layers, and analyze how coherence resonance is induced in this deterministic layer by multiplexing (Sect. 4.2.4). It is important to note that we consider the case of weak multiplexing, i.e., the coupling between the layers is smaller than that inside the layers. Nevertheless, it has a significant effect on the dynamics of the network. We conclude in Sect. 4.2.5 with the summary of the results.

4.2.1 Model

We consider a two-layer multiplex network, where each layer represents a locally coupled ring of FitzHugh-Nagumo (FHN) systems in excitable regime in the presence of Gaussian white noise:

$$\begin{aligned}
\varepsilon \frac{du_{1i}}{dt} &= u_{1i} - \frac{u_{1i}^3}{3} - v_{1i} + \frac{\sigma_1}{2} \sum_{j=i-1}^{i+1} (u_{1j} - u_{1i}) + \\
&\quad \sigma_{12}(u_{2i} - u_{1i}), \\
\frac{dv_{1i}}{dt} &= u_{1i} + a + \sqrt{2D_1}\xi_i(t), \\
\varepsilon \frac{du_{2i}}{dt} &= u_{2i} - \frac{u_{2i}^3}{3} - v_{2i} + \frac{\sigma_2}{2} \sum_{j=i-1}^{i+1} (u_{2j} - u_{2i}) + \\
&\quad \sigma_{12}(u_{1i} - u_{2i}), \\
\frac{dv_{2i}}{dt} &= u_{2i} + a + \sqrt{2D_2}\eta_i(t),
\end{aligned} \tag{4.1}$$

where u_{1i} and v_{1i} are the activator and inhibitor variables in the first(upper) layer, respectively, $i = 1, ..., N$ with N being the total number of elements in the network. All indices are modulo N. In the present section, every layer consists of $N = 500$ elements. In a similar way, u_{2i} and v_{2i} represent the activator and inhibitor variables in the second(lower) layer, respectively. The strength of the coupling within the layer (intra-layer coupling) is given by σ_1 for the first layer and σ_2 for the second layer. Since we consider local coupling for each layer, every node within a layer has only one nearest neighbor in each direction on a ring. The coupling between the layers (inter-layer coupling) is bidirectional, diffusive, and its strength is characterized by σ_{12}. A small parameter responsible for the timescale separation of fast activator and slow inhibitor is given by $\varepsilon > 0$. Here we fix $\varepsilon = 0.01$. Parameter a defines the excitability threshold. For an individual FHN element, it determines whether the system is in the excitable ($|a| > 1$), or oscillatory ($|a| < 1$) regime. In the present study, we assume that all elements are in the excitable regime close to the threshold ($a = 1.05$). Further, $\xi_i(t) \in \mathbb{R}$ is Gaussian white noise with intensity D_1 introduced into the first layer, i.e., $\langle \xi_i(t) \rangle = 0$ and $\langle \xi_i(t)\xi_j(t') \rangle = \delta_{ij}\delta(t - t')$, $\forall i, j$. Similarly, $\eta_i(t) \in \mathbb{R}$ is Gaussian white noise with intensity D_2 introduced into the second layer. These noise sources are uncorrelated. In the present section, we consider multiplex networks. They represent a special class of multilayer networks where the layers contain the same number of nodes, and the inter-layer links are allowed only for replica nodes, i.e., there is a one-to-one correspondence between the nodes in different layers (Fig. 4.1).

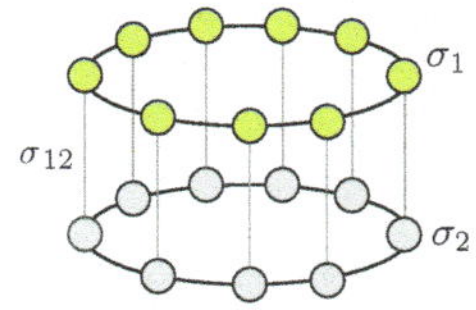

Fig. 4.1 Schematic diagram showing a multiplex network consisting of two layers where each layer is represented by a locally coupled ring network [142]

4.2.2 Dynamics of Isolated Layers

Coherence resonance characterizes the emergence of relatively coherent noise-induced oscillations occurring for an optimal noise intensity. There exist several different measures for quantifying coherence resonance, such as the normalized standard deviation of the interspike interval, the correlation time, and the signal-to-noise ratio [134, 135, 139]. Since in the present work we deal with a neural model demonstrating spiking behavior, it is convenient to use the first one. It is defined as $R_T = \frac{\sqrt{\langle t_{ISI}^2 \rangle - \langle t_{ISI} \rangle^2}}{\langle t_{ISI} \rangle}$, where t_{ISI} is the time between two subsequent spikes and $\langle \cdots \rangle$ indicates the average over the time series. A system undergoing coherence resonance will show a pronounced minimum in the value of R_T [135]. The above definition of R_T is limited to characterizing coherence resonance for a single FitzHugh-Nagumo oscillator. For a network of oscillators, coherence resonance can be measured by redefining R_T as follows [141]:

$$R_T = \frac{\sqrt{\langle \overline{t_{ISI}^2} \rangle - \langle \overline{t_{ISI}} \rangle^2}}{\langle \overline{t_{ISI}} \rangle}, \tag{4.2}$$

where the over-bar indicates the additional average over nodes. Since we are interested in the impact of coupling strength on coherence resonance, we measure R_T for varying noise intensity D at different values of coupling strength. We find that for the considered network topology and the chosen parameter setting $a = 1.05$, $\varepsilon = 0.01$, $N = 500$, the optimal coupling strength value is $\sigma = 0.2$ (Fig. 4.2). We obtain a non-monotonic dependence of R_T on D with a well-pronounced deep minimum, and the network shows coherence resonance for a very narrow interval of the noise intensity values around $D = 0.002$. If we decrease the coupling strength, the depth of the local minimum becomes negligibly small, the dependence of R_T on D becomes rather monotonic, and coherence resonance disappears. Furthermore, lower coupling strength results in much higher values of R_T indicating less coherent oscillations: the whole curve for $\sigma = 0.1$ lies above the optimal dependence (Fig. 4.2). For values of σ larger than the optimal, we can still detect coherence resonance. However, it becomes less pronounced: the depth of the minimum becomes smaller and it is observed for much broader interval of noise intensity (see curve for $\sigma = 0.4$

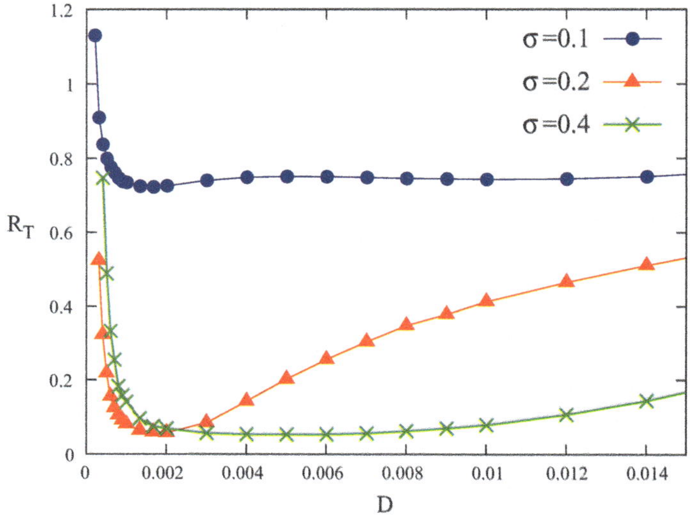

Fig. 4.2 Normalized standard deviation of the interspike interval R_T for varying noise intensity D for an isolated ring with local coupling and three different values of coupling strength: $\sigma = 0.1$ (circles), $\sigma = 0.2$ (triangles), and $\sigma = 0.4$ (crosses). The results are obtained by integrating over 2000 time units and then averaging over time, oscillators and realizations (for 10 simulations each). Other parameters: $a = 1.05$, $\varepsilon = 0.01$, $N = 500$ [142]

in Fig. 4.2) if compared with the optimal case (see curve for $\sigma = 0.2$ in Fig. 4.2). Therefore, there is a clear resonance for a certain optimal coupling strength value corresponding to the most pronounced coherence resonance. In Sect. 4.2.3, we will concentrate on the two values: $\sigma = 0.1$ and $\sigma = 0.2$.

4.2.3 Multiplex Network: Intra-layer Coupling Strength Mismatch

Now we consider a two-layer multiplex network with non-identical layers where the upper(first) and the lower(second) rings have different intra-layer coupling strengths $\sigma_1 = 0.1$ and $\sigma_2 = 0.2$, respectively. The noise intensity is the same in both layers $D_1 = D_2 = D$. As discussed in Sect. 4.2.2, an isolated one-layer network exhibits optimal coherence resonance for $\sigma = 0.2$ while at $\sigma = 0.1$ coherence resonance is not observed. We investigate the impact of multiplexing on the coherence of noise-induced oscillations and address the question whether coherence resonance can be induced in the first layer. The multiplexing introduces an additional parameter σ_{12} that characterizes the coupling between the layers. In the present work, we consider weaker coupling between the layers than that inside the layers, i.e., we focus on weak multiplexing. We find that even for small inter-layer coupling strength $\sigma_{12} = 0.01$

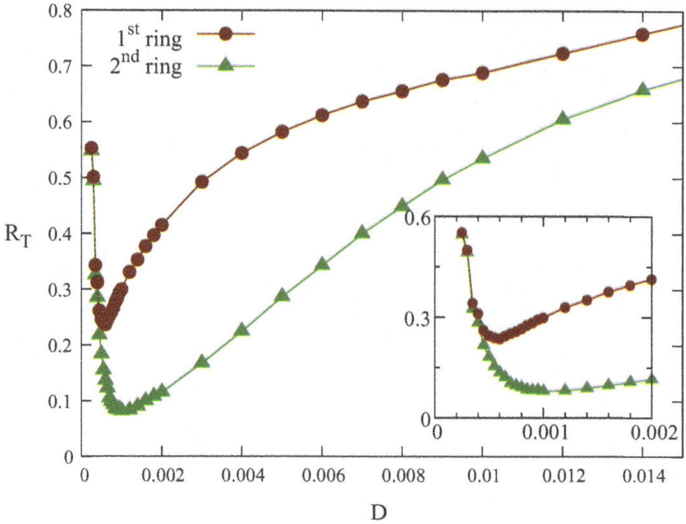

Fig. 4.3 Normalized standard deviation of the interspike interval R_T for varying noise intensity D for a multiplex network of two coupled rings at $\sigma_1 = 0.1$, $\sigma_2 = 0.2$ and $\sigma_{12} = 0.04$. Other parameters: $a = 1.05$, $\varepsilon = 0.01$, $N = 500$ [142]

multiplexing induces coherence resonance in the first ring. For increasing inter-layer coupling strength $\sigma_{12} = 0.04$, coherence resonance becomes more pronounced (Fig. 4.3). At the same time, coherence resonance in the second ring is preserved. Interestingly, the curve corresponding to the multiplex-induced coherence resonance lies above the one displaying coherence resonance in the second ring (Fig. 4.3). This indicates the fact that the oscillations induced by multiplexing in the first ring are not as coherent as those in the second ring. Therefore, coherence resonance manifests itself differently for the layers of the multiplex network. To confirm these conclusions, we further analyze the power spectral density for the two layers of the multiplex network and calculate full spectral width at half maximum $\Delta\omega$ (Fig. 4.4). Indeed, both layers undergo coherence resonance: for an intermediate optimal value of noise intensity D_{opt} (which is different for the two layers), we observe the narrowest spectrum of the noise-induced oscillations (Fig. 4.4). Moreover, the spectral width at the optimal noise intensity for the first ring $\Delta\omega = 0.055$ is smaller than that for the second ring $\Delta\omega = 0.008$ indicating lower coherence of the oscillations in the first ring. It turns out that stronger multiplexing (larger values of the inter-layer coupling strength σ_{12}) allows to increase the coherence of these oscillations, i.e., the optimal value of R_T in the first ring decreases (Fig. 4.5). However, in this case, stronger noise is required to achieve coherence resonance (Fig. 4.5).

Therefore, for increasing inter-layer coupling, multiplex-induced coherence resonance becomes more pronounced and its R_T-curve shifts down towards lower values and approaches the R_T-dependence corresponding to the second ring. Hence, the noise-induced oscillations in the two layers reach the same level of coherence. This

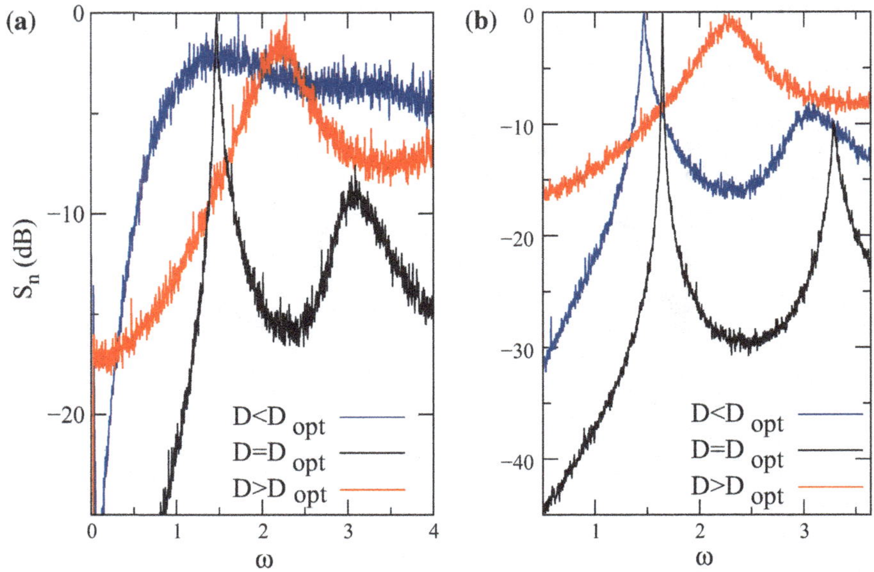

Fig. 4.4 Power spectral density for three different values of the noise intensity for one selected node $i = 200$ from the first (**a**) and the second (**b**) ring of the multiplex network: **a** $D = 0.0003$; $D_{\text{opt}} = 0.0006$ ($\Delta\omega = 0.055$); $D = 0.05$ ($\Delta\omega = 0.43$) and **b** $D = 0.0006$ ($\Delta\omega = 0.06$); $D_{\text{opt}} = 0.001$ ($\Delta\omega = 0.008$); $D = 0.05$ ($\Delta\omega = 0.43$). The coupling parameters are as in Fig. 4.3: $\sigma_1 = 0.1$, $\sigma_2 = 0.2$ and $\sigma_{12} = 0.04$. Other parameters: $a = 1.05$, $\varepsilon = 0.01$, $N = 500$ [142]

can be explained by the fact that larger inter-layer coupling makes the influence of the second ring on the first one stronger. Consequently, the rings start to behave in a similar way. Nevertheless, the inter-layer coupling strength remains smaller than that within the layers. Therefore, we demonstrate that even weak multiplexing can have a significant impact on the noise-induced dynamics. To stress this point, we show in one plot the R_T-curve for the first ring in isolation and in the presence of weak multiplexing (Fig. 4.6). The isolated ring demonstrates a monotonic dependence of R_T on D while for the multiplexed layer the whole curve shifts towards lower values of R_T and shows a well-pronounced deep minimum indicating coherence resonance (Fig. 4.6).

4.2.4 A Deterministic Layer Multiplexed with a Noisy Layer

Further, we investigate the case when a deterministic ring network $D_2 = 0$ is multiplexed with a noisy network $D_1 \neq 0$. The main question we address in this section is whether coherence resonance can be induced by multiplexing in the deterministic network. We now fix the intra-layer coupling strength $\sigma_1 = \sigma_2 = 0.1$ that corresponds to the absence of coherence resonance in an isolated noisy ring (see Fig. 4.2).

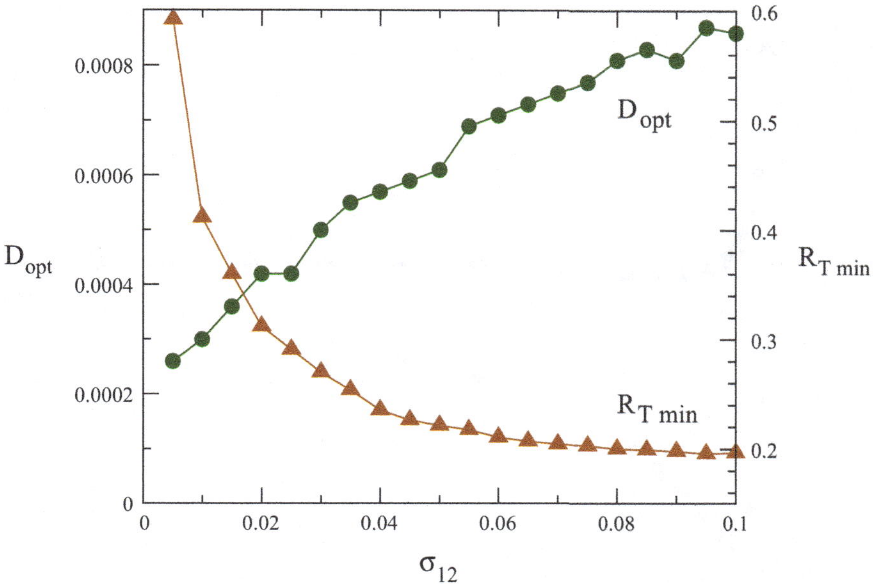

Fig. 4.5 Optimal values of noise intensity D_{opt} (circles) and corresponding optimal values of R_T (triangles) for varying inter-layer coupling strength σ_{12} at $\sigma_1 = 0.1, \sigma_2 = 0.2$. The results are shown for the first ring of the multiplex network. Other parameters: $a = 1.05$, $\varepsilon = 0.01$, $N = 500$ [142]

Intriguingly, once the coupling between the ring networks is introduced, coherence resonance occurs in the second layer, even for a rather weak multiplexing $\sigma_{12} = 0.01$ (Fig. 4.7). The R_T-curve displays a clear deep minimum (Fig. 4.7a), and the power spectral density demonstrates the smallest width for the optimal intermediate value of noise intensity (Fig. 4.7b) indicating coherence resonance for the noise-induced oscillations.

To investigate the impact of inter-layer coupling strength on multiplex-induced coherence resonance, we calculate the R_T value in the (D_1,σ_{12}) parameter plane for the first (Fig. 4.8a) and second (Fig. 4.8b) layers of the multiplex network. It turns out that both layers undergo coherence resonance. Moreover, the increase of the inter-layer coupling strength shifts the minimum of R_T to higher values of D_1. Surprisingly, coherence resonance in the second layer ($D_2 = 0$) is more pronounced compared to the first layer ($D_1 \neq 0$). This is clearly indicated by the lower optimal values of R_T in the second ring (dark gray(blue) color in Fig. 4.8b) if compared to the first ring (light gray(blue) color in Fig. 4.8b). It is important to note that multiplexing induces coherence resonance in the deterministic ring network only for rather small values of σ_{12} (Fig. 4.8b). For larger values of the inter-layer coupling strength $\sigma_{12} > 0.04$, the R_T values grow monotonically with increasing noise intensity D_1 and, therefore, the effect of coherence resonance disappears (Fig. 4.8b).

A similar effect can be observed for larger values of intra-layer coupling strength $\sigma_1 = \sigma_2 = 0.4$. In this case, the isolated noisy ring network demonstrates coherence

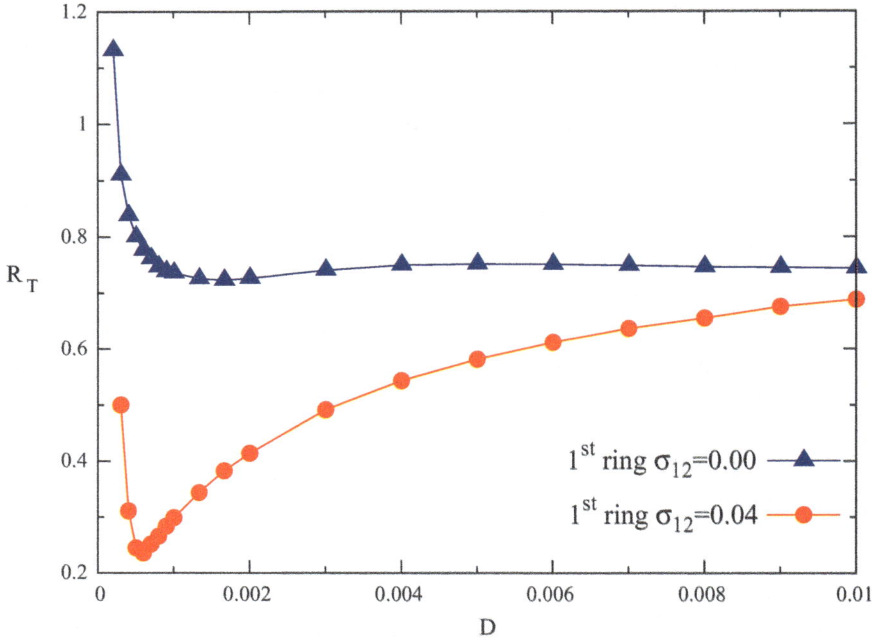

Fig. 4.6 Normalized standard deviation of the interspike interval R_T for varying noise intensity D for the first ring of the multiplex network at $\sigma_1 = 0.1$, $\sigma_2 = 0.2$ and two different values of the inter-layer coupling strength: $\sigma_{12} = 0.0$ (isolated ring) and $\sigma_{12} = 0.04$ (weak multiplexing). Other parameters: $a = 1.05$, $\varepsilon = 0.01$, $N = 500$ [142]

resonance not for a particular value of noise intensity, but for a broad interval of noise intensity values (see Fig. 4.2). Once the deterministic ring network is multiplexed with a noisy one, coherence resonance is induced in both layers and appears to be more pronounced in the deterministic layer (see Fig. 4.9).

4.2.5 Summary

In conclusion, we have investigated a two-layer multiplex network of coupled FitzHugh-Nagumo neurons in the excitable regime in the presence of Gaussian white noise. We have shown that noise-induced dynamics can be controlled by multiplexing even if the coupling between the layers is weaker than that within the layers. The control strategy we describe here is based on multiplexing and allows to induce coherence resonance in parameter regimes where it is absent for isolated networks. We have applied our control scheme to two different networks. First, we have considered a network with interactions between the excitable units being non-optimal. It means that coherence resonance in an isolated network of this type is not observed since

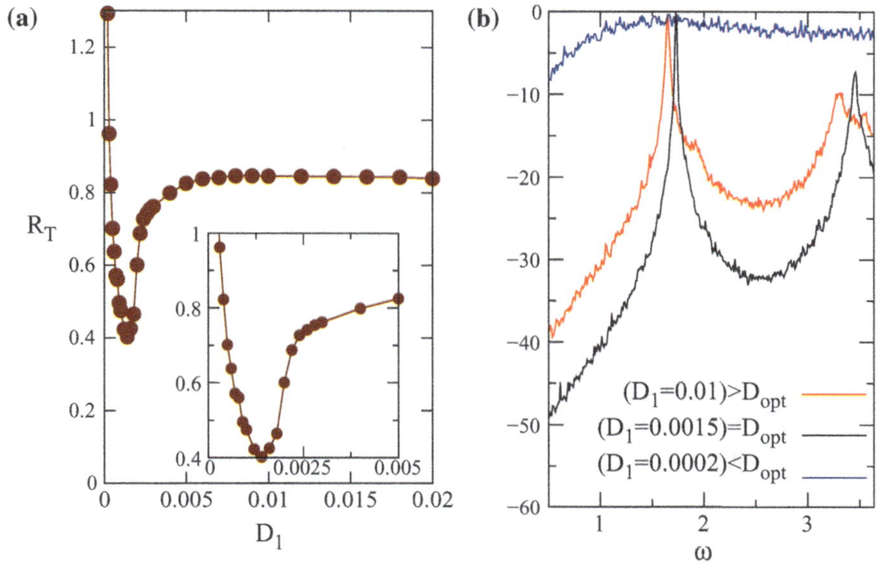

Fig. 4.7 Multiplex-induced coherence resonance in the second layer for $\sigma_1 = \sigma_2 = 0.1$, $\sigma_{12} = 0.01$ and $D_2 = 0$ (noise is added only to the first ring). **a** Normalized standard deviation of the interspike interval R_T for varying noise intensity D_1. **b** Power spectral density for one selected node $i = 300$ at $D_1 = 0.0002$; $D_{1opt} = 0.0015$ ($\Delta\omega = 0.018$); $D_1 = 0.01$ ($\Delta\omega = 0.028$). Other parameters: $a = 1.05$, $\varepsilon = 0.01$, $N = 500$ [142]

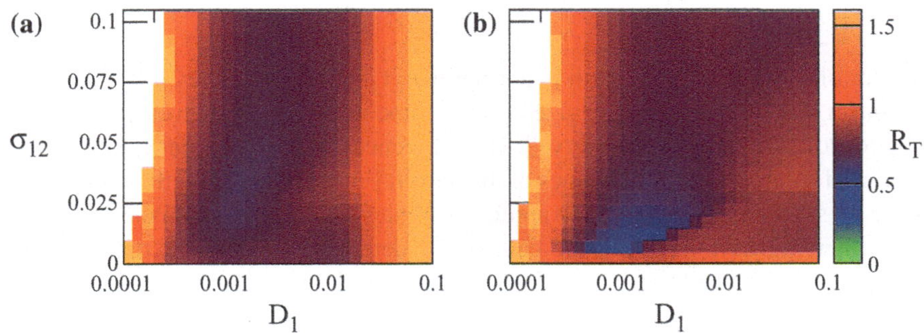

Fig. 4.8 Normalized standard deviation of the interspike interval R_T for varying noise intensity D_1 and inter-layer coupling strength σ_{12} for the first (**a**) and the second (**b**) ring of the multiplex network. Parameters as in Fig. 4.7: $\sigma_1 = \sigma_2 = 0.1$, $D_2 = 0$ (noise is added only to the first ring). Other parameters: $a = 1.05$, $\varepsilon = 0.01$, $N = 500$ [142]

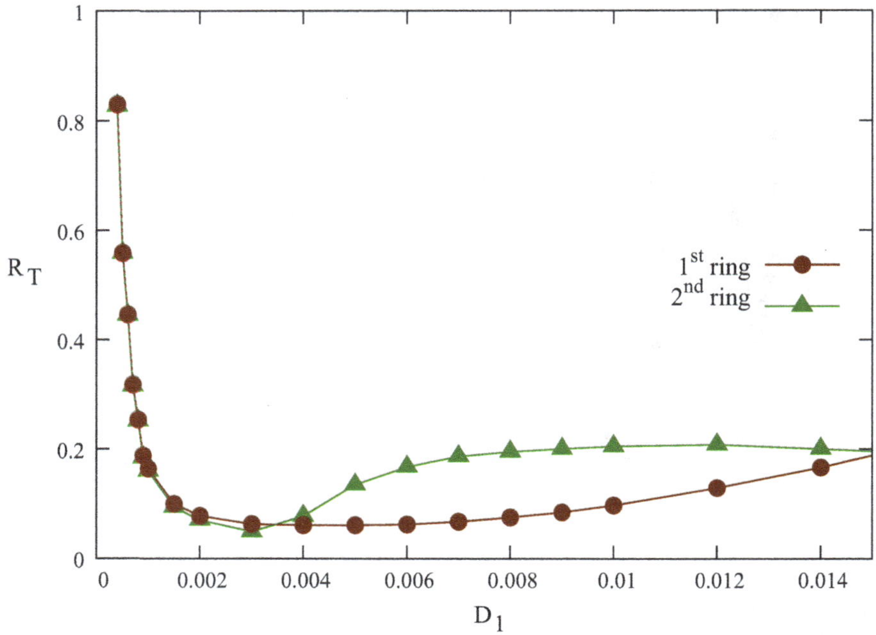

Fig. 4.9 Normalized standard deviation of the interspike interval R_T for varying noise intensity D_1 for the multiplex network at $\sigma_1 = \sigma_2 = 0.4$, $\sigma_{12} = 0.01$ and $D_2 = 0$ (noise is added only to the first ring). Other parameters: $a = 1.05$, $\varepsilon = 0.01$, $N = 500$ [142]

the intra-layer coupling strength is not optimal. Once we multiplex it with another one-layer network of the same size and the same noise intensity, well-pronounced coherence resonance can be observed. The coherence of the oscillations can be further increased by choosing larger values of the inter-layer coupling strength. However, in this case stronger noise should be applied to achieve coherence resonance.

Further, we have analyzed the case of the deterministic network that does not demonstrate coherence resonance in isolation. Multiplexing with a noisy network of the same size and the same intra-layer coupling strength leads to the occurrence of coherence resonance. Moreover, coherence resonance induced by multiplexing in the deterministic layer is better pronounced than that in the noisy layer. Multiplexing induces coherence resonance in the deterministic ring network only for rather small values of the inter-layer coupling strength.

4.3 Coherence-Incoherence Patterns in Multiplex Networks

Coupled oscillatory units can demonstrate various types of collective behavior, including completely synchronized states, partially synchronized patterns, oscillation suppression, and desynchronized dynamics. The transition from completely synchronized to completely irregular behavior can occur via different mechanisms involving special types of partial synchronization patterns. Chimera states represent an intriguing scenario, where the system spontaneously splits into coexisting domains of coherent (e.g., synchronized) and incoherent behavior, which are localized in space [61, 62]. An alternative scenario involves solitary states where individual "solitary" oscillators leave the synchronous cluster at random positions in space [127, 128].

Chimera states and solitary states have been found in multilayer networks only very recently [20, 23, 126, 218, 235, 250, 351, 356, 357, 357, 358]. In particular, the impact of strong multiplexing, when the strength of the coupling between the layers is comparable with that inside each layer, has been investigated for coupled chaotic maps [124, 125]. In more detail, it has been shown that strong multiplexing can be used to control chimeras in networks of coupled chaotic maps [125]. The impact of strong multiplexing with an uncoupled layer has been investigated for a network of Hindmarsh-Rose neurons in [217], where the strong inter-layer links represent chemical connections and weak intra-layer couplings model electrical synapses. In the real-world networks, it is often the case that the nodes form layers (communities or populations) where the coupling within a layer is much stronger than that across the layers. This property of community structure is common in many social and biological networks [359], and in particular in neural networks [187].

Multilayer networks not only allow for a better representation of the topology and dynamics for natural and man-made systems in comparison with isolated one-layer structures, but also open up new possibilities of control. Although the phenomenon of synchronization [23, 346–349] and formation of partial synchronization patterns [23, 124, 125, 218, 223, 250, 350, 351, 360] have been recently considered in multilayer networks, the challenging problem of controlling chimeras by weak multiplexing, in particular, in neuronal networks has been much less investigated [126].

There occurs a question whether weak inter-layer coupling, i.e., weak multiplexing, can have a significant effect on the dynamics of the network. In particular, in this section we address the following questions: Can weak multiplexing be used to control the spatiotemporal patterns? Does it allow to achieve desired dynamic regimes, i.e., induce, design, and suppress chimera states in neural networks? Are other scenarios such as solitary states possible in the presence of weak multiplexing? What are in this situation efficient control strategies?

In the present section, we investigate a multilayer network of coupled FitzHugh-Nagumo oscillators. We demonstrate that weak multiplexing has an essential impact on the dynamical patterns and can be used for controlling. We show that the desired states can be achieved in a particular layer when the coupling between the layers is rather weak. Different types of chimera states can be induced and suppressed. Moreover, we report the occurrence of solitary states for small intra-layer coupling strength mismatch between the layers. Therefore, by weak multiplexing we can switch from chimera to solitary patterns. The advantage of multiplexing control we discuss here is that it allows to achieve the desired state in a certain layer without manipulating its parameters, and it works for weak coupling between the layers.

The section is organized as follows. We first introduce the model of a multiplex network consisting of two layers where each layer is represented by a nonlocally coupled ring of FitzHugh-Nagumo (FHN) oscillators (Sect. 4.3.1). Next, the dynamics of isolated layers is discussed (Sect. 4.3.2). After that, we study a multiplex network composed of two non-identical layers. In particular, we investigate the following two cases: (i) the layers are characterized by different coupling ranges, while the coupling strength within the layers is the same (Sect. 4.3.3); (ii) the layers have an intra-layer coupling strength mismatch, while the coupling range is fixed and the same for both layers (Sect. 4.3.4). Finally, we demonstrate that small intra-layer coupling strength mismatch leads to the occurrence of solitary states throughout the whole network (Sect. 4.3.5). The results on chimera and solitary states in a two-layer network of coupled FHN systems are summarized in Sect. 4.3.6.

4.3.1 Model

We investigate a multiplex network consisting of two layers where each layer is represented by a nonlocally coupled ring of FitzHugh-Nagumo (FHN) oscillators.[2] As discussed in Chap. 3, this two-dimensional system is a paradigmatic model for neural excitability [141]. Previously, chimera states have been found in one-layer networks consisting of coupled oscillatory [79] and excitatory [131–133] FHN systems. In Sect. 4.2, we have shown that weak multiplexing plays a significant role for coherence resonance in a network of coupled excitable FHN units under the influence of noise [142]. Here, we consider the occurrence of chimera patterns and solitary states in the presence of weak multiplexing focusing on oscillatory FHN neurons [126].

We consider a two-layer multiplex network, where each layer is given by a ring of N nonlocally coupled FHN oscillators:

$$
\begin{aligned}
\varepsilon \frac{du_{1i}}{dt} &= u_{1i} - \frac{u_{1i}^3}{3} - v_{1i} + \frac{\sigma_1}{2R_1} \sum_{j=i-R_1}^{i+R_1} [b_{uu}(u_{1j} - u_{1i}) + \\
&\quad + b_{uv}(v_{1j} - v_{1i})] + \sigma_{12}(u_{2i} - u_{1i}), \\
\frac{dv_{1i}}{dt} &= u_{1i} + a_i + \frac{\sigma_1}{2R_1} \sum_{j=i-R_1}^{i+R_1} [b_{vu}(u_{1j} - u_{1i}) + \\
&\quad + b_{vv}(v_{1j} - v_{1i})], \\
\varepsilon \frac{du_{2i}}{dt} &= u_{2i} - \frac{u_{2i}^3}{3} - v_{2i} + \frac{\sigma_2}{2R_2} \sum_{j=i-R_2}^{i+R_2} [b_{uu}(u_{2j} - u_{2i}) + \\
&\quad + b_{uv}(v_{2j} - v_{2i})] + \sigma_{12}(u_{1i} - u_{2i}), \\
\frac{dv_{2i}}{dt} &= u_{2i} + a_i + \frac{\sigma_2}{2R_2} \sum_{j=i-R_2}^{i+R_2} [b_{vu}(u_{2j} - u_{2i}) + \\
&\quad + b_{vv}(v_{2j} - v_{2i})],
\end{aligned}
\tag{4.3}
$$

[2]Portions of the following text have been re-printed from [126] with the permission of AIP Publishing.

where u_{1i} and v_{1i} are the activator and inhibitor variables in the first(upper) layer, respectively, $i = 1, ..., N$ with N being the total number of elements in the network. All indices are modulo N. In a similar way, u_{2i} and v_{2i} represent the activator and inhibitor variables in the second (lower) layer, respectively. The strength of the coupling within the layer (intra-layer coupling) is given by σ_1 for the first layer and σ_2 for the second layer. The parameters R_1 and R_2 indicate the number of nearest neighbors in each direction on a ring for the first and second layers, respectively. The coupling between the layers (inter-layer coupling) is bidirectional, diffusive, and its strength is characterized by σ_{12}. As in Sect. 4.2 here, we are also interested in the impact of weak multiplexing, i.e., when the inter-layer coupling σ_{12} is smaller than the strength of the intra-layer connections σ_1 and σ_2. We also introduce a coupling range for both layers. It is represented by the normalized number of nearest neighbors for the first (upper) layer $r_1 = R_1/N$ and for the second (lower) layer $r_2 = R_2/N$. A small parameter responsible for the timescale separation of fast activator and slow inhibitor is given by $\varepsilon > 0$ and a_i defines the excitability threshold. For an individual FHN element, it determines whether the system is in the excitable ($|a_i| > 1$), or oscillatory ($|a_i| < 1$) regime. In the present study, we assume that all elements are in the oscillatory regime ($a_i \equiv a = 0.5$). A schematic representation of a multiplex network consisting of two layers with different coupling ranges is given in Fig. 4.10.

Equation (4.3) contains not only direct, but also cross-couplings between activator (u) and inhibitor (v) variables, which is modeled by a rotational coupling matrix [79]:

$$\boldsymbol{B} = \begin{pmatrix} b_{\mathrm{uu}} & b_{\mathrm{uv}} \\ b_{\mathrm{vu}} & b_{\mathrm{vv}} \end{pmatrix} = \begin{pmatrix} \cos\phi & \sin\phi \\ -\sin\phi & \cos\phi \end{pmatrix}, \tag{4.4}$$

where $\phi \in [-\pi; \pi)$. As in Chap. 3 here we also fix the parameter $\phi = \pi/2 - 0.1$. As discussed in Chap. 3, chimera states have been found for this value of ϕ in both the

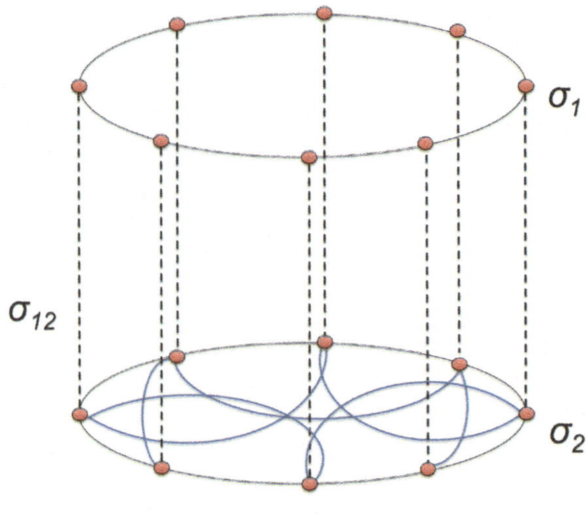

Fig. 4.10 Schematic diagram showing a multiplex network consisting of two layers with different coupling ranges [126]

deterministic oscillatory [79] and the noisy excitable regime [131, 132]. Moreover, it has been shown that chimera states occurring in the excitable regime [131, 132] are different from those detected in the oscillatory regime [79].

4.3.2 Dynamics of Isolated Layers

First, we consider the dynamics of the disconnected non-identical layers $\sigma_{12} = 0$. Both of them are represented by a nonlocally coupled ring of $N = 300$ identical elements, and the intra-layer coupling strength is fixed $\sigma_1 = \sigma_2 = 0.1$. We introduce the coupling range mismatch by choosing $r_1 = 0.2$ for the first (upper) layer and $r_2 = 0.35$ for the second (lower) layer. Therefore, the second layer is characterized by the higher link density compared with the first layer. The isolated layer with the smaller coupling range exhibits desynchronized dynamics (Fig. 4.11a–c). This becomes evident from the space-time plot that shows incoherent pattern (Fig. 4.11a). In the other layer, chimera states are observed (Fig. 4.11d–f). Coexistence in space of well-separated synchronized and desynchronized groups of oscillating FHN neurons is clearly seen from the space-time plot (Fig. 4.11d) and the snapshot (Fig. 4.11e). The mean phase velocity profile has a typical arc-shaped profile, a characteristic signature of chimera states (Fig. 4.11f). The map of regimes in the (r, σ) parameter plane for an isolated nonlocally coupled ring of oscillatory FHN elements has been analyzed in detail in [79].

4.3.3 Multiplex Network: Coupling Range Mismatch

Next, we introduce the coupling between the non-identical layers and investigate the impact of the inter-layer coupling strength σ_{12} on the dynamics of the network. Interestingly, even if the inter-layer coupling $\sigma_{12} = 0.01$ is much smaller than that within the layers $\sigma_1 = \sigma_2 = 0.1$, chimera states are observed for both rings (Fig. 4.12). One can clearly distinguish coherent and incoherent groups in the snapshots (Fig. 4.12a, b) and identify the typical arc shape of the mean phase velocity profiles (Fig. 4.12c, d). Interestingly, the location in space of coherent and incoherent domains of the chimera pattern coincides in the two layers of the multiplex network (Fig. 4.12).

Therefore, weak multiplexing with a denser layer allows to induce chimera states in the layer with the lower link density that does not demonstrate chimera states in isolation. Consequently, in a weakly multiplexed neural network, the control of the dynamics in one of the layers can be realized without manipulating the internal parameters of its elements or the couplings between them. The control is achieved by adjusting the topology (coupling range in this particular case) of the other layer. This is relevant for the applications, since it is not always possible to directly access the desired layer while the network it is multiplexed with may be adaptable. It is important to note that the same effect can be achieved in the presence of strong

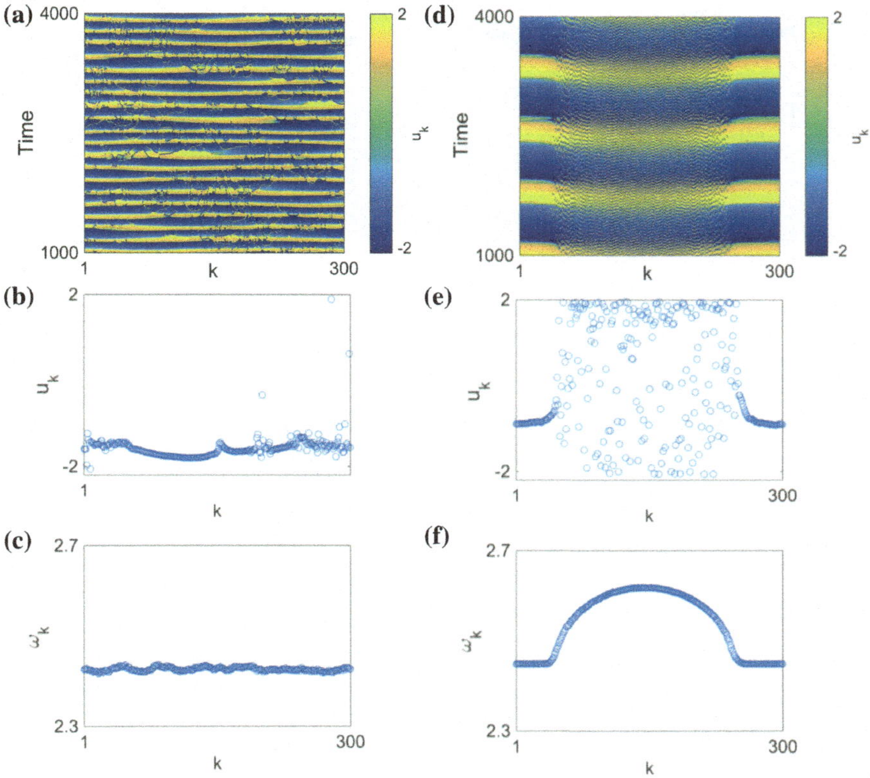

Fig. 4.11 Two disconnected layers ($\sigma_{12} = 0$) with different coupling ranges. **a**, **b**, **c** first (upper) layer, $r_1 = 0.2$; **d**, **e**, **f** second (lower) layer $r_2 = 0.35$; **a**, **d** space-time plots for the variable u_k; **b**, **e** snapshots for the variable u_k; **c**, **f** mean phase velocity profiles. Initial conditions: randomly distributed on the circle $u^2 + v^2 = 4$. Other parameters: $N = 300$, $\varepsilon = 0.05$, $\phi = \pi/2 - 0.1$, $a = 0.5$, $\sigma_1 = \sigma_2 = 0.1$ [126]

multiplexing. In particular, when the strength of the coupling between the layers is equal to the inter-layer coupling $\sigma_{12} = \sigma_1 = \sigma_2 = 0.1$, multiplexing with a denser layer induces chimera patterns in the sparser layer (Fig. 4.13).

To get an overall view on the dynamics of the network, we calculate the map of regimes in each layer for varying coupling range r_2 and inter-layer coupling strength σ_{12} and keeping all other parameters fixed (Fig. 4.14). In other words, we only manipulate the second (lower) layer while keeping the parameters of the first (upper) layer unchanged.

It turns out that even very weak coupling between the layers forces them to behave the same way ($\sigma_{12} > 0.007$). We detect different types of chimera states (points A, B, C, D in Fig. 4.14a, b) and incoherent patterns (white region in Fig. 4.14a, b) depending on coupling parameters. Interestingly, chimeras can be induced in the first (upper) layer by multiplexing it with not exclusively denser layer as shown in

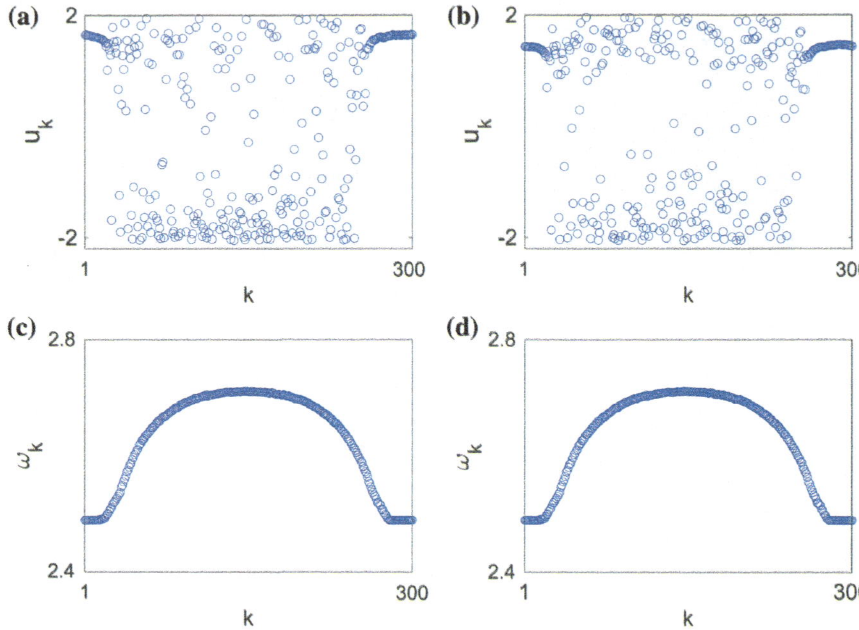

Fig. 4.12 Two weakly multiplexed layers ($\sigma_{12} = 0.01$) with different coupling ranges. **a**, **c** first (upper) layer, $r_1 = 0.2$; **b**, **d** second (lower) layer, $r_2 = 0.35$; **a**, **b** snapshots of variable u_k; **c**, **d** mean phase velocity profiles. Other parameters: $N = 300$, $\varepsilon = 0.05$, $\phi = \pi/2 - 0.1$, $a = 0.5$, $\sigma_1 = \sigma_2 = 0.1$ [126]

Figs. 4.12, 4.13 and 4.14 (see point D in Fig. 4.14a, b and panel D in (Fig. 4.14c), but also with the layer characterized by lower density (point A in Fig. 4.14a, b and panel A in (Fig. 4.14c). Depending on the coupling range in the second (lower) layer r_2, we can observe different chimera patterns. For small coupling range r_2 (point A in Fig. 4.14a, b and panel A in (Fig. 4.14c), the chimera pattern is characterized by a smaller incoherent domain compared with those observed for larger values of r_2 (points B, C, D in Fig. 4.14a, b and panels B, C, D in (Fig. 4.14c). Moreover, for the patterns in region A (Fig. 4.14a, b) the difference between the maximum frequency from the incoherent domain and that of coherent domain is smaller. The latter property also applies to the patterns in region B (Fig. 4.14a, b). Furthermore, the mean phase velocity of the chimeras in region B (panel B in Fig. 4.14c) has a plateau-shaped profile, while chimeras in regions A and D (Fig. 4.14a, b) demonstrate a classical arc-shaped profile (panels A and D in Fig. 4.14c). A distinguishing feature of chimeras in region C (panel C in (Fig. 4.14c) is a step-like structure of the mean phase velocity profile. The formation of such a structure can be explained by the complex synchronization cascade mechanisms recently analyzed in [242]. It is important to note that the coupling range where chimeras are observed in the second (lower) layer in the multiplex network is shifted towards higher values if compared with the isolated second (lower) layer (Fig. 4.14b). Therefore, weak multiplexing induces

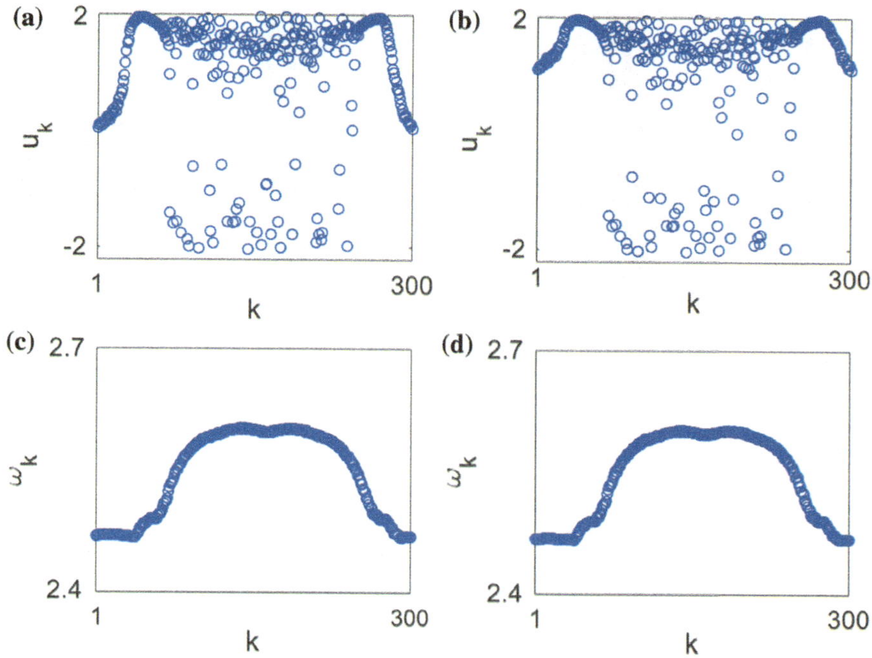

Fig. 4.13 Two strongly multiplexed layers ($\sigma_{12} = \sigma_1 = \sigma_2 = 0.1$) with different coupling ranges. **a, c** first (upper) layer, $r_1 = 0.2$; **b, d** second (lower) layer, $r_2 = 0.35$; **a, b** snapshots of variable u_k; **c, d** mean phase velocity profiles. Other parameters: $N = 300$, $\varepsilon = 0.05$, $\phi = \pi/2 - 0.1$, $a = 0.5$ [126]

chimeras not only in the first (upper) layer, but also in the parameter range of the second (lower) layer where no chimeras are observed in isolation.

4.3.4 Multiplex Network: Coupling Strength Mismatch

Further, we consider a different type of a multiplex network with non-identical layers. We fix the link density, i.e., the coupling range r, in both rings $r_1 = r_2 = 0.35$ and introduce a mismatch in the intra-layer coupling strength $\sigma_1 \neq \sigma_2$. In more detail, the elements in the first (upper) layer are coupled more weakly than the nodes in the second (lower) layer $\sigma_1 < \sigma_2$. Without multiplexing $\sigma_{12} = 0$, the first (upper) layer exhibits a chimera state (Fig. 4.15a, g). In contrast, the strongly coupled layer ($\sigma_2 = 0.4$) demonstrates coherent behavior: both the snapshot of the variable u_k (Fig. 4.15b) and the flat mean phase velocity profile (Fig. 4.15h) indicate synchronization.

Once the layers are connected, chimera states in the first (upper) layer become less pronounced if compared with the isolated case. In more detail, from the snapshots of variable u_k it can be seen that the size of the incoherent domain decreases

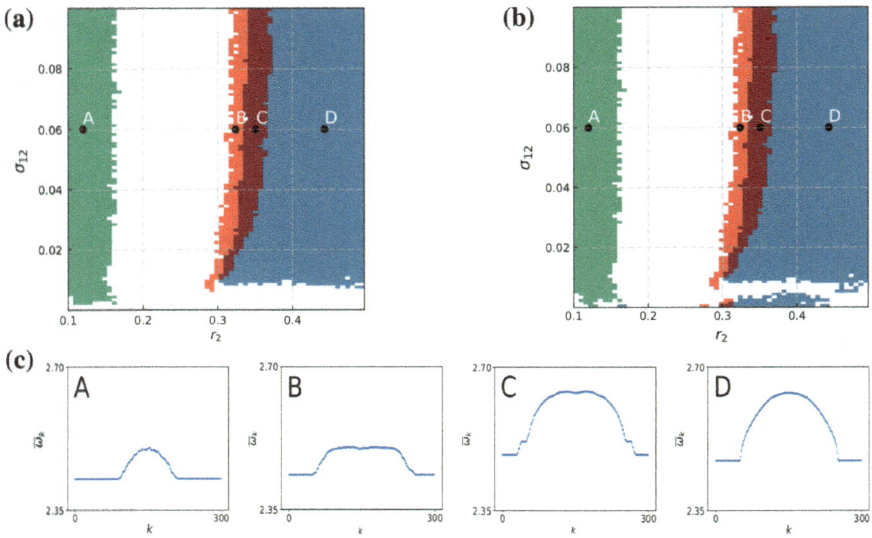

Fig. 4.14 Map of regimes in the case of coupling range mismatch for **a** the first (upper) layer and **b** the second (lower) layer in the (r_2,σ_{12}) parameter plane. The parameters $[r_2, \sigma_{12}]$ for the selected points: A = [0.12, 0.06], B = [0.323 ,0.06], C = [0.35, 0.06], D = [0.44, 0.06]. The colors stand for different chimera types; white region indicates incoherent patterns. **c** Mean phase velocity profiles for the upper layer (lower layer behaves similarly) for points A, B, C, D. Other parameters: $N = 300$, $\varepsilon = 0.05$, $\phi = \pi/2 - 0.1$, $a = 0.5$, $\sigma_1 = \sigma_2 = 0.1$,$r_1 = 0.2$ [126]

(see Fig. 4.15a) for the isolated case and (Fig. 4.15c) for the multiplexed case). In the mean phase velocity profile, the difference between the maximum frequency of incoherent domain and that of coherent domain decreases and at the same time a dip is formed in the middle of the incoherent domain (Fig. 4.15i). The occurrence of the dip can be explained by the fact that in the isolated nonlocally coupled ring the increase of the coupling strength for the fixed coupling range leads to the formation of multichimera states. In other words, the transition from a classical chimera state with one incoherent domain to a chimera pattern with two incoherent domains occurs through the formation of the dip in the incoherent domain of the mean phase velocity profile [79]. Here, such a modification of the mean phase velocity profile is caused by weak multiplexing with the strongly coupled layer. The increase of the inter-layer coupling (within the weak multiplexing range) destroys chimera states in the first (upper) layer and induces synchronization throughout the whole network: both within and across the layers (Fig. 4.15e, f, k, l). Therefore, weak multiplexing allows not only to induce chimeras (as shown in Sect. 4.3.3), but also to suppress them.

To get an overall view on the dynamics of the network, we calculate the map of regimes in each layer for varying coupling strength σ_2 and inter-layer coupling strength σ_{12} and keeping all other parameters fixed (Fig. 4.16). Therefore, we again manipulate only the second (lower) layer while keeping the parameters of the first

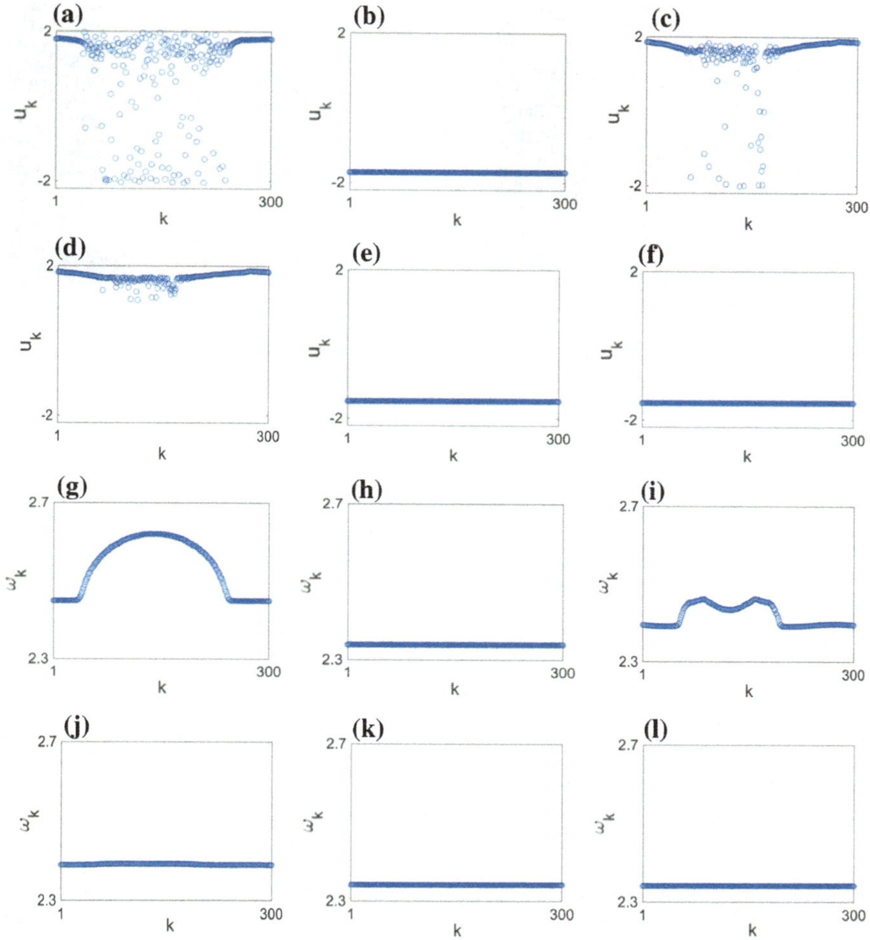

Fig. 4.15 Two-layer multiplex network with intra-layer coupling strength mismatch for different values of inter-layer coupling strength σ_{12}. Snapshots of variable u_k (top row) and mean phase velocity profiles (bottom row). **a, b, g, h** two disconnected layers $\sigma_{12} = 0$ (**a, g**—first (upper) layer and **b, h**—second (lower) layer); **c, d, i, j** weakly coupled layers $\sigma_{12} = 0.01$ (**c, i**—first (upper) layer and **d, j**—second (lower) layer); **e, f, k, l** weakly coupled layers $\sigma_{12} = 0.05$ (**e, k**—first (upper) layer and **f, l**—second (lower) layer). Other parameters: $N = 300$, $\varepsilon = 0.05$, $a = 0.5$, $\phi = \pi/2 - 0.1$, $r_1 = r_2 = 0.35$, $\sigma_1 = 0.1$, $\sigma_2 = 0.4$ [126]

(upper) layer unchanged. The difference from Sect. 4.3.3 is that here we adjust the strength of the intra-layer coupling and the range. In comparison with the previous case, where the coupling range was varied (see Fig. 4.14), here we observe a more complex structure of regimes (Fig. 4.16). Depending on coupling parameters, the layers may behave the same way or differently (hatched region in Fig. 4.16a, b). For a fixed value of the inter-layer coupling strength ($\sigma_{12} = 0.03$), by changing the intra-layer coupling strength (σ_2), we observe a change in the mean phase velocity

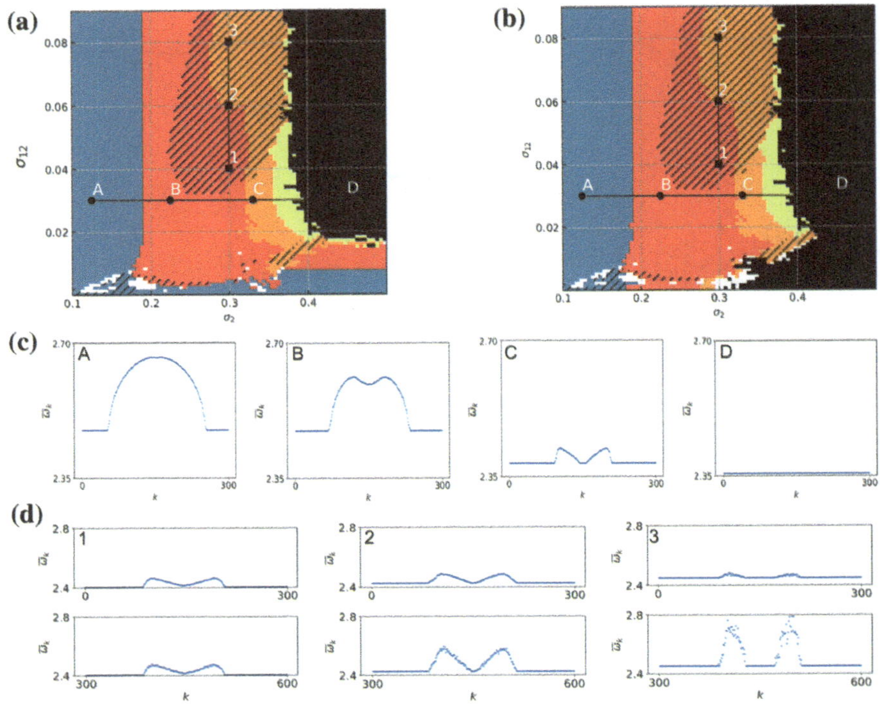

Fig. 4.16 Map of regimes in the case of coupling strength mismatch for **a** the first (upper) layer and **b** the second (lower) layer in the (σ_2, σ_{12}) parameter plane. The parameters $[\sigma_2, \sigma_{12}]$ for the selected points: A = [0.125, 0.03], B = [0.225, 0.03], C = [0.330, 0.03], D = [0.450, 0.03], 1 = [0.3, 0.04], 2 = [0.3, 0.06], 3 = [0.3, 0.08]. Hatching marks the regions where the layers behave differently; colors stand for different chimera types; white region indicates incoherent patterns. **c** Mean phase velocity profiles for the first (upper) layer (lower layer behaves similarly) for points A, B, C, D. **d** Mean phase velocity profiles for the first (upper) layer (top row) and the second (lower) layer (bottom row) for points 1, 2, 3. Other parameters: $N = 300$, $\varepsilon = 0.05$, $\phi = \pi/2 - 0.1$, $a = 0.5$, $\sigma_1 = 0.1$, $r_1 = r_2 = 0.35$ [126]

profile of chimeras (points A, B, C, D in Fig. 4.16a, b). For relatively small values of σ_2, we observe classical chimeras with one incoherent domain (for example, point A in Fig. 4.16a, b and panel A in Fig. 4.16c). The increase of σ_2 leads to the formation of a dip in the mean phase velocity profile (panel B in Fig. 4.16c). When we increase σ_2 even further, the dip reaches the frequency level of the coherent domain resulting in the formation of two-headed chimera (panel C in Fig. 4.16c). Further increase of σ_2 leads to an in-phase synchronized regime (panel D in Fig. 4.16c). Therefore, we can suppress chimeras in the first (lower) layer not only by changing the inter-layer coupling strength (σ_{12}) (as shown in Fig. 4.15, but also by increasing the intra-layer coupling strength in the second (lower) layer (σ_2) for a fixed value of σ_{12} (Fig. 4.16a).

Compared with Sect. 4.3.3, where both layers show similar behavior, here the layers may behave differently depending on the coupling parameters. For the fixed value of intra-layer coupling ($\sigma_2 = 0.3$), the increase in inter-layer coupling strength σ_{12} makes the layers demonstrate different mean phase velocity profiles (point 1, 2, 3 in Fig. 4.16a, b and panels 1, 2, 3 in Fig. 4.16d). Although the position in space for coherent and incoherent domains in both layers stays the same and the coherent domains are synchronized, the incoherent domains start demonstrating different behaviors. For example, the maximum value of mean phase velocity becomes higher in the second (upper) layer (panel 2 in Fig. 4.16d). Further increase of the inter-layer coupling strength σ_{12}, when the two-headed chimera is formed, results in a better pronounced incoherent domain in the second (lower) layer, while in the first (upper) layer, the chimera is almost suppressed (panel 3 in Fig. 4.16d). This result is intriguing since the two-headed chimeras are better pronounced in the second (lower) layer for the range of σ_2 values that correspond to no chimera (in-phase synchronization) for this layer in isolation. Therefore, weak multiplexing can have a dramatic effect on the dynamic regime of the network.

4.3.5 Multiplex Network: Switching to Solitary States

A solitary state is a partial synchronization pattern whose formation mechanism is different from that of a chimera state. In what follows we show that weak multiplexing allows not only for control of chimeras, but can also lead to the occurrence of solitary states. In particular, we demonstrate that for small intra-layer coupling strength mismatch, multiplexing induces solitary states in nonlocally coupled rings, which do not show these patterns in isolation.

In more detail, the isolated layers of nonlocally coupled FHN oscillators for the chosen set of coupling parameters exhibit synchronization. Once the two rings are coupled (and rather weak inter-layer coupling is enough), solitary states are observed in both layers (Fig. 4.17). We can see from the space-time plots that solitary nodes are distributed randomly along the network, a characteristic signature of solitary states (4.17a, d). The snapshots indicate the presence of two groups of FHN elements: the coherent cluster and solitary nodes split from the synchronized group (4.17b, e). Therefore, multiplexing allows to achieve solitary states throughout the whole network without manipulating the intra-layer coupling parameters.

It is important to point out that the solitary states we observe here for FHN neurons are different from those observed in networks of Kuramoto phase oscillators. They have a rather flat mean phase velocity profile indicating that the frequency of the solitary nodes is close to that of the synchronized cluster (Fig. 4.17c, f). Further, the solitary nodes and the synchronized elements have different amplitudes (Fig. 4.18a). Another distinctive feature of the solitary patterns in FHN model is that the solitary nodes and the synchronized cluster are characterized by a small phase shift (Fig. 4.18b). The occurrence of the phase shift has been previously reported for globally coupled networks combining repulsive and attractive interactions [127]. In contrast to that, in our case, the solitary states are observed in the network where the interactions are all of the same type.

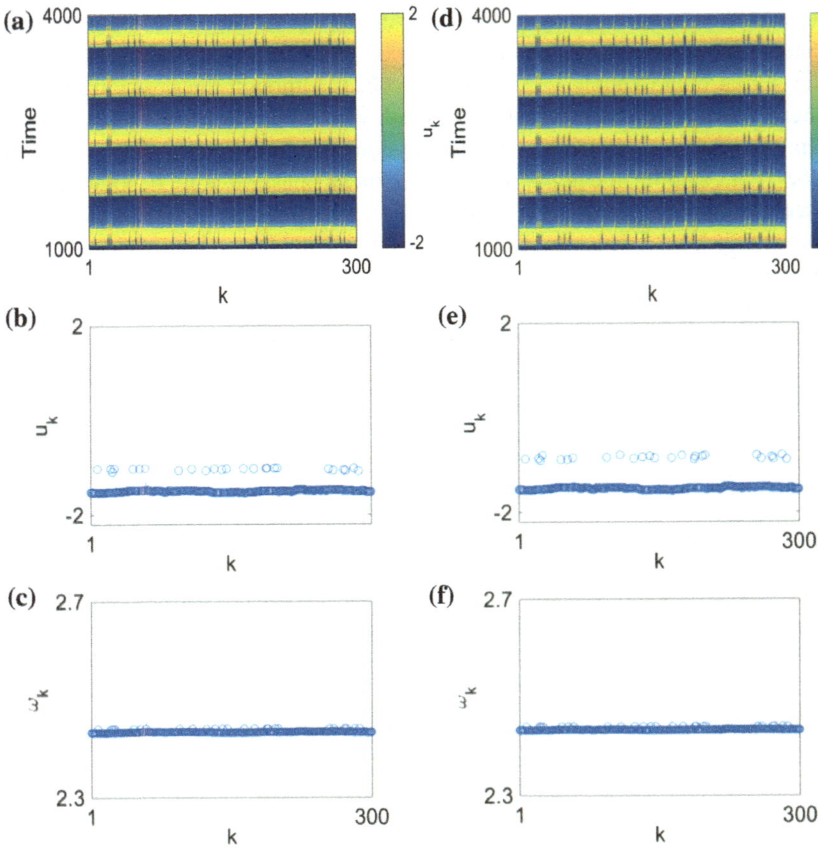

Fig. 4.17 Two weakly coupled layers ($\sigma_{12} = 0.05$) with small intra-layer coupling strength mismatch. **a**, **b**, **c** first (upper) layer, $\sigma_1 = 0.4$; **d**, **e**, **f** second (lower) layer $\sigma_2 = 0.3$; **a**, **d** space-time plots for the variable u_k; **b**, **e** snapshots for the variable u_k; **c**, **f** mean phase velocity profiles. Other parameters: $N = 300$, $\varepsilon = 0.05$, $\phi = \pi/2 - 0.1$, $a = 0.5$, $r = 0.35$ [126]

4.3.6 Summary

In conclusion, weak multiplexing can have a dramatic effect on the behavior of the network. One possible explanation for this fact can be provided by drawing an analogy with classical synchronization theory of periodic oscillations [20, 358]. For example, mutual synchronization of two bidirectionally coupled periodic oscillators can be achieved for a weaker connection between them if compared with external synchronization of unidirectionally coupled oscillating systems. Here, we deal with two layers that are coupled mutually; therefore, weak multiplexing is enough to make them essentially influence the dynamics of each other.

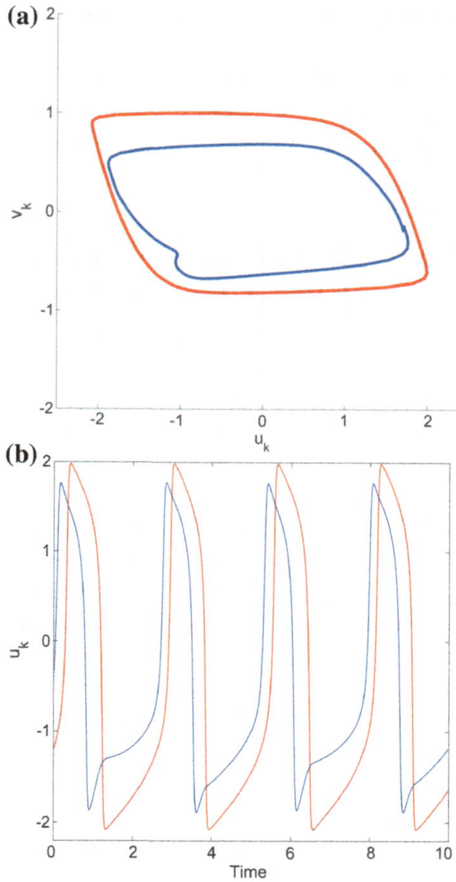

Fig. 4.18 Solitary state in first (upper) layer of the multiplex network for $\sigma_{12} = 0.05$. **a** phase portrait and **b** time series for two selected nodes: solitary node (blue) and a node from the synchronized cluster (red). Other parameters: $N = 300$, $\varepsilon = 0.05$, $\phi = \pi/2 - 0.1$, $a = 0.5$, $r = 0.35$ $\sigma_1 = 0.4$, $\sigma_2 = 0.3$ [126]

Weak multiplexing represents a powerful tool for controlling dynamic patterns in neural networks. It allows to adjust the dynamics of one layer without manipulating its parameters. In more detail, we demonstrate two control strategies: (i) by tuning the coupling range in one layer, chimeras with desired mean phase velocity profile can be induced in the other layer. Moreover, the same dynamical patterns across the layers are achieved for even very weak coupling between them. (ii) By tuning the intra-layer coupling strength, we can suppress chimera states with one incoherent domain and induce a variety of other regimes, including in-phase synchronization and two-headed chimeras. Furthermore, we can make the layers behave differently. Interestingly, for small intra-layer coupling strength mismatch between the layers, we can switch from a chimera to a solitary state.

In both cases, the control of the dynamics for one of the layers can be realized without manipulating the internal parameters of its elements or the connections between them. The control is achieved by adjusting the coupling parameters (coupling range or coupling strength) of the other layer and varying the coupling strength between the layers within the weak multiplexing range. This is important from the point of view of applications, since it is not always possible to directly access the desired layer while the network it is multiplexed with may be adaptable.

4.4 Coherence-Incoherence Patterns in Networks with Power-Law Coupling

In the long-range coupling, each node in the network is connected to all other nodes with an effective interaction strength that decreases with increasing distance according to a power law. Thus, the oscillators are subject to a long-range interaction whose strength is controlled by the power-law exponent. Previously, the long-range interaction obeying a power law has been considered in ferromagnetic spin models [361], hydrodynamic interaction of active particles [362, 363], coupled-map lattices (see [364] and references therein), and phase oscillators [365–367] in the context of synchronization. In particular, Rogers and Wille [365] numerically showed that a one-dimensional ring of coupled nonlinear phase oscillators with frequency mismatch undergoes a phase transition from a synchronized to a completely desynchronized state as the range of interaction is decreased.

In the present section, we show that long-range interaction with distance-dependent power-law coupling can induce various spatiotemporal patterns (like chimera states, tree-like patterns, or solitary states), and at the same time tuning of the power-law exponent can mediate transitions between them. While previously the power-law coupling has been considered for chimera states in ecological oscillators [84], here we consider a network consisting of neural oscillators modeled by the FitzHugh-Nagumo system. We explore and demonstrate the influence of the coupling scheme and map all the spatiotemporal patterns including chimera states. We identify the possible transitions between the spatiotemporal patterns that arise due to the interplay of coupling strength and coupling topology characterized by the power-law exponent of the long-range interaction.

The section is organized as follows. First, we introduce the model (Sect. 4.4.1). Then, we investigate the map of regimes in the FitzHugh-Nagumo network with power-law coupling kernel for varying the power-law exponent and the coupling strength (Sect. 4.4.2). Further, we study the detected spatiotemporal patterns in more detail: chimera states (Sect. 4.4.3), tree-like perturbations (Sect. 4.4.4), and solitary states (Sect. 4.4.5). Moreover, we study transitions between various patterns in the network (Sect. 4.4.6). Finally, we summarize the results in Sect. 4.4.7.

4.4.1 Model

We consider a network of coupled oscillatory FitzHugh-Nagumo (FHN) systems. As described in Chap. 3, this two-dimensional model is a paradigmatic for neural excitability. Here, we investigate the impact of power-law interactions on the spatiotemporal patterns occurring in the network given by the following equations:

$$\begin{aligned} \varepsilon \frac{du_i}{dt} &= u_i + \frac{u_i^3}{3} - v_i + \frac{\sigma}{\eta} \sum_j^N G_{ij}(b_{uu}(u_j - u_i) + b_{uv}(v_j - v_i)), \\ \frac{dv_i}{dt} &= u_i + a_i + \frac{\sigma}{\eta} \sum_j^N G_{ij}(b_{vu}(u_j - u_i) + b_{vv}(v_j - v_i)), \end{aligned} \tag{4.5}$$

where u_i and v_i are the activator and inhibitor variables, respectively, $j = 1, ..., N$ and all indices are modulo N, $\varepsilon > 0$ is a small parameter responsible for the timescale separation of fast activator and slow inhibitor, a_i defines the excitability threshold. For an individual FHN element, it determines whether the system is excitable ($|a_i| > 1$), or oscillatory ($|a_i| < 1$). For $|a_i| < 1$, the system is in the oscillatory regime where the steady state is unstable and self-sustained oscillations are observed. For $|a_i| > 1$, the system is in the excitable regime and characterized by a locally stable steady state. In the present section, we focus on oscillatory regime ($a_i \equiv a = 0.5$). Equation (4.5) contains not only direct, but also cross-couplings between activator (u) and inhibitor (v) variables, which is modeled by a rotational coupling matrix [79]:

$$\boldsymbol{B} = \begin{pmatrix} b_{uu} & b_{uv} \\ b_{vu} & b_{vv} \end{pmatrix} = \begin{pmatrix} \cos\phi & \sin\phi \\ -\sin\phi & \cos\phi \end{pmatrix}, \tag{4.6}$$

where $\phi \in [-\pi; \pi)$. In the present section (as in Chap. 3), we fix the parameter $\phi = \pi/2 - 0.1$ for which chimeras have been found in the deterministic oscillatory regime [79]. The coupling strength is characterized by σ. Finally, the coupling topology is given by a distant-dependent power-law kernel of the form

$$\begin{aligned} G_{ij} &= \frac{1}{\mid j - i \mid^{\alpha}} \qquad i \neq j \\ G_{ii} &= -\sum_{j \neq i} \frac{1}{\mid j - i \mid^{\alpha}} \end{aligned} \tag{4.7}$$

with the diagonal elements being the negative sum of all non-diagonal elements in a row, due to diffusive coupling. We introduce a normalization constant

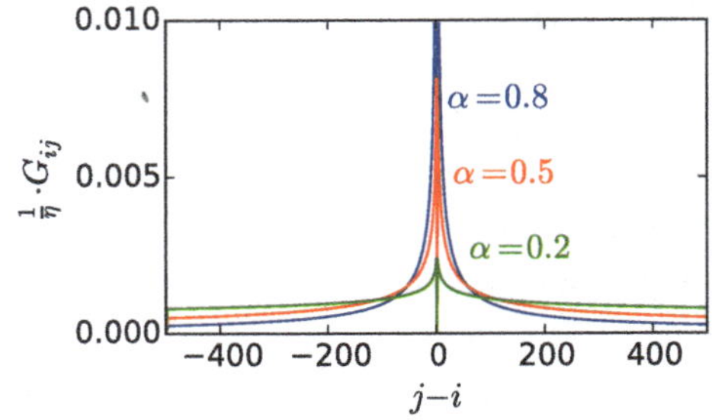

Fig. 4.19 Long-range power-law interaction kernels [201]

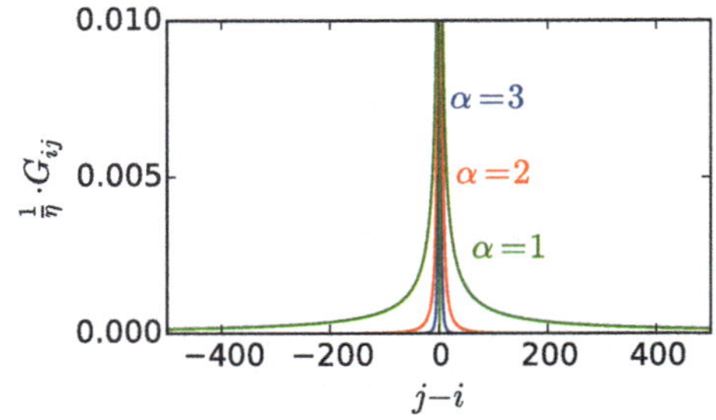

Fig. 4.20 Short-range power-law interaction kernel [201]

$$\eta = 2 \sum_{p=1}^{(N-1)/2} p^{-\alpha} \qquad \text{if } N \text{ odd}$$
$$\eta = 2 \sum_{p=1}^{(N-1)/2} p^{-\alpha} + \left(\frac{N}{2}\right)^{-\alpha} \qquad \text{if } N \text{ even} \tag{4.8}$$

that is given by the area underneath the graph of the kernel. This choice of the coupling kernel and the normalization constant allows to compare the results for different power-law exponents α without changing the maximum input for a given node. We vary the interaction from long to short range by adjusting the power-law exponent. In our case, i.e., in a one-dimensional ring, the most interesting changes happen in the vicinity of $\alpha \sim 1$. The examples of power-law interaction kernels for different α values are given in Figs. 4.19 and 4.20. In the limit $\alpha \to 0$, the coupling becomes global, i.e., all nodes are connected to all other nodes with an equal strength, while in the limit $\alpha \to \infty$ the coupling reduces to local nearest neighbor coupling.

4.4.2 Dynamic Regimes: Impact of Coupling Parameters

We investigate dynamic regimes in system (4.5) in the plane of two parameters that characterize interactions between the network elements: the coupling strength σ and the power-law exponent α of the coupling kernel.

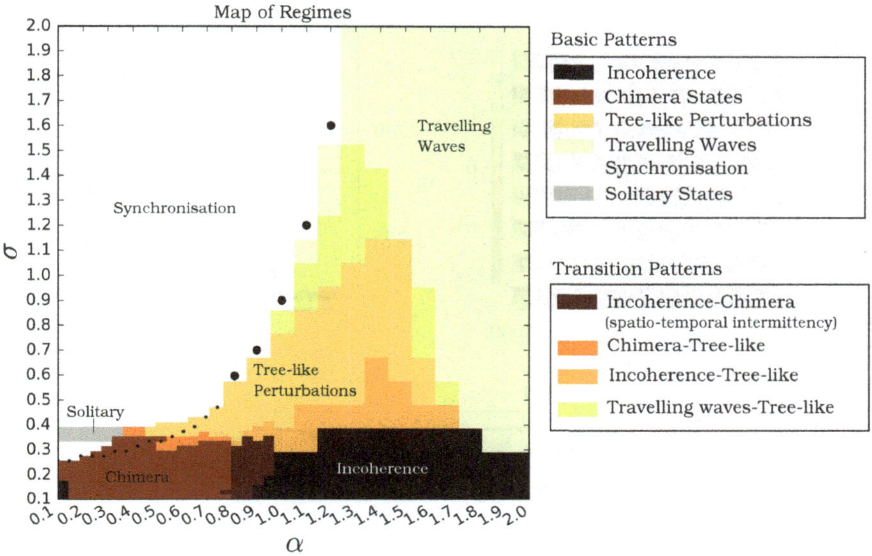

Fig. 4.21 Map of regimes in (α, σ) parameter plane. Other parameters: $\varepsilon = 0.05$, $a = 0.5$, $N = 1000$, $\phi = \pi/2 - 0.1$. Initial conditions: randomly distributed on a circle $u^2 + v^2 = 4$ [201]

Besides a fully synchronized regime, incoherence, and traveling waves, we detect three different partial synchronization patterns (Fig. 4.21). For sufficiently small coupling strength ($\sigma \leq 0.35$) and long-range interactions ($\alpha \leq 0.8$), we find *chimera states*. If the coupling becomes stronger, chimera states disappear and the so-called *tree-like perturbations* are observed. These states also exist for steeper coupling kernels, i.e., larger values of α. Another remarkable network regime we detect in a narrow stripe at around $\sigma \approx 0.35$ is the *solitary states* (gray region in Fig. 4.21). This coupling strength value $\sigma = 0.35$ also marks the border between chimera states and tree-like perturbations. In what follows we discuss chimera states, tree-like perturbations, solitary states as well as various transitions between these patterns in more detail.

4.4.3 Chimera States

Previously, chimera states in the oscillatory regime of FHN system have been found for nonlocally coupled ring networks with rectangular coupling kernel [79]. In such networks, every node has P nearest neighbors in each direction on a ring. Additionally, a normalized number of nearest neighbors $r = P/N$ (also called coupling range), where N is the total number of elements in the network, is used to compare networks of different sizes. By tuning P, one can interpolate between local (nearest neighbor, $P = 1$) coupling and global coupling (all-to-all, $P = N/2$). It has been

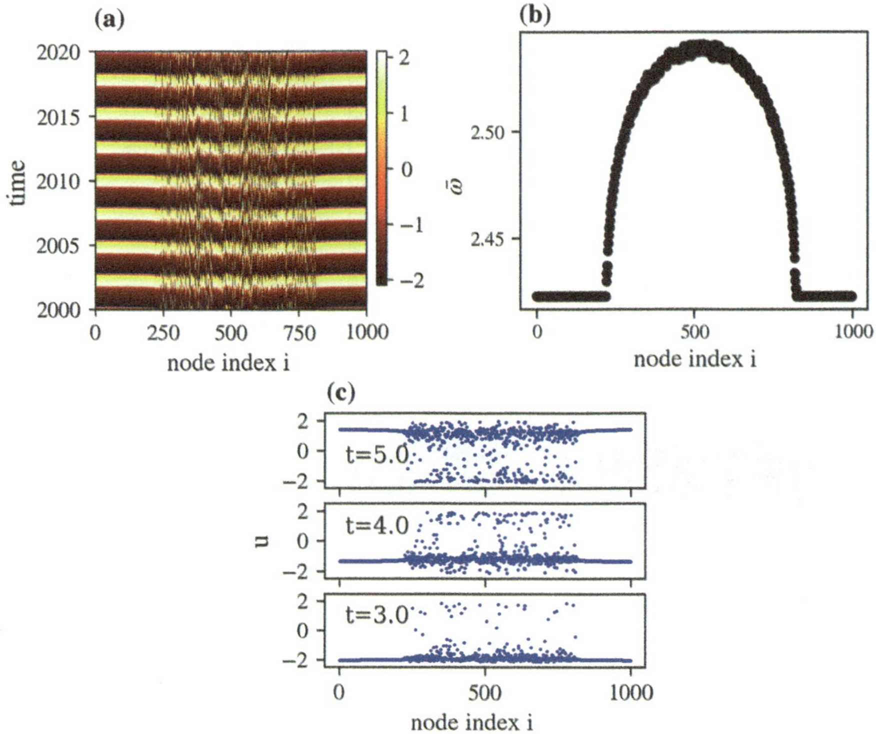

Fig. 4.22 Chimera state in a network with power-law coupling for $\alpha = 0.3$ and $\sigma = 0.1$: **a** Space-time plot for the u-variable; **b** mean phase velocity profile; **c** a sequence of snapshots for the u-variable. Other parameters: $\varepsilon = 0.05$, $a = 0.5$, $N = 1000$, $\phi = \pi/2 - 0.1$ [201]

shown that chimera states in nonlocally coupled rings occur for intermediate coupling range and sufficiently small coupling strength [79]. In the case of power-law coupling that we investigate in this section, chimera states are observed for similar values of coupling strength $\sigma < 0.35$ and relatively small values of power-law exponent $\alpha < 0.8$ (Fig. 4.21). The space-time plot in Fig. 4.22a shows coexistence in space of a coherent domain (to the left and to the right) and incoherent nodes (in the middle). Note that there is only one coherent domain, due to periodic boundary condition. Figure 4.22b displays the typical arc-shaped mean phase velocity profile, a characteristic signature of chimera states. The sequence of snapshots (Fig. 4.22c) further confirms the separation in space of two domains with different behaviors. While the nodes from the coherent domain spike at the same time, the FHN neurons from the incoherent domain spike in a random fashion.

In contrast to the case of a nonlocally coupled ring, we do not detect chimera states with more than one incoherent domain for a network with power-law coupling kernel. Even if we use a two-headed chimera state as initial condition (i.e., a chimera state with two incoherent domains), the system transforms into a one-headed chimera state similarly to the one depicted in Fig. 4.22. This can be explained by taking a closer look at the power-law coupling kernel. All nodes of the network contribute to the coupling input defined in (4.5) and in the case of long-range interaction (4.7) the kernel does not converge to zero within the network size. Even though the coupling for a particular node in the network is strongly influenced by its nearest neighbors, the contribution from the nodes on the opposite side of the ring cannot be neglected in the long-range interaction case. In contrast to that, for the nonlocally coupling ring with a rectangular kernel, the nodes on the opposite side of the ring have no direct influence, as they are not connected. Apparently, this allows the network to form two or more distinct incoherent domains, while this is not possible if all nodes can interact with each other.

There is strong numerical evidence that the basin of attraction of the chimera states in system (4.5) is surprisingly large. For the map of regimes (Fig. 4.21), we use initial conditions randomly distributed on a ring with radius 2. However, the chimera region shown in this map can be obtained also from any other initial condition except the following two cases. First, if a synchronized state is taken as an initial condition, the network will remain in-phase synchronized for all times, as the coupling vanishes. Second, in the case of initial conditions randomly distributed on a ring with a large radius (e.g., 10), the coupling outweighs the intrinsic dynamics and the network diverges quickly.

4.4.4 Tree-Like Patterns

For larger coupling strength, chimera states transform into a pattern that is characterized by tree-like shaped perturbations (see Fig. 4.23). They are observed for the coupling strength $\sigma > 0.35$. In contrast to chimera states, they are not limited to the regime of long-range interactions, where $\alpha < 1$ but can also be detected up to the value of the power-law exponent $\alpha = 1.8$ (see Fig. 4.21). The perturbations in this pattern undergo changes in space and time. They are growing locally, splitting, traveling in space and time via propagating phase flips, decaying, disappearing, and eventually occurring again (Fig. 4.23). Figure 4.24 displays an example of this peculiar motion. We start the observation with three initial perturbations (see Fig. 4.24b bottom panel, $t = 127.5$). In the next step, the perturbation in the middle gets enhanced, while the two perturbations to the left and to the right decay ($t = 132.5$). The middle perturbation grows further ($t = 137.5$) and emits phase flips in both directions ($t = 142.5$), which spread the perturbation in space and time. Eventually, the phase flips stop traveling ($t = 147.5$) and form the seeds of perturbations similarly to the initial ones ($t = 152.5$). Finally, we are back to the initial situation ($t = 162.5$). The alternation of synchronization and desynchronization patterns via propagating phase flips leads to a tree-like shape of the perturbation in the space-time plot (Fig. 4.24a). Which initial perturbation decays and which one grows is random as well as the motion in the network space. This can be seen in the space-time plot over a longer time interval

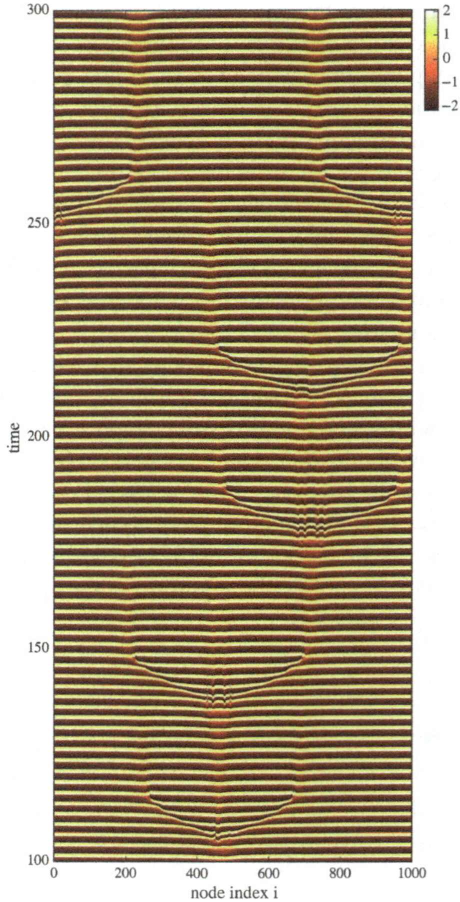

Fig. 4.23 Tree-like perturbation pattern for $\alpha = 0.8, \sigma = 0.5$. Other parameters: $\varepsilon = 0.05$, $a = 0.5$, $N = 1000$, $\phi = \pi/2 - 0.1$ [201]

(see Fig. 4.23) and is also reflected in the indistinct shape of the mean phase velocity profile depicted in Fig. 4.25.

For weaker coupling (i.e., smaller σ) and smaller coupling range (i.e., larger α), the patterns become more complex. The scenario described above occurs simultaneously in multiple parts of the network. The perturbations interfere with each other, merging and splitting again. Therefore, decreasing σ leads to a more turbulent pattern. With increasing α, successively emitted propagating phase flips dominate and a smooth transition to the regime of traveling waves takes place (see Sect. 4.4.6).

The results presented in this section are obtained from random initial conditions. Our numerical simulations show that a perturbation of one node is not enough to induce a tree-like perturbation in the network that is otherwise fully synchronized. Instead, the perturbation decays and the node is entrained. With a larger number of randomly perturbed nodes, the probability to obtain the tree-like perturbations increases. We conclude that these patterns coexist with fully synchronized regime.

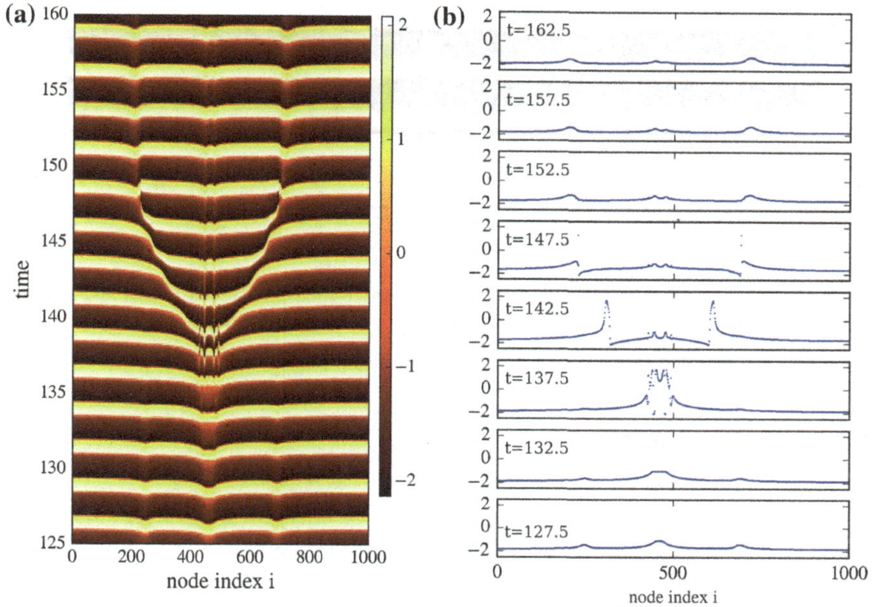

Fig. 4.24 Tree-like perturbation pattern for $\alpha = 0.8$ and $\sigma = 0.5$. **a** Space-time plot for the u-variable; **b** Successive snapshots for the u-variable. Other parameters: $\varepsilon = 0.05$, $a = 0.5$, $N = 1000$, $\phi = \pi/2 - 0.1$ [201]

Fig. 4.25 Mean phase velocity profile for $\alpha = 0.8$ and $\sigma = 0.45$ (averaged over 2000 time units). Other parameters: $\varepsilon = 0.05$, $a = 0.5$, $N = 1000$, $\phi = \pi/2 - 0.1$ [201]

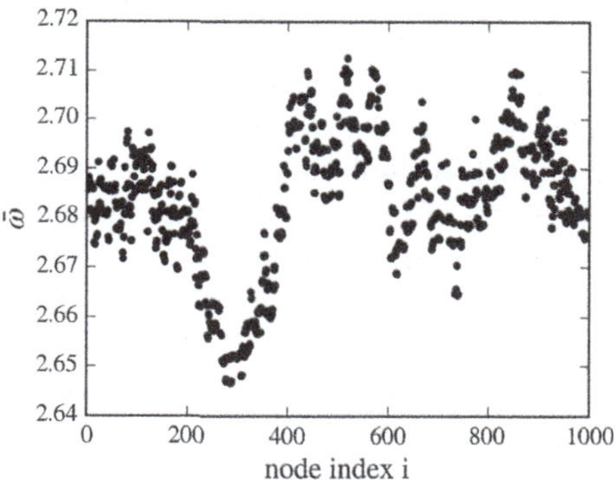

Starting with a snapshot of a chimera state (obtained with $\alpha = 0.8$, $\sigma = 0.15$) as initial conditions in the parameter regime of tree-like perturbations ($\alpha = 0.8$, $\sigma = 0.45$), the clear separation between coherent and incoherent domain breaks up and tree-like perturbations form, indicating that there is no multistability and chimera states do not coexist with tree-like perturbations. This will be explained in more detail in Sect. 4.4.6.

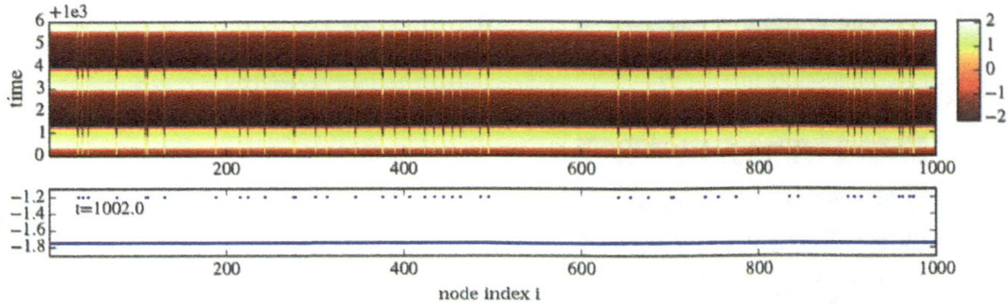

Fig. 4.26 Solitary states for $\alpha = 0.1$ and $\sigma = 0.35$: space-time plot (upper panel) and snapshot at $t = 1002$ (lower panel). Number of solitary nodes: 44. Other parameters: $\varepsilon = 0.05$, $a = 0.5$, $N = 1000$, $\phi = \pi/2 - 0.1$ [201]

4.4.5 Solitary States

For a certain range of coupling strength $0.32 \leq \sigma \leq 0.37$ and long-range interactions with $\alpha \leq 0.3$ we find solitary states (Fig. 4.26). As discussed in Sect. 4.3.5, a solitary state is a partial synchronization pattern whose formation mechanism is different from that of a chimera state. In the case of power-law coupling kernel that we consider in this section, the nodes from the synchronized cluster of the solitary pattern are synchronized with respect to both the frequency and the phase. At the same time, they are shifted in phase with respect to the solitary nodes that start spiking earlier. Therefore, we also call them precursor nodes. The precursor or solitary nodes are phase synchronized with respect to each other and are randomly distributed over the network. In the example given in Fig. 4.26, a small group of 44 nodes spike in a precursor fashion, while the rest of the network (956 elements) is frequency and phase locked. Both the space-time plot and the snapshot indicate the presence of two groups of FHN elements: the coherent cluster and the solitary nodes split from the synchronized group (Fig. 4.26). Moreover, we can clearly see that the precursor nodes are distributed randomly along the network.

Next, we calculate the phase difference between the precursor nodes and the synchronized cluster. The geometric phase is given by

$$\phi_i(t) = \arctan\left(\frac{v_i(t)}{u_i(t)}\right). \tag{4.9}$$

It is important to note that due to the slow-fast dynamics of the FitzHugh-Nagumo oscillator, the motion of a node along the limit cycle is highly inhomogeneous in time. Hence, the phase difference between two oscillators becomes small if both of them are located on the same slow branch, but increases erratically, if one of them reaches the fast branch and "jumps" to the other slow branch. To overcome this problem, we introduce the dynamic phase. We calculate the function $t(\phi_0)$ (which uses the geometric phase of an uncoupled oscillator ϕ_0 with period T as reference) numerically by assigning a value of t from the interval $0 < t(\phi_0) < T$ to each value

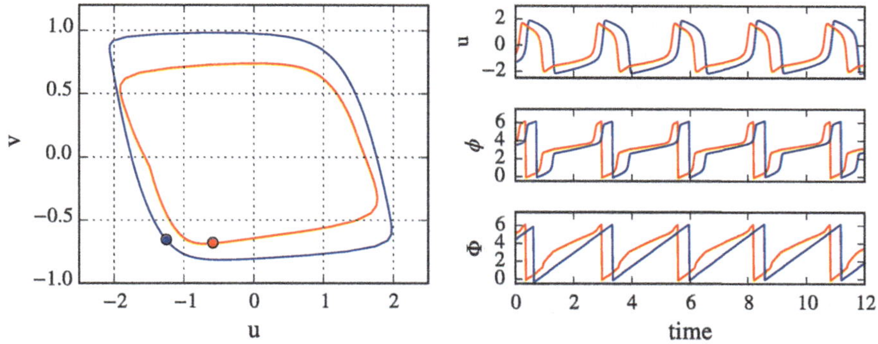

Fig. 4.27 Solitary states for $\alpha = 0.1$ and $\sigma = 0.35$: phase portrait (left panel) for the nodes from the synchronized cluster (blue) and the solitary nodes (red); time series for the variable u (top right panel), the geometric phase ϕ (middle right panel), and the dynamic phase Φ (bottom right panel). Other parameters: $\varepsilon = 0.05$, $a = 0.5$, $N = 1000$, $\phi = \pi/2 - 0.1$ [201]

of the geometric phase (that is equidistant with respect to the system's intrinsic time). The dynamic phase is then defined by

$$\Phi_i = 2\pi \frac{t(\phi_i)}{T}, \tag{4.10}$$

leading to a constant (dynamic) phase velocity $\dot{\Phi}_0$ in the uncoupled case. For the calculation of both the geometric and the dynamic phases, we neglect the amplitude dynamics of the FitzHugh-Nagumo system.

The limit cycle of the synchronized cluster almost coincides with the limit cycle of an uncoupled FHN oscillator (blue cycle in Fig. 4.27 right panel). The limit cycle of the precursor nodes is typically smaller than that of the synchronized cluster and can vary depending on the coupling parameters (red cycle in Fig. 4.27 right panel). Since the coupling is diffusive, the nodes in the synchronized cluster are effectively only influenced by a small number of nodes (44 out of 1000 in this particular example) that spike in a precursor fashion. Hence, the coupling terms are rather small and the nodes in the synchronized cluster behave very similar to the uncoupled system. On the other hand, the precursor nodes get significant inputs from all the nodes belonging to the synchronized cluster leading to a large coupling term and, thus, a strong distortion from the uncoupled limit cycle.

Time series for the variable u (Fig. 4.27 top right panel), the geometric phase ϕ (Fig. 4.27 middle right panel), and the dynamic phase Φ (Fig. 4.27 bottom right panel) clearly demonstrate that the synchronized elements and the solitary nodes are shifted in phase. Moreover, the results obtained for the dynamic phase show that this phase shift can vary (Table 4.1).

Table 4.1 Dynamic phase shift for $\alpha = 0.1$ and $\sigma = 0.35$. Other parameters: $\varepsilon = 0.05$, $a = 0.5$, $N = 1000$, $\phi = \pi/2 - 0.1$

Maximum phase difference	$\Delta\Phi = 0.76$
Minimum phase difference	$\Delta\Phi = 0.26$
Mean phase difference	$\bar{\Delta\Phi} = 0.482$
Normalized phase difference	$\Delta\Phi/T = 0.185$

4.4.5.1 Formation of Solitary States from Random Initial Conditions

Previously, it has been shown that solitary states strongly depend on initial conditions. In particular, for varying initial conditions the solitary nodes can appear at any place of the network space and in any combination [334]. Therefore, we further analyze the role of initial conditions for solitary states in system (4.5). To obtain the results on solitary states shown in Figs. 4.26 and 4.27, we have used initial conditions randomly distributed on the circle $u^2 + v^2 = 4$ (Fig. 4.28). The nodes that subsequently form the synchronized cluster are marked in blue, while the precursor nodes are shown in red (Fig. 4.28). Therefore, initially the oscillators are uniformly distributed along the ring. However, as the system evolves, the dynamics becomes highly heterogeneous along the limit cycle. The nodes that are initially located in the vicinity of the slow branch $u < 0$ form the synchronized domain within a short time. The elements located above the few precursor nodes join the synchronized cluster by catching up the synchronized nodes. The FHN elements positioned below the precursor nodes have to slow down to become a part of the synchronized cluster (to indicate this schematically we use arrows in Fig. 4.28). The precursor elements neither speed up nor slow down, forming in such a way the solitary domain.

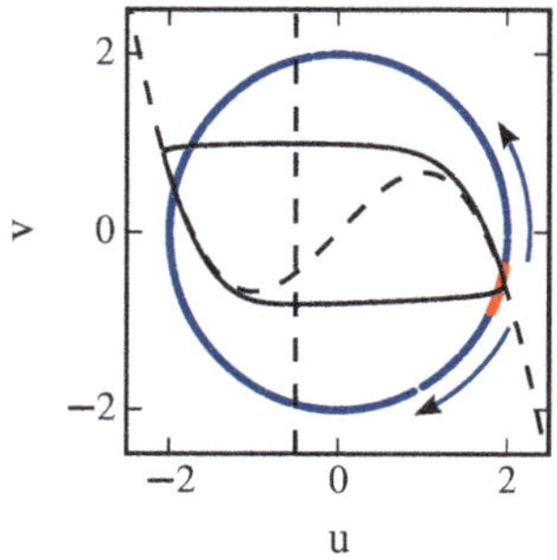

Fig. 4.28 Initial conditions randomly distributed on the circle $u^2 + v^2 = 4$. Blue dots indicate the nodes that form the synchronized cluster and the red dots mark the elements that form the precursor (solitary) nodes. Solid black line denotes the limit cycle of the uncoupled system and the dashed black lines show the corresponding nullclines. The arrows are used to indicate the fact that the elements located above the precursor nodes join the synchronized cluster by speeding up and those positioned below the precursor nodes slow down to become a part of the synchronized cluster [201]

The role of power-law exponent and the case of global coupling: In the network of FHN oscillators with power-law coupling solitary states occur for long-range interactions $0 \leq \alpha < 0.3$ and do not vary qualitatively within this interval. They are also observed for the limit case of $\alpha = 0$, i.e., for global coupling, and disappear for $\alpha \geq 0.35$. This can be explained by the fact that with increasing α the coupling between oscillators at distant positions in the network becomes too weak, i.e., the coupling for a given node is primarily determined by its nearest neighbors and the synchronized domain of the solitary pattern cannot form from random initial conditions. On the contrary, for sufficiently small power-law exponents, the node's neighborhood is no longer well defined and the coupling between the elements at distant positions in the network is strong enough for a synchronized cluster to form. It becomes even easier in the limit case of global coupling when every node is connected to any other element in the network.

The role of the coupling strength: Solitary states occur for a narrow interval of coupling strength values $0.32 \leq \sigma \leq 0.37$. Further, we investigate the phase portrait of solitary states within this interval (Fig. 4.29). It turns out that for increasing coupling strength the number of the solitary nodes decreases. Further, for $\sigma = 0.32$ and $\sigma = 0.34$ a precursor node can have two possible trajectories, leading to two distinct precursor branches. Analogously, there are four possible precursor branches for $\sigma = 0.33$. Which trajectory a solitary node follows is determined by the initial conditions. This conclusion is supported by the following. If we manipulate the solitary state (e.g., with parameters $\alpha = 0.1$, $\sigma = 0.32$, see Fig. 4.29) and manually allocate all solitary nodes on the same precursor branch, the solitary nodes will follow this trajectory. The other precursor branch will not be populated. Further, we discuss the case of specially prepared initial conditions.

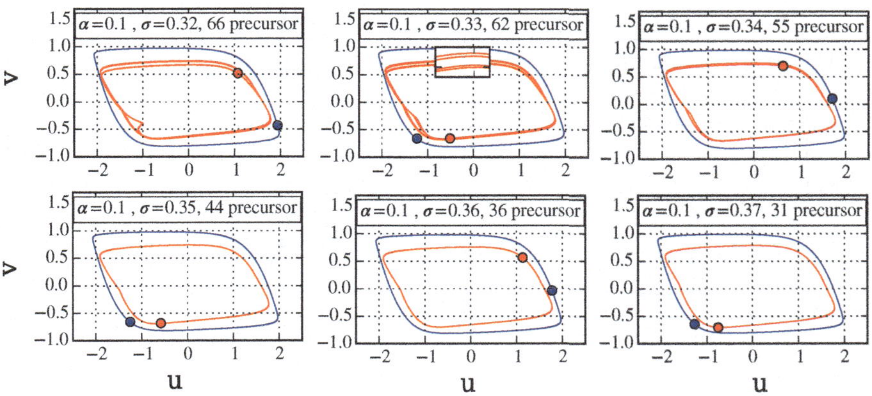

Fig. 4.29 Phase portraits of solitary states for $\alpha = 0.1$ and different values of coupling strengths σ. The values of the coupling strength σ and the number of solitary nodes are indicated in the panels. Initial conditions: random but the same for all panels. The dots display the position of the synchronized cluster (blue) and the solitary nodes (red) at an arbitrarily chosen time step. Other parameters: $\varepsilon = 0.05$, $a = 0.5$, $N = 1000$, $\phi = \pi/2 - 0.1$ [201]

4.4.5.2 Formation of Solitary States from Specially Prepared Initial Conditions

In the solitary states, we report here, the network is divided into two domains with different dynamics. The majority of the elements is synchronized while a few solitary nodes, which are split from the synchronized cluster, are randomly distributed along the network. In what follows we address a question, whether one can manually distribute the nodes of the network between these two domains.

Global coupling: In the case of global coupling, the node's indices do not play any role. Therefore, we start our investigation with the global coupling case to eliminate the influence of the nodes' positions in the network space, i.e., their respective indices.

From a solitary state, formed from random initial conditions, we extract the positions of the two domains in phase space. Further, we manually allocate the nodes of the network to either of these positions. It turns out that in a network of $N = 1000$ globally coupled oscillators the maximum number of precursor nodes is 125. Surprisingly, the solitary state with a single precursor node and a synchronized cluster of 999 nodes and the solitary state involving 125 precursor nodes and 875 synchronized nodes look very similar (see Fig. 4.30).

Power-law coupling: Next, we consider the influence of specially prepared initial conditions on the formation of solitary states. Similarly to the globally coupled case, we initially allocate one part of the network on the precursor trajectory and the rest on the trajectory of the synchronized cluster. It turns out that the introduction of a small neighborhood in the network enlarges the number of nodes that follow the trajectory of the solitary nodes. Since initially we have a well-defined neighborhood and a clear spatial separation between the two domains, the resulting pattern is a cluster state or partially synchronized pattern.

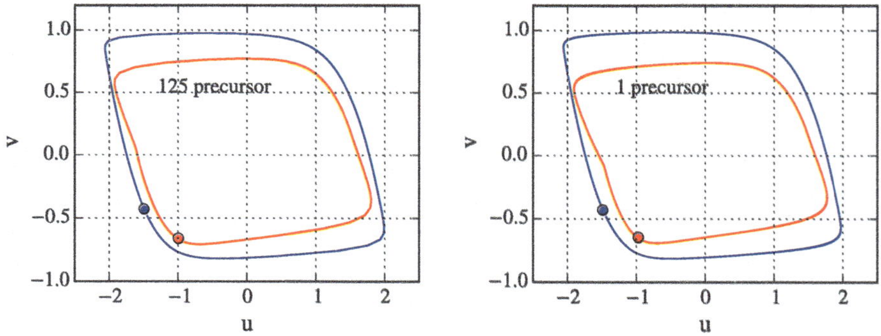

Fig. 4.30 Solitary states with different numbers of precursor nodes: 125 (left panel) and 1 (right panel) in a globally coupled network ($\alpha = 0$) for $\sigma = 0.35$ obtained from specially prepared initial conditions. Other parameters: $\varepsilon = 0.05$, $a = 0.5$, $N = 1000$, $\phi = \pi/2 - 0.1$ [201]

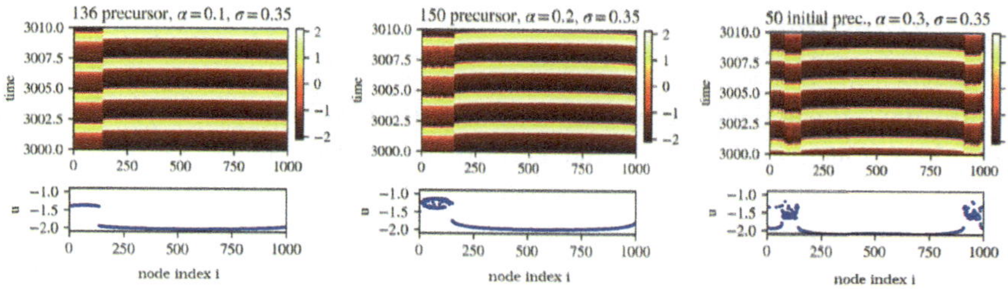

Fig. 4.31 Cluster states for different power-law exponents obtained from specially prepared initial conditions: $\alpha = 0.1$ (left panel), $\alpha = 0.2$ (middle panel), $\alpha = 0.3$ (right panel). Top panels display space-time plots for the u-variable and bottom panels show snapshots of the u-variable. Other parameters: $\varepsilon = 0.05$, $a = 0.5$, $N = 1000$, $\phi = \pi/2 - 0.1$ [201]

Our numerical simulations show that for a power-law exponent $\alpha = 0.1$ a cluster state with 136 nodes on the precursor trajectory can evolve from the specially prepared initial conditions (see Fig. 4.31 left panel). Increasing the initial number of precursor nodes does not lead to the formation of the fully synchronized solution. Instead, more precursor nodes join the synchronized cluster, i.e., the number of solitary nodes decreases. The larger the number of precursor nodes in the initial conditions, the more nodes join the synchronized cluster. Finally, for 200 nodes in the initial precursor cluster, the system reaches the synchronized state.

For the power-law exponent $\alpha = 0.2$, the nodes from the synchronized cluster are synchronized with respect to their frequency as well as the amplitude. The elements in the precursor cluster are only frequency synchronized but have slightly different amplitudes (see Fig. 4.31 middle panel). The number of nodes, which can be manually allocated in the incoherent precursor cluster, is increased. The maximum number of solitary nodes is between 150 and 160. For larger initial number of precursor elements, the network transforms into the synchronized state, similar to the case of $\alpha = 0.1$. If we further increase the power-law exponent, the elements in the precursor cluster differ in the amplitudes even more resulting in a more complex pattern (see Fig. 4.31 right panel).

Therefore, in the detected solitary state, the nodes from the synchronized cluster behave almost like an uncoupled FitzHugh-Nagumo oscillator. Solitary nodes, on the contrary, can follow slightly different trajectories depending on initial conditions allowing for two or even four different coexisting precursor branches. Too many precursor nodes or too strong coupling strength lead to the collapse of the solitary states, and the network becomes fully synchronized. Solitary states can be found for global coupling as well as for power-law coupling with small exponents. We can manually allocate the nodes to either of the precursor branches constructing various cluster states.

4.4.6 Transition Patterns

Next, we investigate the patterns that occur as a result of transitions from chimera states towards incoherence or tree-like perturbations. Furthermore, we analyze how tree-like perturbations arise from incoherence and transform into traveling waves.

From chimera states to incoherence: For the fixed value of coupling strength σ by increasing the power-law exponent α, i.e., reducing the effective coupling range, we observe a transition from chimera states to complete incoherence via the regime of spatiotemporal intermittency (Fig. 4.32). Occurring for $0.8 < \alpha < 1.0$, this pattern is characterized by the coexistence of an ordered structure with incoherence in space and time (Fig. 4.32 middle panel). While in a chimera state we observe a clear separation of coherent and incoherent domains and an arc-shaped mean phase velocity profile (Fig. 4.32 left panel), in the regime of spatiotemporal intermittency, coherent domains exist only temporally. They emerge, drift, and disappear in a random fashion, and the mean phase velocity loses its arc-shaped profile (Fig. 4.32 middle panel). In the regime of complete incoherence, the synchronized domains disappear (Fig. 4.32 right panel).

From chimera states to tree-like perturbations: For the fixed interaction range (e.g., $\alpha = 0.6$) by increasing the coupling strength σ, we observe a transition from chimera states to the tree-like perturbations (Fig. 4.33). It occurs via a pattern that is characterized by the coexistence in space of coherent and incoherent domains, where the incoherent domain exhibits the typical features of tree-like perturbations (Fig. 4.33 middle panel).

From incoherence to tree-like perturbations: The transition from incoherence to tree-like perturbations can be observed in the network by fixing the interaction range (e.g., $\alpha = 1.2$) and varying the coupling strength σ (Fig. 4.34). While there are no synchronized domains in the regime of complete incoherence (Fig. 4.34 left panel), the increase of σ leads to the formation of coherent structures and results

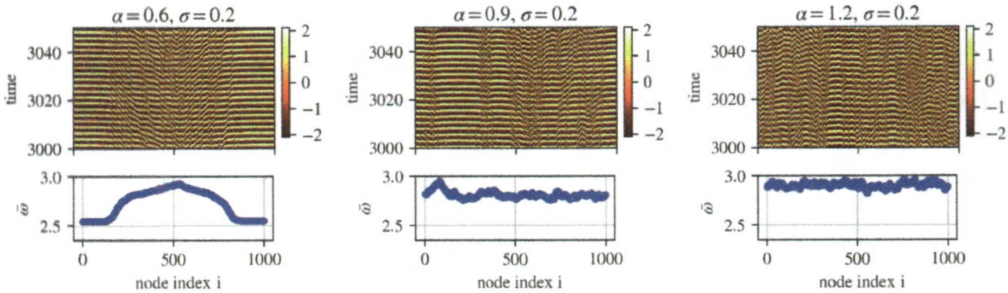

Fig. 4.32 Transition from chimera states at $\alpha = 0.6$ (left panel) to incoherence at $\alpha = 1.2$ (right panel) via spatiotemporal intermittency at $\alpha = 0.9$ (middle panel) for the fixed value of the coupling strength $\sigma = 0.2$. The top row shows space-time plots for the u-variable and the bottom row displays mean phase velocity profile. Other parameters: $\varepsilon = 0.05, a = 0.5, N = 1000, \phi = \pi/2 - 0.1$ [201]

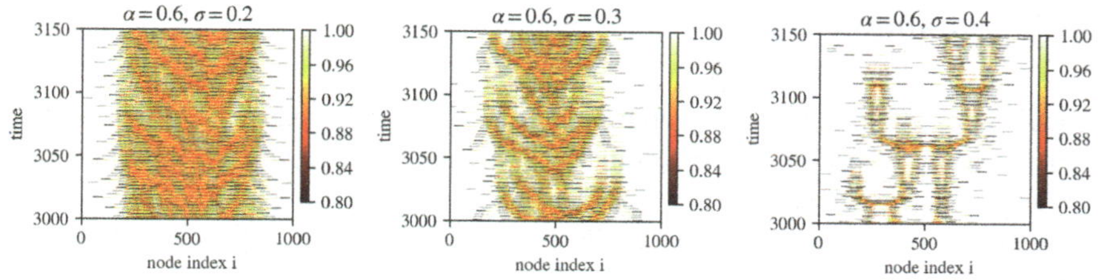

Fig. 4.33 Transition pattern at $\sigma = 0.3$ (middle pattern) from chimera states at $\sigma = 0.2$ (left panel) to tree-like perturbations at $\sigma = 0.4$ (right panel) for the fixed value of the power-law exponent $\alpha = 0.6$. All three panels display space-time plots for the local order parameter. Other parameters: $\varepsilon = 0.05$, $a = 0.5$, $N = 1000$, $\phi = \pi/2 - 0.1$ [201]

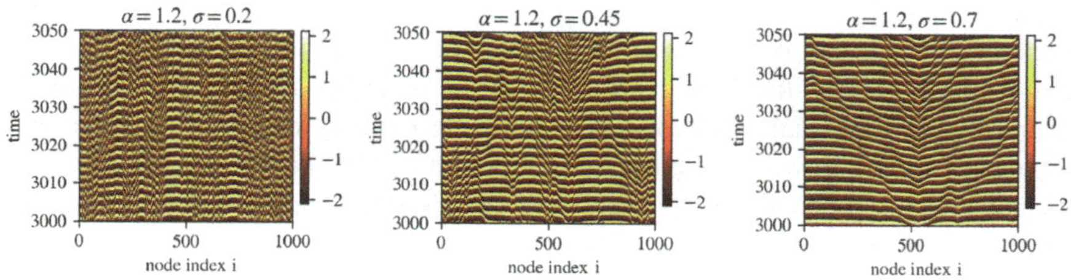

Fig. 4.34 Transition from incoherence at $\sigma = 0.2$ (left panel) to tree-like perturbations at $\sigma = 0.7$ (right panel) via spatiotemporal intermittency at $\sigma = 0.45$ (middle panel) for the fixed value of the power-law exponent $\alpha = 1.2$. All three panels display space-time plots for the u-variable. Other parameters: $\varepsilon = 0.05$, $a = 0.5$, $N = 1000$, $\phi = \pi/2 - 0.1$ [201]

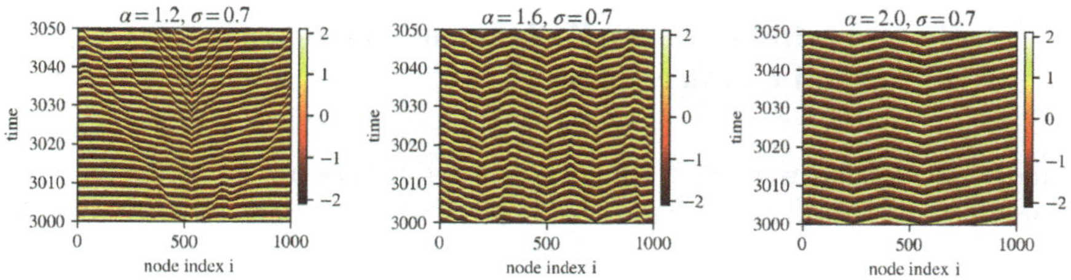

Fig. 4.35 Transition pattern at $\alpha = 1.6$ (middle panel) from tree-like perturbations at $\alpha = 1.2$ (left panel) to traveling waves at $\alpha = 2.0$ (right panel) for the fixed value of the coupling strength $\sigma = 0.7$. All three panels display space-time plots for the u-variable. Other parameters: $\varepsilon = 0.05$, $a = 0.5$, $N = 1000$, $\phi = \pi/2 - 0.1$ [201]

in the regime of spatiotemporal intermittency (Fig. 4.34 middle panel). Finally, for $\alpha = 1.2$ and $\sigma = 0.7$ the tree-like perturbation pattern occurs (Fig. 4.34 right panel).

From tree-like perturbations to traveling waves: By increasing the power-law exponent α for the fixed coupling strength σ, we observe how traveling waves evolve from the tree-like perturbations (Fig. 4.35).

4.4.7 Summary

In a network of FitzHugh-Nagumo oscillators with power-law coupling, we find a large variety of peculiar spatiotemporal patterns. Besides the fully synchronized solution, complete incoherence, and traveling waves, we observe various partial synchronization patterns. We detect classical chimera states, characterized by the coexistence in space of coherent and incoherent oscillations. Above the critical value of coupling strength of $\sigma \geq 0.35$, chimera states disappear and there occurs another partial synchronization pattern—tree-like perturbations. Further, we also observe the regime of solitary states, where the majority of the nodes in the network is synchronized and performs oscillations along the limit cycle similar to that of the single FHN system. The rest of the network forms a smaller group of nodes distributed randomly in space, spiking in a precursor manner and moving along a limit cycle with a smaller amplitude. The distribution of the nodes over these two domains is determined by the initial conditions. Therefore, it is random for randomly distributed initial conditions. However, the initial distribution can be manipulated to form two distinct clusters localized in space. These cluster states can be found for global coupling as well.

Moreover, for varying coupling parameters, we disclose different transitions between the basic patterns observed in the network. These transitions involve regimes exhibiting features of more than one basic pattern.

4.5 Coherence-Incoherence Patterns in Networks with Fractal Connectivities

In the present section, we analyze chimera states in ring networks of Van der Pol oscillators with hierarchical coupling topology (Sect. 4.5.1). In more detail, we investigate the stepwise transition from a nonlocal to a hierarchical topology and propose the network clustering coefficient as a measure to establish a link between the existence of chimera states and the compactness of the initial base pattern of a hierarchical topology. Moreover, we show that a large clustering coefficient promotes the occurrence of chimeras. Depending on the level of hierarchy and base pattern, we obtain chimera states with different numbers of incoherent domains. We investigate the chimera regimes as a function of coupling strength and nonlinearity parameter of the individual oscillators. The analysis of a network with larger base pattern resulting in larger clustering coefficient reveals two different types of chimera states and highlights the increasing role of amplitude dynamics. Further, we study chimera states in ring networks of FitzHugh-Nagumo oscillators with fractal connectivities (Sect. 4.5.2). In particular, we focus on the interplay of time delay in the coupling term and the network topology. In the parameter plane of coupling strength and delay time, we find tongue-like regions of existence of chimera states alternating with regions of coherent dynamics. We show analytically and numerically that the period of the synchronized dynamics as a function of delay is characterized by a sequence of piecewise linear branches. In between these branches, various chimera states and other partial synchronization patterns are induced by the time delay. By varying the time delay, one can deliberately choose and stabilize desired spatiotemporal patterns. Finally, for coupled FitzHugh-Nagumo neurons, we compare two topologies: a mathematically constructed network with 2D modular fractal connectivity and an empirical structural neural connectivity derived from diffusion-weighted magnetic resonance imaging (Sect. 4.5.3). We analyze the properties of chimeras and partially synchronized states, and obtain regions of their stability in the parameter planes. Furthermore, we qualitatively simulate the dynamics of epileptic seizures and study the influence of the removal of nodes on the network synchronizability, which can be useful for applications to epileptic surgery.

4.5.1 Ring Networks of Van der Pol Oscillators

Recent results in the area of neuroscience increased the interest in irregular coupling topologies. Diffusion tensor magnetic resonance imaging (DT-MRI) studies revealed an intricate architecture in the neuron interconnectivity of the human and mammalian brain: the connectivity of the neuron axons network represents a hierarchical (quasi-fractal) geometry [341–345]. These works motivate the investigation of networks with hierarchical connectivity and, in particular, the mechanisms of formation of chimera states in such networks. It is important to note that quasi-fractal coupling topologies have hierarchical structure, although they differ from the hierarchical tree-like models widely studied in the network science. The latter network model is a part of the scale-free model family, having many hubs among the nodes, and displaying power-law distributed degrees and clustering coefficients [368]. In contrast, the coupling topologies we consider in the present section have the form of a ring with hierarchical structures of connectivity gaps, and the degree and clustering coefficient are the same for all nodes.

In what follows we uncover characteristic measures of the hierarchical connectivities which determine the emergence of chimeras.[3] As a key measure we identify the clustering coefficient. We focus on the stepwise transition of network topology from nonlocal to hierarchical and analyze different types of chimera states that can arise in systems with different numbers of hierarchical steps.

The dynamics of individual units in the network is governed by the Van der Pol oscillator [369], which has a long history of being used in both the physical and biological sciences, and allows for a continuous transition between sinusoidal and strongly nonlinear relaxation oscillations by tuning a single parameter.

[3]Portions of the following text have been re-printed from [118] with the permission of AIP Publishing.

4.5.1.1 Model: Local Dynamics and Algorithm to Construct Hierarchical Connectivities

We consider a ring of N identical Van der Pol oscillators with different coupling topologies, which are given by the respective adjacency matrix $\boldsymbol{G}$. While keeping the periodicity of the ring, and the circulant structure of the adjacency matrix, we vary the connectivity pattern of each element. The dynamical equations for the two-dimensional phase space variable $\mathbf{x}_k = (u_k, \dot{u}_k)^T = (u_k, v_k)^T \in \mathbb{R}^2$ are as follows:

$$\dot{\mathbf{x}}_i(t) = \mathbf{F}(\mathbf{x}_i(t)) + \frac{\sigma}{g} \sum_{j=1}^{N} G_{ij} \boldsymbol{H}(\mathbf{x}_j - \mathbf{x}_i) \tag{4.11}$$

with $i \in \{1, ..., N\}$. The dynamics of each individual oscillator is governed by

$$\mathbf{F}(\mathbf{x}) = \begin{pmatrix} v \\ \varepsilon(1 - u^2)v - u \end{pmatrix}, \tag{4.12}$$

where ε denotes the bifurcation parameter. The uncoupled Van der Pol oscillator has a stable fixed point at $\mathbf{x} = 0$ for $\varepsilon < 0$ and undergoes an Andronov-Hopf bifurcation at $\varepsilon = 0$. Here, only $\varepsilon > 0$ is considered. The parameter σ denotes the coupling strength, and $g = \sum_{j=1}^{N} G_{ij}$ is the number of links for each node (corresponding to the row sum of $\boldsymbol{G}$). The interaction is realized as diffusive coupling with coupling matrix

$$\boldsymbol{H} = \begin{pmatrix} 0 & 0 \\ b_1 & b_2 \end{pmatrix} \tag{4.13}$$

and real interaction parameters b_1 and b_2. In accordance with Omelchenko et al. [78], throughout the section we fix parameters $b_1 = 1.0$ and $b_2 = 0.1$.

Hierarchical topologies can be generated using a classical Cantor construction algorithm for a fractal set [210, 211]. This iterative hierarchical procedure starts from a *base pattern* or initiation string b_{init} of length b, where each element represents either a link ('1') or a gap ('0'). The number of links contained in b_{init} is referred to as c_1. In each iterative step, each link is replaced by the initial base pattern, while each gap is replaced by b gaps. Thus, each iteration increases the size of the final bit pattern, such that after n iterations the total length is $N = b^n$. Since the hierarchy is truncated at a finite n, we call the resulting pattern hierarchical rather than fractal. Using the resulting string as the first row of the adjacency matrix $\boldsymbol{G}$, and constructing a circulant adjacency matrix $\boldsymbol{G}$ by applying this string to each element of the ring, a ring network of $N = b^n$ nodes with hierarchical connectivity is generated [80, 83, 119]. Here, we will slightly modify this procedure by adding an additional zero in the first instance of the sequence, which corresponds to the self-coupling. Note that the diffusive coupling scheme in (4.11) cancels out any instantaneous self-coupling. Therefore, there is no net effect of the diagonal elements of the adjacency matrix G_{ii} on the network dynamics, and hence the first link in the clockwise sense from the reference node is effectively removed from the link pattern. Without our modification, this would lead to a breaking of the base pattern symmetry, i.e., if the base pattern is symmetric, the resulting coupling topology would not be so, since the first link to the right is missing from the final link pattern. Our procedure, in contrast, ensures the preservation of an initial symmetry of b_{init} in the final link pattern, which is crucial for the observation of chimera states, since asymmetric coupling leads to a drift of the chimera [145, 146]. Thus, a ring network of $N = b^n + 1$ nodes is generated.

While fully hierarchical topologies can be generated using this classical construction algorithm, a further modification allows us to study systematically the transition between nonlocal and hierarchical topologies via a stepwise iteration process. Nonlocal coupling schemes have widely been used in the context of chimera state research and do therefore provide a good reference point to compare hierarchical networks with.

A nonlocal topology can be generated from a base pattern b_{init} of length b, which contains an equal number of links only at its beginning and end (for instance (101) or (110011)). The link pattern is then expanded to a predetermined system size $N = b^n + 1$, corresponding to the final size of a fully hierarchical connectivity, by replacing each element with $\frac{N-1}{b}$ copies of itself and again adding the additional zero in first instance of the final pattern. Thus, a suitable base pattern b_{init} can be used to construct either a fully hierarchical or a nonlocal topology. The stepwise transition between these two types of topologies is realized as follows: First, b_{init} is iterated m times according to the Cantor construction process, generating a pattern of size b^m. Afterwards, this pattern is expanded to the final size N by replacing each element with $\frac{N-1}{b^m}$ copies of itself. The initial base pattern $b_{\text{init}} = (101)$ and a predetermined system size of $N = 27 + 1$ ($b = 3, n = 3$) provide a simple example. The resulting coupling topologies are illustrated in Fig. 4.36. The number of Cantor iterations of *b before* the expansion is defined as the mth hierarchical step, with $m \in \{1, 2, ..., n\}$ such that

$m = 1$: $b^1_{\text{init}} = (101)$, the initial base pattern is expanded to a 1-step hierarchical connectivity by replacing each element with $\frac{27}{3} = 9$ copies of itself. This corresponds to nonlocal coupling with coupling radius $r = \frac{R}{N} = \frac{9}{27} = 0.333$. See Fig. 4.36a.

$m = 2$: $b^2_{\text{init}} = (101000101)$, the once iterated initial base pattern is expanded to a 2-step hierarchical connectivity by replacing each element with $\frac{27}{9} = 3$ copies of itself. This marks the first step in the transition from a nonlocal ($m = 1$) to a fully hierarchical ($m = n = 3$) topology. See Fig. 4.36b.

$m = n = 3$: $b^3_{\text{init}} = (101000101000000000101000101)$, this link pattern is of size N for $m = n = 3$ and corresponds to the fully hierarchical or $n-$step hierarchical connectivity. See Fig. 4.36c.

Thus, taking up to $m = n$ steps in the hierarchical expansion it is possible to tune a suitable initial base pattern between a nonlocal and a fully hierarchical connectivity. In the following, an $m-$step hierarchical connectivity is denoted as $(b_{\text{init}})^m$, where

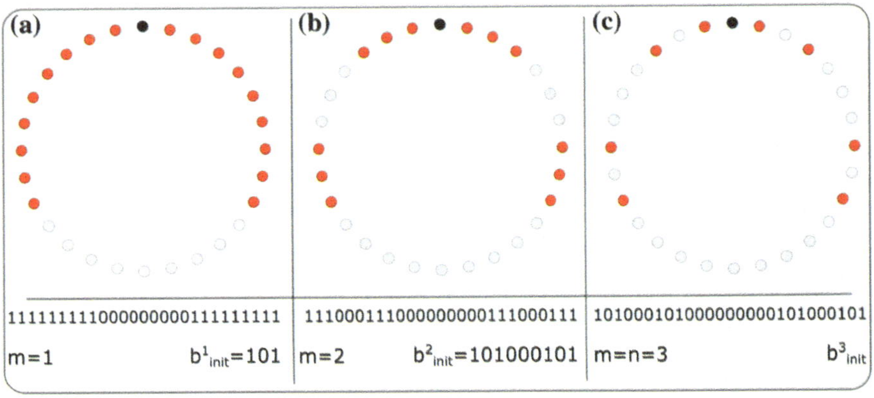

Fig. 4.36 Transition from nonlocal to hierarchical connectivity via hierarchical steps m. The reference node is colored in black, linked nodes in red, and unconnected nodes (gaps) in gray. The initial base pattern of all panels is $b_{\text{init}} = (101)$, the level of hierarchy is $n = 3$, $N = b^n + 1 = 28$ nodes. **a** $m = 1$, each element in the initial base pattern is expanded by $\frac{27}{3} = 9$ elements and the final 1-step hierarchical system corresponds to nonlocal coupling where each element is coupled to its $R = 9$ nearest neighbors in both directions. Clustering coefficient $C(101, 3, 1) = 0.705882$ and link density $\rho = 0.64$; **b** $m = 2$, expansion by $\frac{27}{9} = 3$ elements to a 2-step hierarchical system. $C(101, 3, 2) = 0.409091$ and $\rho = 0.428$; **c** $m = n = 3$, fully hierarchical or $n-$step hierarchical system without further expansion of the base pattern. $C(101, 3, 3) = 0$ and $\rho = 0.286$. With each $m-$step, the total number of links for each node and the clustering coefficient decrease [118]

b_{init} is the underlying base pattern, n is the level of hierarchy, and m is the hierarchical step in a transition topology.

4.5.1.2 Clustering Coefficients

Besides the bifurcation parameter ε and the coupling strength σ, the topological quantities b_{init}, c_1, n and the resulting link density $\rho = \frac{c_1^n}{N}$ (for $m = n$) or $\rho = \frac{c_1^m b^{(n-m)}}{N}$ (for $m \neq n$), and fractal dimension $d_f = \ln c_1 / \ln b$ are important parameters in the study of chimera states in hierarchical systems. However, since there are several distributions of links for a given set of b and c_1 that result in unique topological structures, the arrangement of links in b_{init} has to be accounted for. Therefore, we propose to consider the local clustering coefficient introduced by Watts and Strogatz [370], which, for a network containing a set of nodes V and edges E, describes the number of links in the neighborhood $N_i = \{v_j : e_{ij} \in E \vee e_{ji} \in E\}$ relative to the maximum number of links possible. If k_i is the number of neighbors for a node v_i, then maximum number of links is given by $k_i \cdot (k_i - 1)$ and the clustering coefficient C_i for a node v_i is defined as

$$C_i = \frac{|\{e_{jk} : v_j, v_k \in N_i, e_{jk} \in E\}|}{k_i(k_i - 1)}. \tag{4.14}$$

While common nonlocal coupling schemes can rely on the coupling range (or variations thereof [80]) as a definite measure, this is not the case in hierarchically coupled networks. Different arrangements of links change the compactness of the base pattern, and the final compactness of the system strongly depends on the distribution properties of the c_1 links in a base pattern b_{init}. The clustering coefficient is such a suitable measure since it directly expresses the compactness of links in the final system. In the following, $C(b_{\text{init}}, n, m)$ denotes the clustering coefficient of an m−step hierarchical connectivity with base pattern b_{init}. In the transition scenario from nonlocal to hierarchical topologies, the system size and the symmetry properties stay the same throughout. However, each hierarchical step modifies the compactness and total number of links of the final topology which changes the clustering coefficient C as well as the link density ρ. Since hierarchical coupling introduces irregular and long-ranging links and gaps, the clustering coefficient will decrease when transiting towards a hierarchical connectivity. Out of all the 2^b possible base patterns for a given length b, only a fraction is relevant for discussion in the context of hierarchical networks. For example, mirror symmetric base patterns (such as (110101) and (101011)) result in topologies with identical clustering coefficients, while patterns containing only one link would result in a network with only a single link to each node. In the case of $b = 6$, out of 64 possible patterns, only 31 will yield unique hierarchical topologies. Figure 4.37 demonstrates clustering coefficients $C(b_{\text{init}}, n, m)$ as a function of the hierarchical step m for a selected number of representative base patterns.

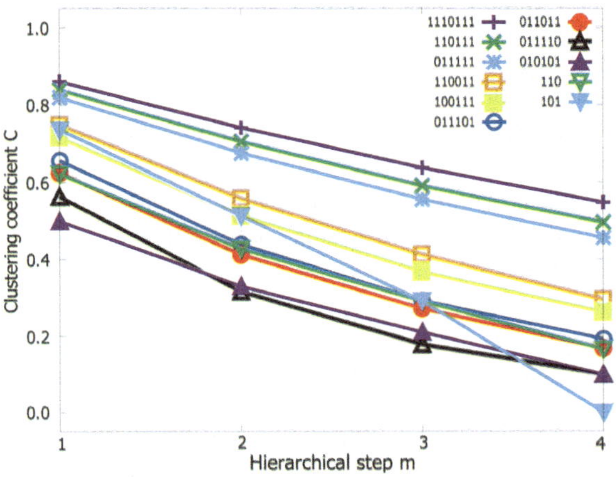

Fig. 4.37 Clustering coefficients C for networks with topologies varied stepwise from nonlocal to hierarchical for different base patterns (see legend). The hierarchical step $m \in \{1, ..., n = 4\}$ is used to tune between nonlocal coupling and fully hierarchical coupling. At higher hierarchical step m, the clustering coefficients decrease due to the disproportional increase of gaps versus links in the base pattern. Non-compact base patterns lead to vanishing clustering coefficients C. This suppresses chimera states in hierarchical networks generated from these base patterns. System sizes (n = 4): $N = 82$ for $b = 3$, $N = 1297$ for $b = 6$, $N = 2402$ for $b = 7$ [118]

These results provide an instructive overview over the dependence of C upon m. As expected, the clustering coefficients decrease across the board when introducing hierarchical components. With each step m, the number of gaps increases disproportionally compared to the number of links, since each gap in b_{init} introduces b new gaps with each iteration, while each link only leads to c_1 new links and adds $b - c_1$ further gaps. Thus, the link density decreases with each iteration of m and the remaining links are distributed in a more irregular, hierarchical manner. A close analysis of C with respect to the underlying base pattern shows that base patterns with gaps at either the end or the beginning of b_{init} (such as (011011), Fig. 4.37, red dots) result in low clustering coefficients, compared to systems of equal link density but different distributions (such as (110011), orange squares). Furthermore, isolated links in b_{init} (such as (100111), yellow squares) further decrease $C(b_{\text{init}}, n = m)$ compared to topologies constructed from more compact base patterns (such as (110011), orange squares). However, more compact base patterns b_{init} that have isolated links (such as (011101), blue circles) *can have* larger $C(b_{\text{init}}, n = m)$, as compared to topologies with same number of links but without isolated links (such as (011011), red dots). This can be explained by the fact that isolated links lead to far-ranging links rather than coupling to those nodes in their close neighborhood. Considering the choice of suitable base patterns b_{init} for the construction of hierarchical topologies which exhibit chimeras, it should be noted that base patterns with larger clustering coefficients are preferable. Therefore, base patterns should be chosen avoiding isolated links, gaps at the beginning or end, and with compact intervals, if possible. This highlights the basic trade-off when studying hierarchical topologies. Compactness is an important requirement for the existence of chimera states and at the same time, networks are less compact if they are more hierarchically structured. If one wants to study dynamic phenomena and maintain a hierarchical connectivity, this has to be balanced in a careful manner, since low n decreases the actual degree of hierarchy of the system, while large n produces highly hierarchical connectivities (fractal, in the case of $n \to \infty$) but with very low compactness. Our observations show that hierarchical topologies resulting from base patterns that contain only isolated links for given b and c_1 have the lowest clustering coefficient, and chimera states are rarely observed in such networks.

4.5.1.3 Chimera States in Networks with Hierarchical Connectivities

The simplest illustrative example of a hierarchical connectivity is constructed from the base pattern (101). We consider this pattern in order to demonstrate the step-by-step transition from nonlocal to hierarchical network topology with $n = 6$. At the initial step $m = 1$, the system is a nonlocally coupled ring with $N = 730$ and coupling radius $r = \frac{243}{730} = 0.333$. The clustering coefficient for this topology is $C(101, 6, 1) = 0.748451$. This system has been studied in [78], where it was shown that depending on the coupling radius and strength, chimera states with different numbers of incoherent domains can be observed. Indeed, Fig. 4.38a shows a 2-chimera state. Snapshots of the variable u_k (upper panels) demonstrate a clear distinction

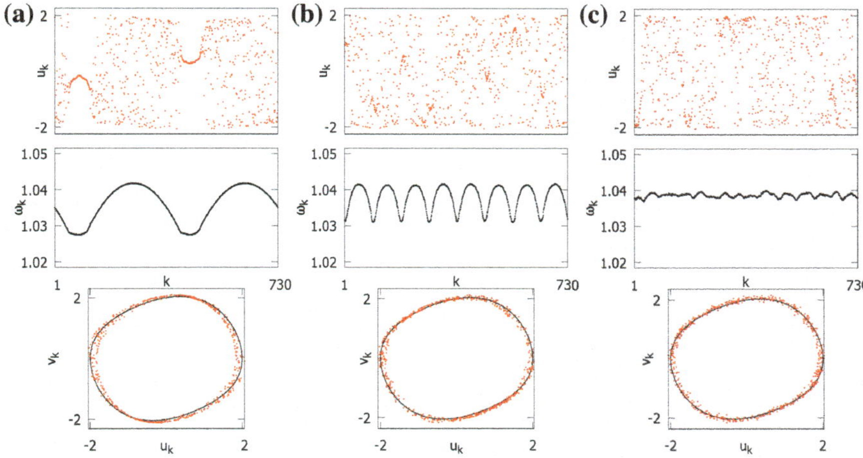

Fig. 4.38 Chimera states in transiting topology with $b_{\text{init}} = (101)$, $n = 6$, $N = 730$, $\sigma = 0.09$, $\varepsilon = 0.2$. Snapshots of variables u_k (upper panels), mean phase velocities ω_k (middle panels), and snapshots in the phase space (u_k, v_k) (bottom panels, limit cycle of the uncoupled unit shown in black). **a** Hierarchical step $m = 1$, corresponding to nonlocal coupling with $r = 0.333$, 2-chimera state, clustering coefficient $C(101, 6, 1) = 0.748451$ and link density $\rho = 0.667$; **b** $m = 2$, 8-chimera state, $C(101, 6, 2) = 0.55727$ and $\rho = 0.444$; **c** $m = 3$, completely incoherent state, $C(101, 6, 3) = 0.408141$ and $\rho = 0.296$. No chimeras are observed for further steps in hierarchical connectivity [118]. Initial conditions as in [78], Fig. 1b

between coherent and incoherent domains, snapshots in the phase space (u_k, v_k) (bottom panels) show that the oscillators are scattered around the limit cycle of the uncoupled element which is shown in black. The middle panels demonstrate the mean phase velocities for each oscillator calculated as $\omega_k = 2\pi M_k/\Delta T$, $k = 1, ..., N$, where M_k is the number of complete rotations around the origin performed by the kth node during the time interval ΔT. Throughout the section, we use $\Delta T = 50000$, which corresponds to several thousand periods. Mean phase velocity profiles are widely used as a criterion to distinguish chimera states: constant ω_k corresponds to coherent domains, where neighboring elements are phase locked. Arc-like parts of the profiles correspond to incoherent domains. Usually, chimera states with different numbers of incoherent domains exhibit high multistability [78], and the choice of initial condition is crucial. We use randomly distributed phases on the circle $u^2 + v^2 = 4$ as initial conditions, as in [78], Fig. 1b. The 2-step hierarchical network (Fig. 4.38b) is characterized by a smaller clustering coefficient $C(101, 6, 2) = 0.557279$, and we observe an 8–chimera state. This fact can be explained by drawing the analogy to nonlocally coupled systems, where a decrease of the coupling radius (the number of neighbors coupled to each element) results in an increase of the number of incoherent domains [78], i.e., the chimera multiplicity.

Already for the 3–step hierarchical network shown in Fig. 4.38c, the chimera state vanishes and is not observed for any further iterations up to $m = n = 6$. This goes along with a decrease of the clustering coefficient to $C(101, 6, 3) = 0.408141$ and

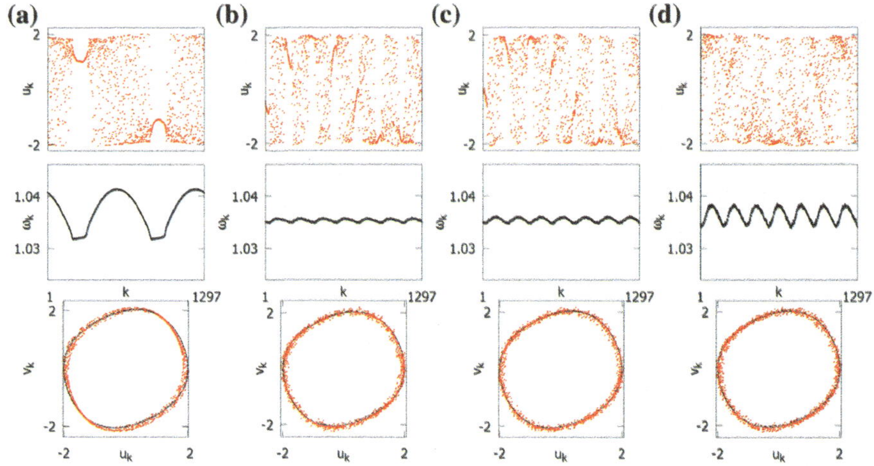

Fig. 4.39 Chimera states in transiting topology with $b_{\text{init}} = (110011)$, $n = 4$, $N = 1297$, $\sigma = 0.09$, $\varepsilon = 0.2$, random initial conditions. Snapshots of variables u_k (upper panels), mean phase velocities ω_k (middle panels), and snapshots in the phase space (u_k, v_k) (bottom panels, limit cycle of the uncoupled element shown in black). **a** $m = 1$, corresponding to nonlocal coupling with $r = 0.333$, $C(110011, 4, 1) = 0.749142$, $\rho = 0.667$, 2-chimera state; **b** $m = 2$, $C(110011, 4, 2) = 0.559569$, $\rho = 0.444$, 7-chimera; **c** $m = 3$, $C(110011, 4, 3) = 0.414161$, $\rho = 0.298$, 7-chimera remains; **d** $m = n = 4$, fully hierarchical network with 7-chimera and more pronounced ω_k profile, $C(110011, 4, 4) = 0.297791$ and $\rho = 0.197$ [118]

further down to $C(101, 6, 6) = 0$ for the fully hierarchical network (see Fig. 4.37). Transiting the system from a nonlocal to a hierarchical connectivity decreases the clustering coefficient as well as the total number of links to such an extent that chimera states do not occur after only three out of six possible hierarchical steps.

Figure 4.39 depicts a similar stepwise transition from nonlocal to hierarchical coupling for the symmetric base pattern (110011) with $n = 4$, which has the same fractal dimension and link density but larger clustering coefficients. The total system size is $N = 1297$ and the clustering coefficients are shown in Fig. 4.37 (orange squares). For $m = 1$ we again observe a 2-chimera, like in Fig. 4.38a. At $m = 2$, the multiplicity of the chimera increases to a 7-chimera which remains stable for further hierarchical steps and is characterized by a more pronounced ω_k profile at the fully hierarchical level $m = n = 4$.

The importance of the clustering coefficient as a measure for hierarchical systems with respect to chimera states is further highlighted by considering several permutations of this base pattern of length $b = 6$, and $n = 4$. We have performed a multitude of scans over wide ranges in (ε, σ)-space, for various less symmetric base patterns and varying initial conditions (specially prepared chimera-like as well as random), but none of them resulted in a chimera state. Among them are topologies with the same $c_1 = 4$ but different link distributions and lower clustering coefficients, such as $(100111)^4$, $(011101)^4$, $(110101)^4$, $(011110)^4$, and $(110110)^4$, but also systems with larger link densities and clustering coefficients $(011111)^4$ and $(110111)^4$. Each base pattern at fixed hierarchical step has a fixed relative coupling radius, i.e., ratio between number of links for each element and size N of the system. Usually, intermediate values of the relative coupling radius promote chimera states, whereas at the same time for small and large relative coupling radius chimeras are rarely observable. The subtle interplay of symmetry and compactness of the network topology with the number of links crucially affects the network dynamics. While it is impossible to draw a definite conclusion from purely empirical studies, this strongly suggests that generally the combination of a symmetric base pattern with large clustering coefficient promotes the existence of chimera states in hierarchical topologies.

4.5.1.4 Stability Regimes of Chimera States

To elaborate the role of the system parameters, we construct the maps of stability regimes in the plane of the nonlinearity parameter ε of the individual oscillators, and the coupling strength σ. For this purpose, we choose a system with base pattern (110011) and $n = 4$, and consider its 3−step and 4−step hierarchical connectivity. We start from specially prepared chimera-like initial conditions for a fixed parameter set, and then use the obtained final state as an initial condition for the subsequent set of parameters, and so forth with a step size $\Delta\varepsilon = 0.02$ and $\Delta\sigma = 0.02$. This procedure represents a kind of continuation method for mapping the chimera regime.

Figure 4.40a depicts the stability regimes for the base pattern $b = (110011)$ with $n = 4$ in a 3-step hierarchical connectivity ($N = 1297$). For small values of the nonlinearity parameter ε, a completely coherent state with wave-like profile is observed (yellow triangles, and snapshot A in Fig. 4.40b). Conversely, for larger ε, when the limit cycle of each individual oscillator starts to transform from sinusoidal to relaxation oscillations, we observe completely incoherent states (red circles, and snapshot C in Fig. 4.40b). Between these two regimes, there is a region where chimera states emerge (green squares, and snapshot B in Fig. 4.40b). Thus, the transition from coherence to incoherence occurs through chimera states. Furthermore, the boundaries between the coherent regime and the chimera state as well as between the chimera state and the completely incoherent regime shift to larger ε with increasing σ. Equation (4.11) shows that an upscaling of the local parameter ε has to be counterbalanced by the coupling strength term which is controlled by σ in order to lead to a similar force for each oscillator. Notably, the boundaries for chimera emergence in the (ε, σ) plane are approximately given by straight lines. In the right panel of Fig. 4.40b, the mean phase velocity profiles are shown for the three selected points A, B, C in the (ε, σ) plane. The chimera state in B shows chimera dynamics where the minima of the profile correspond to the coherent regions.

In contrast, Fig. 4.41a shows the same dynamic regimes in (ε, σ) space for the fully hierarchical system (110011) with $n = 4$. The same qualitative shift and increase of the stable ε range with increasing σ is observed. However, the overall area of stability significantly decreases for the fully hierarchical system, while the character of the chimera state (multiplicity and coherent regions) remains unchanged. The system undergoes the same qualitative change of its dynamic behavior, a transition from a

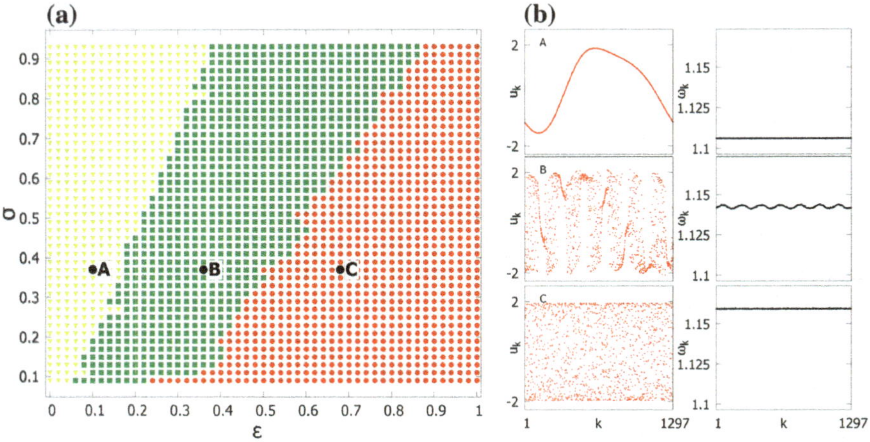

Fig. 4.40 Stability regimes for the system $b_{\text{init}} = (110011)$, $n = 4$ in an $m = 3-$step hierarchical connectivity, $N = 1297$, $C(110011, 4, 3) = 0.414161$, $\rho = 0.296$. **a** Diagram in the parameter space (ε, σ): coherent state (yellow triangles), 7-chimera state (green squares), incoherent state (red circles); **b** exemplary snapshots of variables u_k (left column) and mean phase velocities ω_k (right column) for $\sigma = 0.38$ and $\varepsilon = 0.1$ (A), $\varepsilon = 0.36$ (B), $\varepsilon = 0.68$ (C) [118]

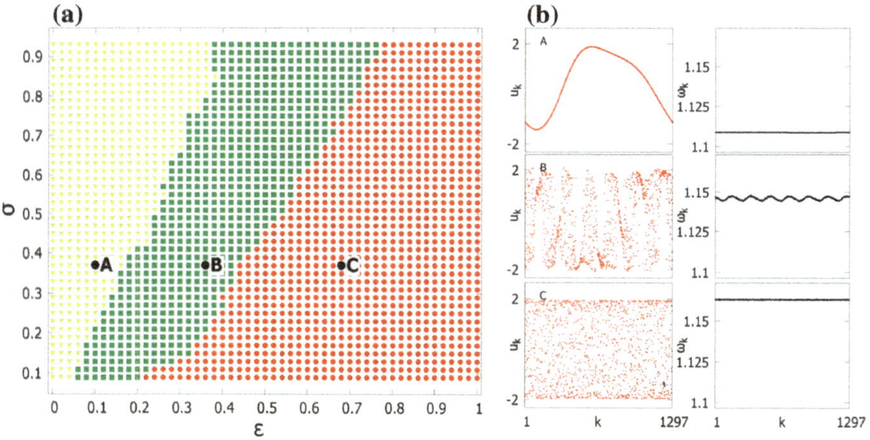

Fig. 4.41 Stability regimes for the system $b_{\text{init}} = (110011)$ in an $m = n = 4$ fully hierarchical connectivity, $N = 1297$, $C(110011, 4, 4) = 0.297791$, $\rho = 0.1974$. **a** Diagram in the parameter space (ε, σ): coherent state (yellow triangles), 7-chimera state (green squares), incoherent state (red circles); **b** exemplary snapshots of variables u_k (left column) and mean phase velocities ω_k (right column) for $\sigma = 0.38$ and $\varepsilon = 0.1$ (A), $\varepsilon = 0.36$ (B), $\varepsilon = 0.68$ (C) [118]

completely coherent state (yellow triangles) to a completely desynchronized state (red circles) via a 7-chimera (green squares) with increasing ε. The only difference between both systems is the decrease in the clustering coefficient C and the link density ρ with increasing hierarchical step m. This indicates that a high clustering coefficients and an increased number of links promote the existence as well as the stability of chimera states in hierarchical connectivities. Stable chimera states are difficult to observe in hierarchical systems that have a very low clustering coefficient.

In this exemplary system, as discussed in the previous section, we have observed only chimera states with seven incoherent domains. For systems with nonlocal coupling topology, we have shown recently that an appropriate choice of the coupling radius determines the multiplicity of the incoherent regions in chimera states, but there may be multistability between different multichimera states [78]. The issue of multistability in networks with hierarchical connectivities still has to be explored systematically. As this requires tremendous computational costs, this task is beyond the scope of the present investigation.

It is remarkable that the regime of chimera states in networks with hierarchical connectivity (Figs. 4.40 and 4.41) extends to much larger values of the coupling strength σ than has been found for nonlocal coupling [78]. This indicates that quasi-fractal connectivities promote chimeras at large coupling strength where in more compact topologies completely coherent states prevail.

4.5.1.5 Larger Base Pattern

Changing the size, symmetry, and number of links in the base pattern b_{init} can lead to various completely different hierarchical connectivities. This strongly influences the resulting clustering coefficient and the link density of the final system. As previous studies of chimera bifurcation scenarios [78] have shown, a change in the effective coupling radius of a nonlocal topology leads to different regimes of chimera states with various multiplicities. The same is true when the link density is increased. Since for the systems with $b = 6, n = 4, c_1 = 4$ we have only found chimera states in some symmetric configurations, we will now consider larger base patterns. This, consequently, allows for more links in the base pattern and generates networks with higher clustering coefficients in the fully hierarchical system $m = n$. However, this dramatically increases the network size, making numerical simulations more expensive.

As an example, we use the base pattern $b_{\text{init}} = (1110111)$ with $n = 4$ to generate a hierarchical connectivity of size $N = 2402$ after full Cantor iteration. The link density is $\rho = \frac{g}{N} = 0.54$ which is considerably larger than for the previously considered examples with base pattern of length $b = 6$. Furthermore, the clustering coefficient of the fully hierarchical network $C(1110111, 4, 4) = 0.548829$ is larger than in all previous examples. To emphasize the aspect of self-organized chimeras, we here use random initial conditions. Of course, then only a small portion of realizations leads to chimera states, because the completely synchronized state is always stable.

We fix the nonlinearity parameter $\varepsilon = 0.1$ and observe the system dynamics for three increasing values of coupling strength $\sigma = 0.25, 0.35, 0.45$, starting from ran-

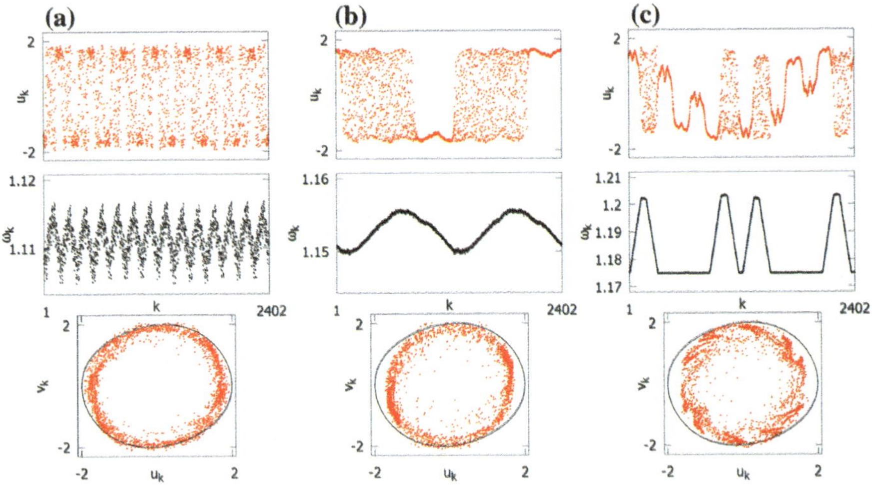

Fig. 4.42 Snapshots of variables u_k (upper panels), mean phase velocities ω_k (middle panels), and snapshots in the phase space (u_k, v_k) (bottom panels, limit cycle of the uncoupled unit shown in black) for fully hierarchical system $b_{\text{init}} = (1110111)$, $n = 4$ with $N = 2402$, $C(1110111, 4, 4) = 0.548829$, $\rho = 0.54$, and $\varepsilon = 0.1$, random initial conditions; **a** $\sigma = 0.25$, 14−chimera; **b** $\sigma = 0.35$, 2-chimera; **c** $\sigma = 0.45$, irregular 4−chimera with nested regions of coherence. Note strong amplitude dynamics in all bottom panels [118]

dom initial conditions. Figure 4.42 depicts the corresponding snapshots, mean phase velocity profiles, and phase portraits. For small coupling strength (Fig. 4.42a), a chimera state with high multiplicity (14−chimera) is obtained. With increasing coupling strength, we move towards chimera states with two incoherent domains, but this chimera performs much stronger amplitude dynamics (bottom panel in Fig. 4.42b). Such two different types of chimera states with strong and weak amplitude dynamics, respectively, were recently observed also in nonlocally coupled networks of Van der Pol oscillators [78]. Further increase of the coupling strength results in a complex 4-chimera state (Fig. 4.42c). Here, the amplitude dynamics is even stronger and the coherent regions exhibit an additional substructure. In hierarchical systems of FitzHugh-Nagumo oscillators [80], nesting effects that appear somewhat similar to the observed structure in the amplitude have been observed in the mean phase velocity ω_k. In Fig. 4.42c, however, the substructure is observed in the amplitude (top panel) and not in the mean phase velocity profile (middle panel). A closer look at the phase portrait (bottom panel) shows strong amplitude dynamics of a peculiar vortex-like structure, caused by the clustering of oscillators in the different coherent regions. Here, clusters of coherent nodes oscillate on smaller cycles in phase space while incoherent nodes roughly follow the limit cycle of the uncoupled system.

Thus, large networks with hierarchical connectivity allow us to observe a variety of chimera states with either weak or strong amplitude dynamics.

4.5.1.6 Summary

In the present section, we have analyzed chimera states in ring networks with hierarchical connectivities. Using a modified iterative Cantor construction algorithm, the network topology is tuned stepwise between nonlocal and hierarchical. We have identified the clustering coefficient and symmetry properties of the base pattern as crucial factors in classifying different topologies with respect to the occurrence of chimera states. We show that symmetric topologies with large clustering coefficients promote the emergence of chimera states, while they are suppressed by slight topological asymmetries or small clustering coefficients. We have determined stability regimes in the plane of coupling strength and nonlinearity parameter of the individual oscillator, which show that chimera states indeed appear on the transition scenario between complete coherence and incoherence. The analysis of an exemplary network with larger base pattern, resulting in larger clustering coefficient and more complex network structure, has revealed two different types of chimera states highlighting the increasing role of amplitude dynamics.

4.5.2 Ring Networks of FitzHugh-Nagumo Oscillators with Time Delay

The topology of the network has been found to play a crucial role in inducing chimera states. While earlier work has focused on simple nonlocal coupling schemes like rings or two-module structures, chimeras have also been found in all-to-all coupled networks [93, 100, 108, 109, 209], as well as in more complex coupling topologies. Of particular interest are networks with hierarchical connectivities, arising in neuroscience as shown by diffusion tensor magnetic resonance imaging analysis, which found a hierarchical (quasi-fractal) connectivity of the neuron axons network [341–345]. Such a network topology can be realized using a Cantor algorithm starting from a chosen base pattern [80, 118, 121, 122] and is in the focus of the present section.

Control of chimera states by extending their lifetime and fixing their spatial position is an important issue [144–147]. A well-known method for stabilization or destabilization of complex patterns in networks is time delay [371–373]. Time-delayed feedback or coupling has been shown to be a versatile method for controlling chimera states [104, 120, 133]. The goal of this section is to study the influence of time delay on chimera states in networks of FitzHugh-Nagumo oscillators with fractal connectivity, and to demonstrate how by varying the time delay one can stabilize chimera states in the network.[4]

---

[4]Portions of the following text have been re-printed from [123] with the permission of The European Physical Journal (EPJ).

4.5.2.1 Model: Local Dynamics and Network Topology

The FitzHugh-Nagumo (FHN) is a paradigmatic model for neural systems [310, 311] but is also used to describe chemical [237] and optoelectronic [315] oscillators and nonlinear electronic circuits [48]. We consider a ring of N identical FHN oscillators with fractal coupling topology, which is given by the adjacency matrix $\boldsymbol{G}$ with a circulant structure. The dynamical equations for the variable $\mathbf{x}_k = (u_k, v_k)^T \in \mathbb{R}^2$, where u_k and v_k are the activator and inhibitor variables, respectively, are

$$\dot{\mathbf{x}}_i(t) = \mathbf{F}(\mathbf{x}_i(t)) + \frac{\sigma}{g}\sum_{j=1}^{N} G_{ij}\boldsymbol{H}[\mathbf{x}_j(t-\tau) - \mathbf{x}_i(t)] \tag{4.15}$$

with $i \in \{1, ..., N\}$ modulo N, and the delay time τ. The dynamics of each individual oscillator is governed by

$$\mathbf{F}(\mathbf{x}) = \begin{pmatrix} \varepsilon^{-1}(u - \frac{u^3}{3} - v) \\ u + a \end{pmatrix}, \tag{4.16}$$

where $\varepsilon > 0$ is a small parameter characterizing a timescale separation, which we fix at $\varepsilon = 0.05$ throughout Sect. 4.5.2. Depending on the threshold parameter a, the FHN oscillator exhibits either oscillatory ($|a| < 1$) or excitable ($|a| > 1$) behavior. Here, we consider the oscillatory regime ($a = 0.5$). The parameter σ denotes the coupling strength, and $g = \sum_{j=1}^{N} G_{ij}$ is the number of links for each node (corresponding to the row sum of $\boldsymbol{G}$). The interaction is realized through diffusive coupling with coupling matrix

$$H = \begin{pmatrix} \varepsilon^{-1}\cos\phi & \varepsilon^{-1}\sin\phi \\ -\sin\phi & \cos\phi \end{pmatrix}. \tag{4.17}$$

In accordance with Omelchenko et al. [79], throughout the section we fix the coupling phase $\phi = \frac{\pi}{2} - 0.1$.

Fractal topologies can be generated using the Cantor construction algorithm for a fractal set [210, 211]. This iterative hierarchical procedure starts from a *base pattern* or initiation string b_{init} of length b, where each element represents either a link ('1') or a gap ('0'). The number of links contained in b_{init} is referred to as c_1. In each iterative step, each link is replaced by the initial base pattern, while each gap is replaced by b gaps. Thus, each iteration increases the size of the final bit pattern, such that after n iterations the total length is $N = b^n$. We call the resulting connectivity fractal or hierarchical. Using the resulting string as the first row of the adjacency matrix $\boldsymbol{G}$, and constructing a circulant adjacency matrix $\boldsymbol{G}$ by applying this string to each element of the ring, a ring network of $N = b^n$ nodes with hierarchical connectivity is generated [80, 83, 119]. Here, we slightly modify this procedure by including an additional zero in the first instance of the sequence [118], which corresponds to the delayed self-coupling. Therefore, there is no net effect of the diagonal elements of the adjacency matrix G_{ij} on the network dynamics. Without our modification, this would lead to a breaking of the base pattern symmetry, i.e., if the base pattern is symmetric, the resulting coupling topology would not be so, since the first link to the right is missing from the final link pattern. Our procedure, in contrast, ensures the preservation of an initial symmetry of b_{init} in the final link pattern, which is crucial for the observation of chimera states, since asymmetric coupling leads to a drift of the chimera [145, 146]. Thus, a ring network of $N = b^n + 1$ nodes is generated.

4.5.2.2 Chimera States

In the following, we consider the network generated with base pattern $b_{\text{init}} = (11011)$ after four iterative steps, resulting in a ring network of $N = 5^4 + 1 = 626$ nodes. Our choice is motivated on the one hand by previous studies of chimera states in nonlocally coupled networks [78, 79], where it has been shown that an intermediate coupling range is crucial for the observation of chimera states, too large and too small numbers of connections make this impossible. On the other hand, it has been demonstrated that hierarchical networks with higher clustering coefficient promote chimera states [80, 118]. For the fractal topology considered here, the clustering coefficient C introduced by Watts and Strogatz [370] is calculated as $C = 0.428$. In our fractal network, we obtain an effective coupling radius $\bar{r} = \frac{c_1^n}{2N} = 0.2$, namely, half the link density, as derived in [118] (see Sect. 4.5.1), which is much smaller than the coupling radius r for which chimeras have been observed in regular nonlocally coupled networks [78, 79].

4.5.2.3 Influence of Time Delay

Figure 4.43 demonstrates a chimera state in the system (4.15) for time delay $\tau = 3.6$, obtained numerically for random initial conditions. We analyze the space-time plot (upper panel), the final snapshot of the activator variables u_k at $t = 50{,}000$ (middle panel), and the phase velocities ω_k of the oscillators (bottom panel). The mean phase velocities of the oscillators are calculated as $\omega_k = 2\pi S_k/\Delta T$, $k = 1, \ldots, N$, where S_k denotes the number of complete rotations realized by the kth oscillator during the time ΔT. Throughout Sect. 4.5.2, we used $\Delta T = 10{,}000$. Oscillators from coherent domains are phase locked and have equal mean frequencies. Arc-like profiles of the mean phase velocities for oscillators from the incoherent domain are typical for chimera states.

To uncover the influence of time delay introduced in the coupling term in system (4.15), we analyze numerically the parameter plane of coupling strength σ and delay time τ. Fixing the network parameters $b_{\text{init}} = (11011)$, $n = 4$, $N = 626$, $a = 0.5$, and $\varepsilon = 0.05$, we choose random initial conditions and vary the values of σ and τ.

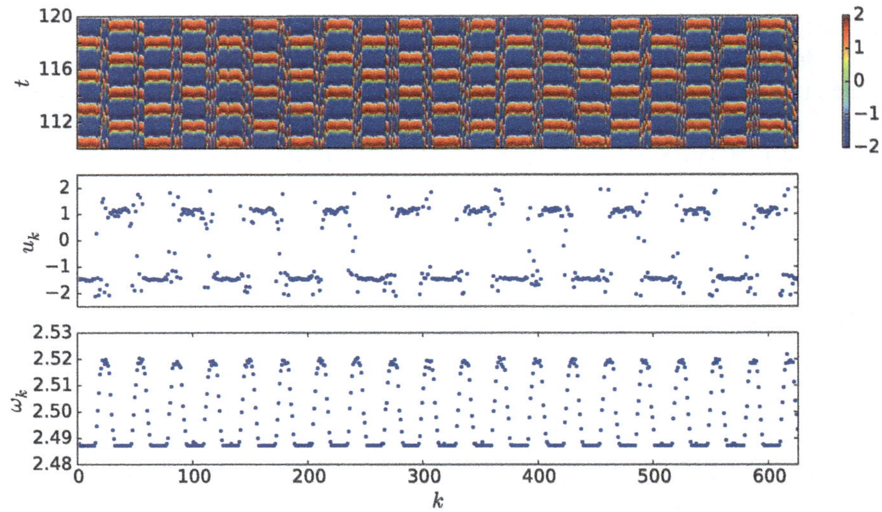

Fig. 4.43 Chimera state in the case $\tau = 3.6$ and $\sigma = 0.05$ for $b_{\text{init}} = (11011), n = 4, N = 626, a = 0.5, \varepsilon = 0.05$, and $\phi = \frac{\pi}{2} - 0.1$. Random initial conditions were used. The three panels correspond to the same simulation: Space-time plot of u_k (upper panel), snapshot of variable u_k at $t = 50{,}000$ (middle panel), and mean phase velocity profile ω_k (bottom panel) [123]

Figure 4.44 shows the map of regimes in the parameter plane (τ, σ). In the undelayed case $\tau = 0$, we cannot observe chimera states for random initial conditions. The introduction of small time delay for weak coupling strength does not change the behavior and the system stays in the completely incoherent regime characterized by chaotic dynamics (gray-dotted region). Nevertheless, for larger values of coupling strength σ chimera states can be observed for small τ. With increasing delay τ, we observe a sequence of tongue-like regions, which are bounded by red curves, on which chimera states occur. These regions appear in between larger areas of coherent structures: fully synchronized states (blue regions with horizontal stripes) alternating with coherent traveling waves, where all nodes oscillate with the same phase velocity (yellow-green regions with diagonal stripes). Inside the tongues, we can observe "salt and pepper" states, which are characterized by strong variations on very short length scales, so that the dynamical patterns have arbitrarily short wavelengths [71, 374] (red-dotted regions). Closer inspection of the chimera tongues shows that increasing τ reduces the size of the tongues and also decreases the maximum values of σ for which chimera states are observed. Moreover, one can easily see that chimera regions appear at τ values close to half-integer multiples of the period of the uncoupled system $T \approx 2.3$.

In many delay systems, one expects resonance effects if the delay is an integer or half-integer multiple of the period of the uncoupled system [329, 375]. The undelayed part of the coupling term in (4.15) is the most important part in case of incoherence

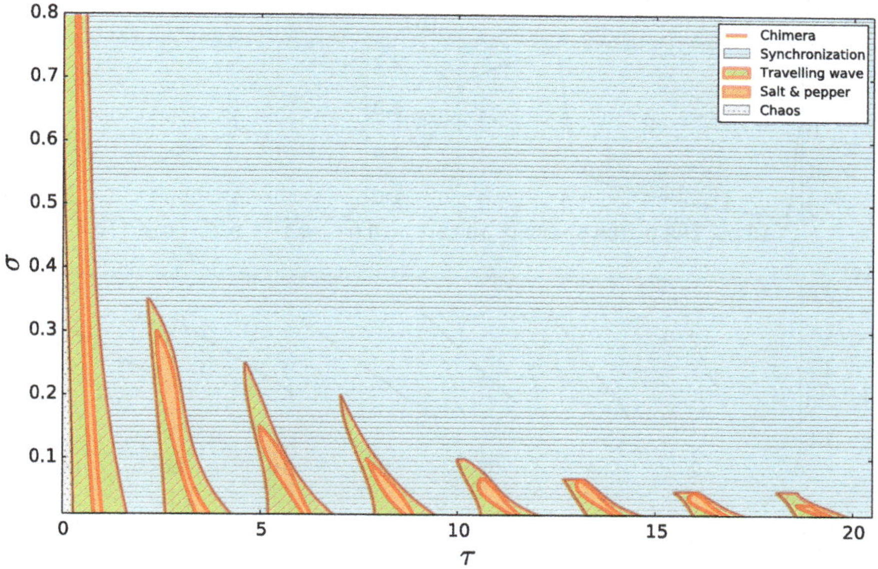

Fig. 4.44 Chimera tongues: Chimeras occur on the boundaries (red curves) between in-phase synchronization (horizontally striped blue region), coherent traveling waves (diagonally striped yellow-green regions), and "salt and pepper" dynamics (dotted red regions) in the parameter plane (τ, σ). Below the first traveling wave region, we can observe chaos (dotted gray region at small τ, σ). Random initial conditions were used for all numerical simulations [123]. Other parameters as in Fig. 4.43

(see tongues in Fig. 4.44) and can be rewritten as follows, neglecting $\cos\phi \ll 1$ and setting $\sin\phi \approx 1$ (it is possible to keep ϕ, but it complicates the algebra, see (4.24)):

$$\begin{aligned} \varepsilon\dot{u} &= u - \frac{u^3}{3} - (1+\sigma)v \\ \dot{v} &= (1+\sigma)u + a. \end{aligned} \tag{4.18}$$

Similar to Brandstetter [376], we employ an analytic approximation for the period of the oscillation defined by (4.18). We consider slow motion on the falling branches of the u-nullcline given by $(1+\sigma)v = u - \frac{u^3}{3}$ and hence $(1+\sigma)\dot{v} = \dot{u}(1-u^2)$, which gives

$$\dot{u} = \frac{(1+\sigma)^2 u + (1+\sigma)a}{1-u^2}. \tag{4.19}$$

It is possible to integrate this equation analytically from $\pm u_+$ to $\pm u_-$, which are approximately the limits of the slow parts of the u-nullcline (see Fig. 4.45a), given

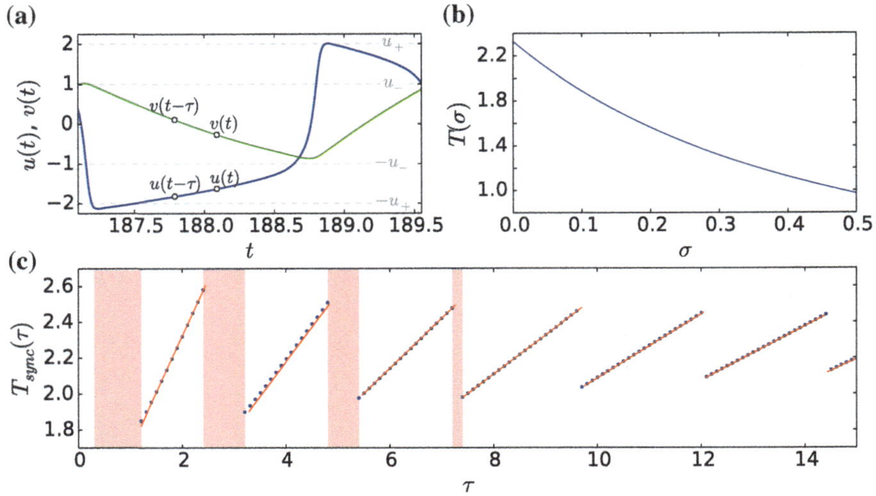

Fig. 4.45 Analytical approximation of the period T for an FHN system with delayed coupling: **a** Limit cycle of the variables $u(t)$ (dark blue line) and $v(t)$ (light green line) of a single FHN oscillator with delayed feedback representing the synchronized state of (4.15) for $\tau = 2.1$ and $\sigma = 0.15$. The dashed gray lines indicate $\pm u_\pm$, respectively, given by $u_+ = 2$ and $u_- = 1$. **b** Period T vs. σ of the FHN oscillator given by ((4.20)), valid for parameters from the incoherent regimes in Fig. 4.44. As the proportionality factor, we assume $1 + e(\varepsilon)$, with $e(0.05) = 0.3$. **c** Period of the synchronized solution T_{sync} versus delay time τ. Comparison of numerics (dots) and analytics (lines), given by ((4.28)) for $\sigma = 0.15$ [123]. The red shaded regions correspond to the tongues in Fig. 4.44. Other parameters for all panels as in Fig. 4.43

by $u_+ = 2$ and $u_- = 1$. With this we obtain a rough approximation of the intrinsic period $T(\sigma)$ of the coupled system, neglecting the fast parts of the trajectory $u(t)$:

$$T(\sigma) \propto (1+\sigma)^{-2}\left[u_+^2 - u_-^2 + \left(1 - \left(\frac{a}{1+\sigma}\right)^2\right)\ln\frac{a^2 - (1+\sigma)^2 u_-^2}{a^2 - (1+\sigma)^2 u_+^2}\right] \tag{4.20}$$

As we can see in Fig. 4.45b, the period T decreases with increasing σ. Therefore, due to the resonance condition of τ with respect to the intrinsic period T, the chimera tongues are shifted to the left with increasing coupling strength σ.

In the case of complete synchronization (blue region in Fig. 4.44), we cannot neglect the delayed terms $v_\tau \equiv v(t-\tau)$ and $u_\tau \equiv u(t-\tau)$ in (4.15):

$$\begin{aligned}\varepsilon\dot{u} &= u - \frac{u^3}{3} - v + \sigma(v_\tau - v)\\ \dot{v} &= u + a - \sigma(u_\tau - u).\end{aligned} \tag{4.21}$$

Due to the almost linear behavior on the slow branches (exemplarily shown by the straight connection between $u(t)$ and $u(t-\tau)$ in Fig. 4.45a), we assume $\mathbf{x}(t) - \mathbf{x}(t-\tau) = \tau\dot{\mathbf{x}}(t)$ for values of τ close to multiples of the period mT with $m \in \mathbb{N}$:

$$\begin{aligned} \varepsilon\dot{u} &= u - \frac{u^3}{3} - v - \sigma\tau\dot{v} \\ \dot{v} &= u + a + \sigma\tau\dot{u}. \end{aligned} \tag{4.22}$$

We can insert the second equation into the first one and analyze the dynamics on the falling branches of the u-nullcline given by $v = u - \frac{u^3}{3} - \sigma\tau(u+a)$:

$$\dot{u} = \frac{u+a}{1 - u^2 - 2\sigma\tau}. \tag{4.23}$$

This is an approximation of the equation which would have been obtained if the phase lag term $\cos\phi$ were not been neglected:

$$\dot{u} = \frac{\frac{u+a}{1+\sigma\tau\cos\phi}}{1 - u^2 - 2\sigma\tau\frac{\sin\phi}{1+\sigma\tau\cos\phi}}. \tag{4.24}$$

In the case of values of τ close to T, we can calculate the period of the synchronized oscillations as

$$T_{sync}(\tau) \propto u_+^2 - u_-^2 + \left(1 - a^2 - 2\sigma\tau\right)\ln\frac{a^2 - u_-^2}{a^2 - u_+^2}. \tag{4.25}$$

As proportionality factor in ((4.20)) and ((4.25)), we assume $1 + e(\varepsilon)$, where $e(0.05) = 0.3$ is a constant parameter, determined by fitting the analytical solution [((4.20)) for $\sigma = 0$] to the numerical simulation [((4.15)) for $\sigma = 0$]. As generally shown in [377], delay systems generically have branches of periodic solutions, which are reappearing for integer multiples of the intrinsic period T of the system. A solution for $\tau = \tau_0 < T$ reappears for all values

$$\tau_m = \tau_0 + mT_{sync}(\tau_0) \tag{4.26}$$

with $m \in \mathbb{N}$, and T_{sync} depends upon τ_0 according to (4.25). The branches T_{sync} of the synchronized solutions are piecewise linear functions of τ, as shown in Fig. 4.45c, where $m = 1, 2, \ldots$ numbers the branches. With increasing m, the branches are stretched by $\frac{\partial\tau_m}{\partial\tau_0}$ and their slope decreases (see [377]). To take into account this mapping for $m > 0$, τ in (4.25) has to be replaced by

$$\tau' = \tau_0\left(\frac{\partial\tau_m}{\partial\tau_0}\right)^{-1} = \tau_0\left(1 - 2\sigma m(1+e)\ln\frac{a^2 - u_-^2}{a^2 - u_+^2}\right)^{-1}, \tag{4.27}$$

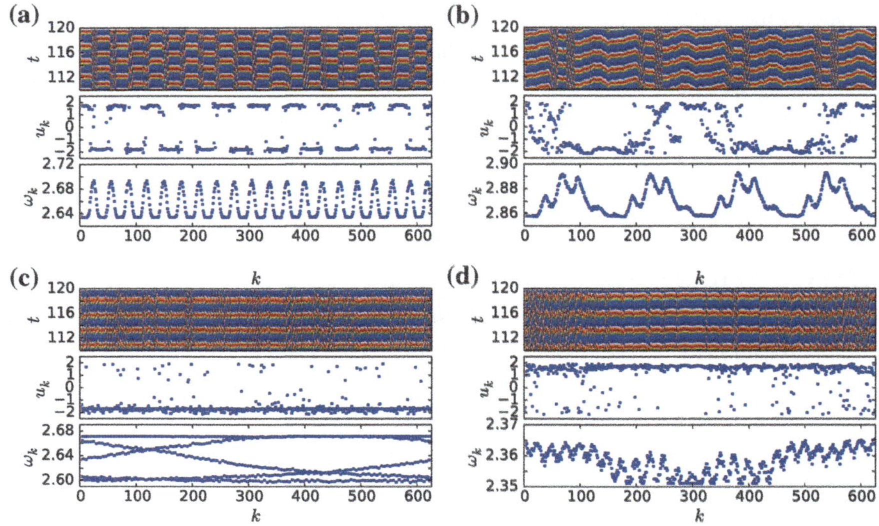

Fig. 4.46 Patterns occurring in the chimera tongues in Fig. 4.44: Space-time plot of u (upper panels), snapshot of variables u_k (middle panels), and mean phase velocity profile ω_k (bottom panels) for **a** $\tau = 1.0$ and $\sigma = 0.1$, **b** $\tau = 1.0$ and $\sigma = 0.15$, **c** $\tau = 1.4$ and $\sigma = 0.05$, and **d** $\tau = 10.7$ and $\sigma = 0.01$ [123]. Other parameters as in Fig. 4.43

where for a given $\tau = \tau_m > T$, τ_0 and m can be calculated from (4.26). Equation (4.25) now reads

$$T_{sync}(\tau') = (1+e)\left[u_+^2 - u_-^2 + \left(1 - a^2 - 2\sigma\tau'\right)\ln\frac{a^2 - u_-^2}{a^2 - u_+^2}\right]. \tag{4.28}$$

A comparison of this analytical result for the period T_{sync} in the synchronized regime with numerical simulations is given in Fig. 4.45c. Depending on the initial conditions, we can find chimera states in the red shaded regions at the boundaries of the piecewise linear branches, which occur if the delay times τ are half-integer multiples of the intrinsic period $T_{sync}(0) = T$. They are marked in Fig. 4.45c (red shaded) for $\sigma = 0.15$. Note that the period is piecewise linear as a function of τ and also of σ (in case of $m = 0$ in (4.27)) in the synchronized regime, whereas it is nonlinear in the non-synchronized regime.

In addition, we can see a decrease of the maximum of the chimera tongues with increasing τ in Fig. 4.44, cf. [378–380]: The maximal value of the coupling strength for which chimera states can be observed decreases for increasing delay.

Let us now take a closer look at the dynamics inside the tongues in Fig. 4.44. For the parameter values chosen inside the first, leftmost and largest, tongue, we find multichimera states (which consist of several coherent and incoherent parts, here 20 each, i.e., we have a 20-chimera) similar to Fig. 4.43 (see Fig. 4.46a) and nested chimera structures (see Fig. 4.46b). These nested structures are slowly shifting in

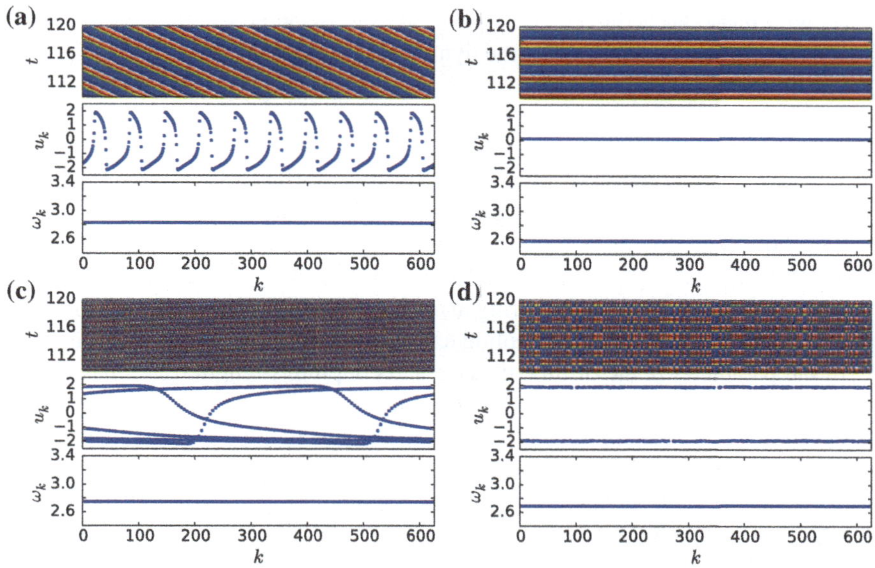

Fig. 4.47 Patterns occurring in non-chimera regimes of Fig. 4.44: Space-time plot of u (upper panels), snapshot of variables u_k (middle panels), and mean phase velocity profile ω_k (bottom panels) for **a** $\tau = 1.1$ and $\sigma = 0.15$, **b** $\tau = 4.5$ and $\sigma = 0.1$, **c** $\tau = 5.1$ and $\sigma = 0.1$, and **d** $\tau = 5.5$ and $\sigma = 0.1$ [123]. Other parameters as in Fig. 4.43

space, so that the mean phase velocity profile (bottom panel of Fig. 4.46b) shows a pyramidal structure instead of an arc-like profile as usually in stationary chimera states. The speed of traveling is sensitive to the coupling strength and delay time. For a pronounced profile of the mean phase velocity, this speed should be small. Otherwise, it is smeared out over time. Figure 4.46c and d shows two examples of the transition region from complete synchronization to chimera states. Also, here we have coherent and incoherent domains. In contrast to the other examples, we can find a complex structure of the mean phase velocity profiles (see bottom panels). In general, the appropriate choice of time delay τ in the system allows one to achieve the desired chimera pattern.

In the parameter plane of delay time τ and coupling strength σ, the region corresponding to coherent states is dominating (blue and yellow regions in Fig. 4.44). On one hand, we observe the in-phase synchronization regime (see Fig. 4.47b) which is enlarged for increasing coupling strength. On the other hand, we also detect a region of coherent traveling waves with wavenumber $k > 1$ (see Fig. 4.47a) and $k < 1$ (see Fig. 4.47c). Varying the delay time τ allows not only for switching between these states, but also for controlling the speed of traveling waves: in the diagonal striped yellow region in Fig. 4.44 the mean phase velocity decreases for increasing delay times. In addition, we can observe *salt and pepper* states (see Fig. 4.47d), where all nodes oscillate with the same phase velocity but they are distributed between states with phase lag π incoherently [374]. As discussed above, the reason for this is arbitrarily short wavelengths of the dynamical patterns.

4.5.2.4 Summary

In conclusion, we have analyzed chimera states in ring networks of FitzHugh-Nagumo oscillators with hierarchical connectivities. For a fixed base pattern, we have constructed a hierarchical connectivity matrix, and numerically analyzed complex spatiotemporal patterns occurring in the network. In particular, we have focused on the role of time delay in the coupling term and its influence on the chimera states.

In the parameter plane of time delay τ and coupling strength σ, we have determined the regimes for different types of chimera states, alternating with regimes of coherent states. An appropriate choice of time delay allows to stabilize several types of chimera states. The interplay of complex hierarchical network topology and time delay results in a plethora of patterns going beyond regular two-population or nonlocally coupled ring networks: we observe chimera states with coherent and incoherent domains of non-identical sizes and non-equidistantly distributed in space. Moreover, traveling and non-traveling chimera states can be obtained for a proper choice of time delay. We also demonstrate that time delay can induce patterns which are not observed in the undelayed case. In addition, we have shown analytically the influence of τ upon the period; i.e., the phase velocity, a piecewise linear dependence in regimes with coherent states, whereas a nonlinear dependence upon τ is found for incoherent states.

Our analysis has shown that networks with complex hierarchical topologies, as arising in neuroscience, can exhibit diverse nontrivial patterns. Time delay can play the role of a powerful control mechanism which allows either to promote or to destroy chimera patterns.

4.5.3 2D Modular Fractal Connectivities in Networks of FitzHugh-Nagumo Oscillators

One of the most promising applications of chimera states in nature is the studies of complex partially synchronized states observed in neural networks. Synchronization and desynchronization of neural activity are essential for explaining both healthy brain function and brain disorders, such as epileptic seizures [158, 160, 161] and Parkinson's disease. It is well known that the synchronization pattern and the related functionality depend both on the structural connectivity and on the type of local dynamics and type and strength of coupling [381–385], making all these parameters potentially relevant for explaining synchronization alterations. For instance, during an epileptic seizure the electrical activity in the brain is excessive and synchronous, and studying chimera states can give further insight in the underlying mechanisms of the initiation or termination of epileptic seizures. Recent studies on the architecture of the neuron interconnectivity of the human and mammalian brain have shown that the connectivity of the neuronal axon network represents a hierarchical, quasi-fractal structure [341, 343]. The existence of chimera states in networks with hierarchical connectivity (fractal) has been recently discovered [33, 80, 118, 120]. Chimera-like states in neural networks were reported under the notion of bump states [386, 387]. They were also observed for nonlocally coupled Hodgkin-Huxley models [185, 388], FitzHugh-Nagumo [79, 80], Hindmarsh-Rose [81, 187], and integrate-and-fire neurons [389].

In the present section, motivated by studies of epileptic seizures, we provide a comparison of the transition from asynchronous behavior to synchrony via chimera states in an empirical structural brain network derived from diffusion-weighted magnetic resonance imaging and in a mathematically constructed network with modular fractal connectivity.[5] Our purpose is to provide insight into the nature of the real brain connectivities by studying an artificial connectivity matrix which is constructed by a well-defined iterative mathematical algorithm generating a hierarchical, nested, quasi-fractal connectivity structure. To describe the dynamics of individual units, we use the paradigmatic FitzHugh-Nagumo oscillator, describing the activation and inhibition dynamics of a spiking neuron. We analyze and compare the network dynamics and complex spatiotemporal patterns in both topologies. Additionally, we analyze the influence of the removal of nodes on the network synchrony, an important aspect for the understanding of the dynamical nature of epileptic seizures [159] and for improving the outcome of epilepsy surgery [390, 391].

4.5.3.1 Model

We consider a system of N identical coupled FitzHugh-Nagumo oscillators:

$$\begin{aligned} \varepsilon\frac{du_k}{dt} &= u_k - \frac{u_k^3}{3} - v_k \\ &\quad + \sigma\sum_{j=1}^{N} G_{kj}\left[b_{uu}(u_j - u_k) + b_{uv}(v_j - v_k)\right] \\ \frac{dv_k}{dt} &= u_k + a \\ &\quad + \sigma\sum_{j=1}^{N} G_{kj}\left[b_{vu}(u_j - u_k) + b_{vv}(v_j - v_k)\right], \end{aligned} \tag{4.29}$$

where u and v are the activator and inhibitor variables and all indices $k = 1, ..., N$ are taken modulo N. The parameter $\varepsilon > 0$ characterizes the timescale separation; it

[5] Portions of the following text have been re-printed from [122] with the permission of AIP Publishing.

is fixed at $\varepsilon = 0.05$ throughout Sect. 4.5.3. The coupling strength is determined by σ, and a is the threshold parameter. Here the oscillatory regime of the FitzHugh-Nagumo oscillators is considered, i.e., $|a| < 1$.

The adjacency matrix G defines the topology of the network. The two investigated topologies are (i) a symmetrized empirical structural neural connectivity G_{emp} obtained from diffusion-weighted magnetic resonance imaging and (ii) a mathematically constructed topology with modular fractal connectivity G_{mod}. In both cases, the adjacency matrix G is weighted and undirected.

The local interaction scheme is characterized by a rotational coupling matrix [79]:

$$\mathbf{B} = \begin{pmatrix} b_{uu} & b_{uv} \\ b_{vu} & b_{vv} \end{pmatrix} = \begin{pmatrix} \cos\phi & \sin\phi \\ -\sin\phi & \cos\phi \end{pmatrix} . \tag{4.30}$$

The coupling phase ϕ is fixed at $\phi = \frac{\pi}{2} - 0.1$, allowing information transfer between the activator and inhibitor variable of the oscillators. It has been shown that the cross-couplings between activator and inhibitor in systems of identical nonlocally coupled FitzHugh-Nagumo oscillators are essential for the observation of chimera states [79].

4.5.3.2 Empirical Structural Topology

The structural brain network of a healthy human subject shown in Fig. 4.48a was obtained from diffusion-weighted magnetic resonance imaging data measured in a randomly selected member of the healthy control group that took part in the early-stage schizophrenia outcome study (ESO). For details of the original study including data acquisition parameters, see [392]. The data were analyzed using probabilistic tractography [393] as implemented in the FMRIB Software Library (FSL) where

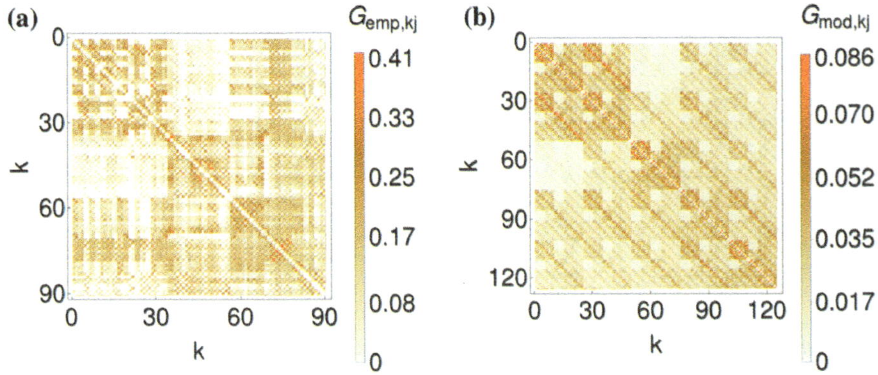

Fig. 4.48 Adjacency matrix of a network topology with **a** empirical structural connectivity and **b** modular fractal connectivity with the base matrix from (4.33) [122]

FMRIB stands for functional magnetic resonance imaging of the brain (www.fmrib.ox.ac.uk/fsl/, Oxford). Note that diffusion-weighted magnetic resonance imaging is a principally "anatomical" method, different from the well-known functional magnetic resonance imaging (fMRI) that provides indirect measurement of neuronal activity over time. The data were acquired during awake resting state with the instruction to stay as still as possible during the scanning. The anatomic network of the cortex and subcortex is constructed using diffusion tensor imaging (DTI) and is divided into 90 cortical and subcortical regions according to the automated anatomical labeling (AAL) atlas [394], each region corresponding to a node in the network. Diffusion magnetic resonance imaging (dMRI) provides information on the distribution of preferred diffusion direction in each volume element (voxel) of the brain, giving thus indirect information for probabilistic estimation of the trajectory of white matter fibers connecting different brain areas. Probabilistic tractography then provides for each voxel a set of N_s streamlines simulating the possible white matter tracts. A coefficient P_{ij} giving the connectivity probability from the ith to the jth region (node) can thus be introduced for all $i, j \in \{1, \ldots, 90\}$. It is defined as the number of probabilistic tracts connecting voxels in the ith to voxels in the jth area, normalized by $N_s = 5000$ times the number of voxels in region i—giving the probability of a streamline entering j, conditional on starting in region i. Because of the high correlation of P_{ij} and P_{ji}, the connectivity probability is approximated with an undirected connectivity probability given by the average of P_{ij} and P_{ji} for each pair of areas (i, j). This way, a weighted and undirected (symmetric) adjacency matrix of size 90×90 is constructed, where each node corresponds to a cortical or subcortical area and each link is defined by the averaged connectivity probability between a pair of areas. The detailed pipeline for constructing the structural network has been adopted from a previous study focusing on the differences between healthy subjects and schizophrenia patients [385].

Note that while the FitzHugh-Nagumo model is a simplified model of a single neuron, it is also often used as a generic model for excitable media on a coarse-grained level. However, studies of FitzHugh-Nagumo networks often lack realistic heterogeneity of the connectivities of the network. While the full connectivity between all neurons of the human brain is not known, as a coarse approximation here we use the matrix of structural connections among 90 well-defined anatomical regions covering the main cortical and subcortical areas.

4.5.3.3 Topology with Modular Fractal Connectivity

In order to provide more insight into the nature of the empirical brain connectivities, here we use an artificial connectivity matrix which has some similar features as the empirical one (Fig. 4.48b). It is constructed by a well-defined iterative algorithm generating a hierarchical, quasi-fractal connectivity structure from the iteration of an initial $b \times b$ base matrix $\mathbf{A}_1$, as described in [33]. The $b^n \times b^n$ adjacency matrix $\mathbf{G}$ for the nth hierarchy level can then be formed by taking $n-1$ Kronecker products of the initial adjacency matrix $\mathbf{A}_1$ with itself, i.e.,

$$\mathbf{G} = \overbrace{\mathbf{A}_1 \otimes \cdots \otimes \mathbf{A}_1}^{n \text{ times}} . \tag{4.31}$$

In the binary case, where $\mathbf{A}_1$ contains only ones and zeros, this is essentially the 2D version of the Cantor construction of a fractal: We start with a $b \times b$ base matrix $\mathbf{A}_1$ of size $m \times m$. If we encounter a nonzero element in the base, we substitute it with the element times the matrix $\mathbf{A}_1$, whereas a zero is replaced by a zero matrix of size equal to the size of $\mathbf{A}_1$. We repeat this substitution procedure n times resulting in the adjacency matrix $\mathbf{G}$ of size $m^{n-1} \times m^{n-1}$ given in (4.31). This matrix $\mathbf{G}$ then defines the coupling topology, but it is no longer a circulant matrix like in a 1D ring topology but describes a modular topology. Note that by the method used to construct the adjacency matrix $\mathbf{G}$, $\mathbf{G}$ has a constant row sum if $\mathbf{A}_1$ has a constant row sum.

To apply this algorithm to generate a similar structure as the empirical topology in Fig. 4.48a, the empirical adjacency matrix $G_{\text{emp,il}}$ is divided into 5^2 equal squares of size $m \times m$, $m = 18$. The sum of links in each square determines the 5×5 elements of the base matrix $\mathbf{A}_1$:

$$A_{1,kj} = \sum_{\substack{i= \\ (k-1)m+1}}^{k \cdot m} \sum_{\substack{l= \\ (j-1)m+1}}^{j \cdot m} G_{emp,il}, \tag{4.32}$$

with $k, j \in \{1, \ldots, 5\}$. For the empirical data of Fig. 4.48a, it reads explicitly:

$$\mathbf{A}_1 = \begin{bmatrix} 5.25677 & 3.22776 & 0.02343 & 1.00899 & 0.86886 \\ 3.22776 & 4.77906 & 0.71110 & 1.58785 & 0.68990 \\ 0.02343 & 0.71110 & 5.39732 & 1.27769 & 1.03968 \\ 1.00899 & 1.58785 & 1.27769 & 3.83577 & 1.92157 \\ 0.86886 & 0.68990 & 1.03968 & 1.92157 & 4.69323 \end{bmatrix} . \tag{4.33}$$

The third iteration of the base matrix gives the investigated modular adjacency matrix $\mathbf{G}_{mod} = \mathbf{A}_1 \otimes \mathbf{A}_1 \otimes \mathbf{A}_1$ with $N = 125$ elements. The size of the base matrix A_1 and the number of iterations are chosen such that the mathematically constructed network topology is of comparable size and visually similar to the empirical topology. The base matrix is weighted and, therefore, a weighted modular fractal connectivity emerges. The link weights are normalized, and the empirical and the mathematically constructed network have equal mean link weights. Furthermore, the self-coupling is set to zero. The degree distributions of the empirical and the mathematically constructed network are compared in Fig. 4.49. It is important to note that since both topologies are undirected, the in-degree is equal to the out-degree.

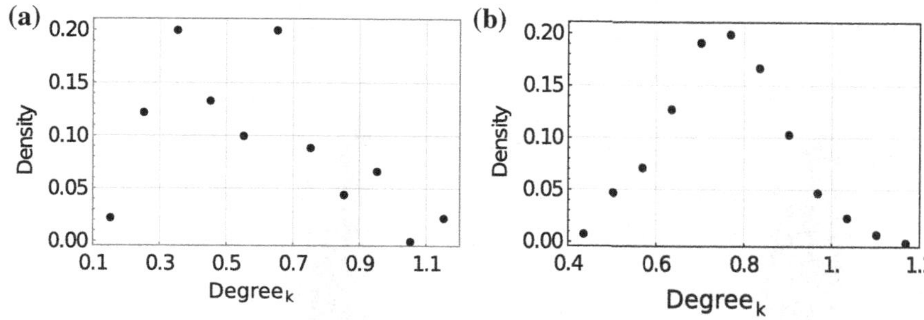

Fig. 4.49 Degree distribution of the nodes indexed by k of a network topology with **a** empirical structural connectivity and **b** modular fractal connectivity with the base matrix from (4.33) [122]

4.5.3.4 Dynamical Regimes

Next, we compare possible dynamical regimes of the system (4.29) with the two network topologies described above. As control parameters, we will consider the threshold parameter a of the individual FitzHugh-Nagumo units, and the coupling strength σ. Starting from random initial conditions, we provide numerical simulations of the underlying networks and observe a plethora of dynamical behaviors.

Together with the coexistence of coherent and incoherent domains in space, the second significant feature of chimera states is the difference of averaged oscillator frequencies. Usually, the oscillators belonging to the coherent domains have identical frequencies, and oscillators from incoherent domains are characterized by higher or lower mean frequencies. If the average time window is sufficiently large and the network is characterized by a regular topology, characteristic arc-like mean phase velocity profiles can be formed. In networks with irregular or all-to-all coupling topologies in some cases, spatial reordering of the nodes is necessary to group the oscillators with equal average frequencies. The mean phase velocities of the oscillators are calculated as $\omega_k = 2\pi M_k/\Delta T$, $k = 1, ..., N$, where M_k denotes the number of complete rotations realized by the kth oscillator during the time ΔT. As a measure for the characterization of the mean phase velocity profiles, we use the standard deviation $\Delta_\omega = \sqrt{(1/N)\sum_{k=1}^{N}(\omega_k - \bar{\omega})^2}$, where $\bar{\omega} = (1/N)\sum_{k=1}^{N}\omega_k$. Larger values of Δ_ω thus correspond to more pronounced frequency differences, giving an opportunity to uncover possible chimera states.

For the characterization of the spatial structure of the obtained patterns, we use the global Kuramoto order parameter $r = \left|\frac{1}{N}\sum_{k=1}^{N} e^{i\phi_k}\right|$, $k = 1, ..., N$, where ϕ_k is the dynamical phase. In the uncoupled case, the geometrical phase is defined as $\tilde{\phi_k}(t) = arctan(v_k(t)/u_k(t))$. The function $t(\tilde{\phi_k})$ is calculated numerically, assigning a value of time $0 < t(\tilde{\phi_k}) < T$ for every value of the geometrical phase, where T is the oscillation period. The dynamical phase is then defined as $\phi_k = 2\pi \cdot t(\tilde{\phi_k})/T$, which

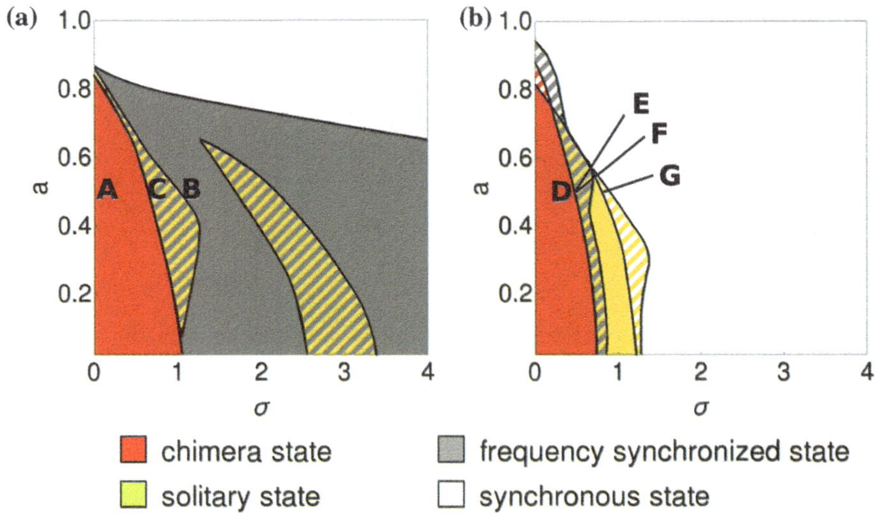

Fig. 4.50 Map of regimes in the parameter space of coupling strength σ and threshold parameter a. **a** Empirical structural connectivity, $N = 90$, **b** modular fractal connectivity, $N = 125$. Hatched regions denote multistability of dynamical states shown with corresponding colors. The letters A, B, C, D, E, F, G mark parameter values corresponding to Figs. 4.52a, 4.53a, 4.54a, 4.52e, 4.53d, g, and 4.54d, respectively [122]

yields constant phase velocity $\dot{\phi}_k$. The global order parameter varies between 1 and 0, and $r = 1$ corresponds to the completely synchronized state in the system. Small values denote spatially desynchronized states.

Figure 4.50 depicts the numerically obtained maps of regimes for empirical and modular fractal networks of FitzHugh-Nagumo oscillators, respectively, in the plane of the coupling strength σ and the threshold parameter of the individual units a. Note that the two systems exhibit a qualitatively similar dynamical picture with cascades of chimera states, solitary states [127], frequency synchronized, and fully coherent states that are presented in detail in the following.

Figure 4.51 shows the numerical evaluation of three quantities: Spatially averaged mean phase velocity $\bar{\omega}$, their standard deviation Δ_ω, and the temporal average of the global Kuramoto order parameter $\langle r \rangle$ for both types of connectivity. The initial conditions for the smallest values of the coupling strength are random. The obtained final state is used as an initial condition for the subsequent set of parameters, and so forth. Additionally, we calculate the Master Stability Function [395] which defines the stability regime for the completely synchronized state. In Fig. 4.51, the orange dots show the stability boundary of the synchronous state, which is stable in the parameter regime above this line, i.e., the Master Stability Function is negative for all eigenvalues of the adjacency matrix.

For small values of the coupling strength σ and the threshold parameter in the approximate range $a \in (0, 0.8)$, both networks exhibit chimera states (shown red in Fig. 4.50). This is in line with earlier observations of chimera states in ring networks

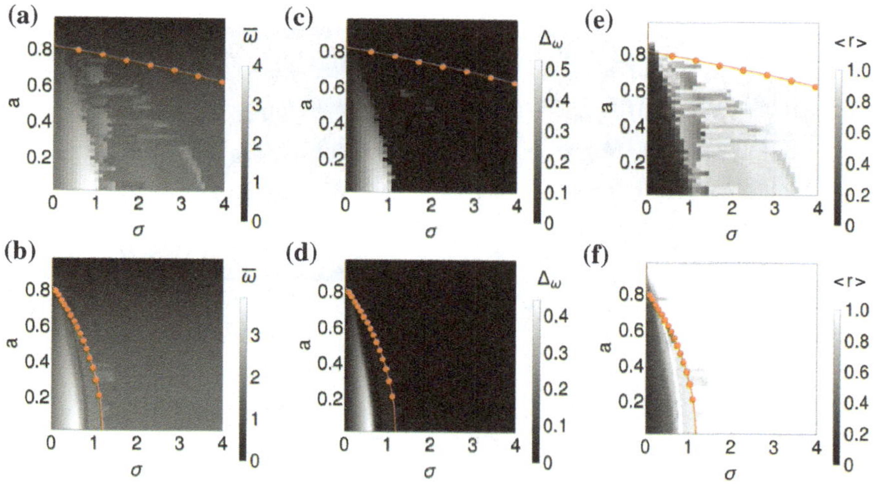

Fig. 4.51 Spatially averaged mean phase velocity $\bar{\omega}$ (left panels), standard deviation of the mean phase velocities Δ_ω (middle panels), and the temporal average of the global Kuramoto order parameter $\langle r \rangle$ (right panels) in the (σ, a) parameter space. The orange dots depict the stability boundary of the synchronous state evaluated from the Master Stability Function; it is stable in the parameter regime above this line. **a**, **c**, **e** empirical structural connectivity, $N = 90$; **b**, **d**, **f** modular fractal connectivity, $N = 125$ [122]

of FitzHugh-Nagumo oscillators with nonlocal coupling [79], where chimera states are usually found for weak coupling strength and a similar range of the threshold parameter.

Examples of chimera states are shown in Fig. 4.52, where panels (a)–(d) correspond to the network with empirical structural connectivity and parameter values marked by **A** in Fig. 4.50a, and panels (e)–(h) correspond to the network with modular fractal connectivity with parameter values marked by **D** in Fig. 4.50b. The upper panels 4.52a, 4.52e in Fig. 4.52 show space-time plots for the activator variables u_k. The middle panels 4.52b, 4.52f depict the mean phase velocity profiles, where the network nodes are reordered by increasing ω_k (index $\widetilde{k}$). We observe well-pronounced differences of the mean phase velocities, which is usually a prominent signature of chimera states. Due to the spatial motion of the incoherent domains, which is common for chimera states in networks of small or moderate size [166], and the two-dimensionality of the considered topologies, the mean phase velocity profile does not allow for a clear distinction of a coherent and an incoherent domain. For both networks, the mean phase velocity profile is independent of the initial conditions. The middle panels 4.52c, 4.52g present the spatial and temporal coherence measures g_0 and h_0, respectively, introduced by Kemeth et al. [168] for a general classification of chimera patterns. Corresponding to this classification scheme, the states presented in Fig. 4.52 are *stationary moving chimeras*. The network's temporal coherence is equal to zero and the spatial coherence is constant, indicating that the incoherent domain of the chimera states is moving in space and has a stable size. The lower pan-

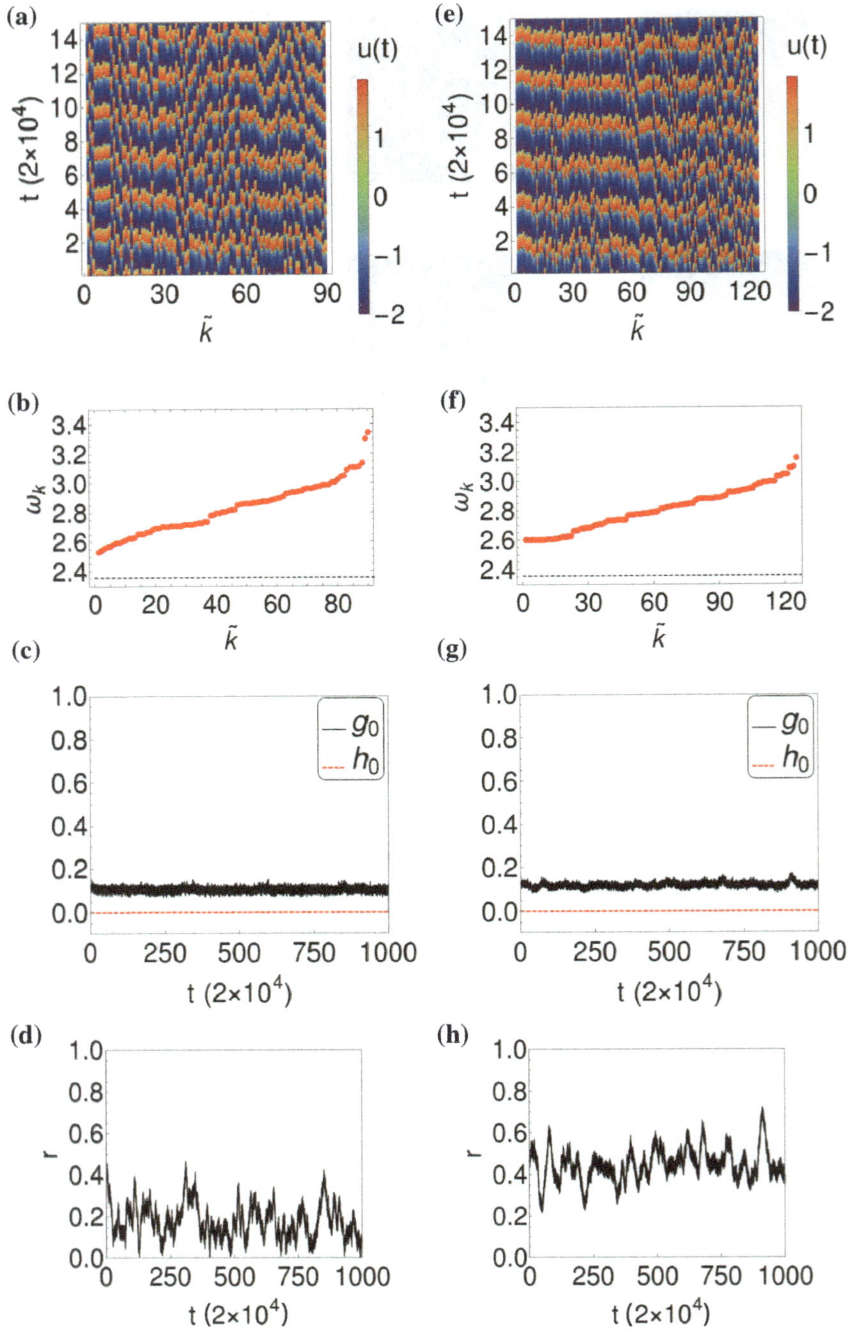
(a)
(b)
(c)
(d)
(e)
(f)
(g)
(h)
u(t)
t (2×10⁴)
$\tilde{k}$
ω_k
g_0
h_0
r

◀**Fig. 4.52** Chimera states: **a–d** Network with empirical structural connectivity, $a = 0.5$, $\sigma = 0.2$, $N = 90$; **e–h** Network with modular fractal connectivity, $a = 0.5$, $\sigma = 0.3$, $N = 125$. **a**, **e** Space-time plots of u_k. Oscillators are reordered from lowest to highest value of ω_k (index $\widetilde{k}$); **b**, **f** Mean phase velocity profile ω_k; the mean phase velocity of an uncoupled oscillator is depicted by a thin black-dotted line; **c**, **g** Spatial and temporal coherence measures g_0 and h_0 versus time; **d**, **h** Dynamics of global Kuramoto order parameter r [122]

els Fig. 4.52d, h demonstrate the dynamics of the global Kuramoto order parameter. Low values indicate the incoherence of the observed chimera patterns, and strong fluctuations of r indicate the random distribution of the incoherent oscillators around the limit cycle in phase space, which changes in time.

Chimera states, which we observe both in networks with empirical structural connectivity, and networks with modular fractal connectivity, have a complex spatial structure due to the complexity of the network topologies and the absence of natural spatial ordering, in contrast to ring networks. This makes their visualization more difficult, but at the same time the spatial and temporal correlation measures clearly indicate a chimera state.

With increasing threshold parameter a and coupling strength σ, in both considered networks chimera states cannot be observed anymore, and two other types of dynamical states occur. One of them is a frequency-synchronized state shown in gray in Fig. 4.50a, b. In contrast to the chimera state, now all the oscillators have equal mean phase velocities. Examples are shown in Fig. 4.53, corresponding to parameter values marked by **B** (empirical structural connectivity, Fig. 4.53a–c), **E** and **F** (modular fractal connectivity, Fig. 4.53d–i). The upper panels in Fig. 4.53 show space-time plots for the activator variables u_k. For better visualization, we reorder the network nodes corresponding to the values of u_k and obtain a coherent traveling wave-like pattern. The middle panels show the mean phase velocities which are the same for all oscillators and, furthermore, they stay constant in time. The lower panels depict the dynamics of the global Kuramoto order parameter, which has large values and exhibits small periodic oscillations on the short timescale due to the amplitude dynamics, i.e., slow-fast motion on the limit cycle. Moreover, in the network with modular fractal connectivity at slightly larger coupling strength, we observe a *breathing* frequency-synchronized state (Fig. 4.53g–i), which exhibits stronger and slower periodic oscillations of the global Kuramoto order parameter on the long timescale due to the cyclic changes of the traveling waves speed.

For increasing values of coupling strength σ and threshold parameter a, we observe bistability of frequency-synchronized states and *solitary states* [127] in both considered networks. *Solitary states* are characterized by frequency synchronization of the majority of oscillators, while single oscillators have a different frequency. The regimes of solitary states are shown in yellow in Fig. 4.50a, b; the dynamics of both networks is multistable, and solitary states coexist with frequency-synchronized states (hatched regions). These patterns might also be classified as *weak chimera states* according to the definition introduced by Ashwin and Burylko [97], where partial frequency synchronization is the main indicator of such states.

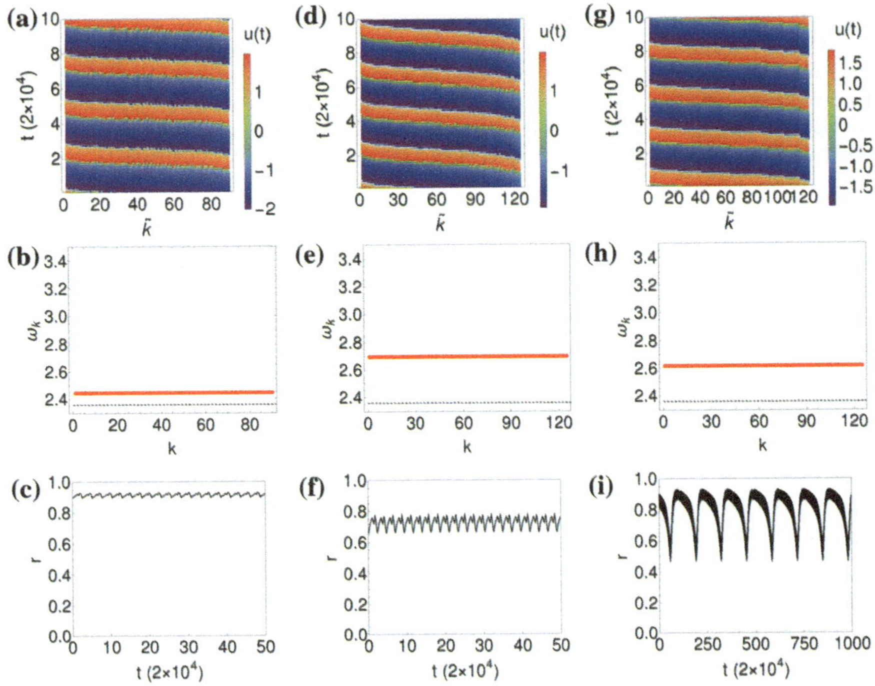

Fig. 4.53 Frequency-synchronized state: **a–c** network with empirical structural connectivity, $a = 0.5, \sigma = 1.3, N = 90$; **d–f** network with modular fractal connectivity, $a = 0.5, \sigma = 0.45, N = 125$. **g–i** Breathing frequency-synchronized state: network with modular fractal connectivity, $a = 0.5$, $\sigma = 0.51$), $N = 125$. **a**, **d**, **g** Space-time plots of u_k. Oscillators are reordered from lowest to highest value of u_k at a fixed time (index $\widetilde{k}$); **b**, **e**, **h** mean phase velocities ω_k; the mean phase velocity of an uncoupled oscillator is shown as thin black-dotted line; **c**, **f**, **i** dynamics of global Kuramoto order parameter r [122]

In the network with empirical structural connectivity, we find two parameter regimes of solitary states, shown in yellow in Fig. 4.50a. In the first regime, usually only one oscillator splits off, exhibiting a different mean phase velocity. This state, corresponding to parameter values marked by **C** in Fig. 4.50a, is shown in Fig. 4.54a–c and has one oscillator with different frequencies (panel b). This solitary incoherent oscillator periodically, after a fixed number of periods, receives a strong input from the network and performs an additional oscillation, indicated by strong oscillations of the global order parameter (panel c), while the rest of the network remains frequency synchronized. A deeper analysis shows that this solitary oscillator has the smallest clustering coefficient, and variable input from different distant parts of the network might be the reason for its incoherent behavior. Note also the regular oscillatory dynamics of the global Kuramoto order parameter depicted in Fig. 4.54c. In the second yellow region (Fig. 4.50a), for larger coupling strength σ, we observe solitary states with three incoherent oscillators, as well as turbulent solitary states with one incoherent oscillator and nonperiodic dynamics.

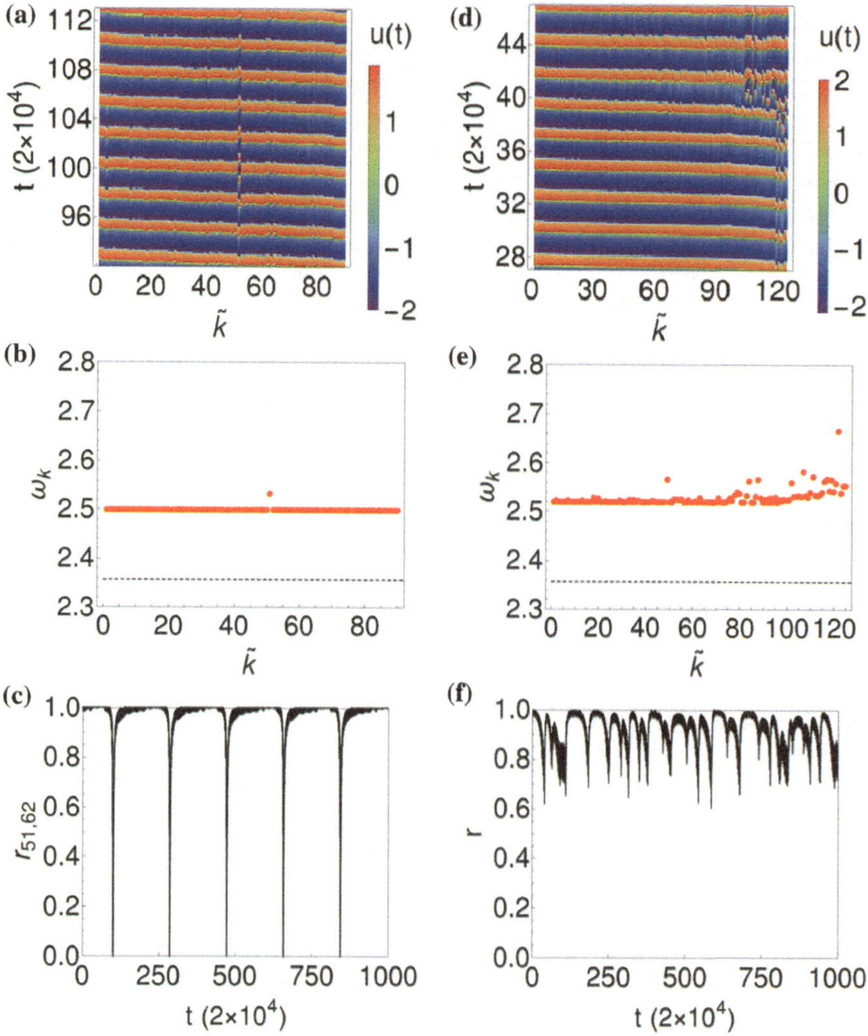

Fig. 4.54 Solitary states: **a**–**c** network with empirical structural connectivity, $a = 0.5$, $\sigma = 0.7$, $N = 90$; **d**–**f** network with modular fractal connectivity, $a = 0.5$, $\sigma = 0.85$, $N = 125$. **a**, **d** Space-time plots of variables u_k, oscillators are reordered from lowest to highest value of u_k at a fixed time (index $\tilde{k}$); **b**, **e** mean phase velocities ω_k, the mean phase velocity of an uncoupled oscillator is shown by a thin black-dotted line; **c** dynamics of Kuramoto order parameter $r_{51,62}$ restricted to the incoherent node $\tilde{k} = 51$ and a reference node $\tilde{k} = 62$; **f** dynamics of global Kuramoto order parameter r [122]

In the network with modular fractal connectivity, we observe similar regimes of solitary states at the transition from chimera states to complete synchronization, as well as a regime of coexistence of solitary and frequency-synchronized states. An example corresponding to parameter values marked by **G** in Fig. 4.50b is shown in Fig. 4.54d–f. In this case, numerous solitary oscillators split off from the coherent domain.

Hence, in both empirical and mathematically constructed networks, we observe similar dynamical scenarios: for weak coupling strength we find chimera states; further increase of the coupling strength leads to a transition to complete synchronization via solitary and frequency-synchronized states. For the empirical network, due to its irregular structure, larger values of coupling strength are needed, while for the modular fractal network, which due to its construction has some topological symmetries, smaller values of the coupling strength σ are sufficient for the stabilization of the completely synchronized pattern.

4.5.3.5 Simulation of Epileptic Seizures

Considering the empirical network, dynamics qualitatively similar to the dynamics of epileptic seizures can be computed. Spontaneous or driven high coherence events occur, indicating pathological seizure behavior. In Fig. 4.55a, an example of such an spontaneous event is given. The temporal evolution of the order parameter for a chimera state at fixed values $(a, \sigma) = (0.5, 0.6)$ is shown. The temporal mean of the order parameter is $\langle r \rangle \approx 0, 5$, and longer events of high coherence spontaneously occur. Before the high coherence event, a drop of the order parameter can be noticed, while the highest value of r is obtained right before its collapse. Both effects have been observed in the study of synchrony in epileptic seizures [158], while a decrease of order preceding complete synchronization has been observed by Andrzejak et al. in a ring of nonlocally coupled phase oscillators [161]. The high coherence events can be controlled by the coupling strength σ; strong coupling in the regime of chimera states increases their probability of occurrence. Furthermore, as shown in Fig. 4.55b, changing the coupling can induce switching between the chimera and the solitary or frequency-synchronized state, which controls the pathological dynamics. It is interesting to note that the system needs some time to respond to modified coupling strength and the high synchronization event persists for some time, even after the coupling is changed back to its initial value.

4.5.3.6 Node Removal and Synchronizability

In studies of epileptic seizures, the focus is usually placed on possible ways to avoid the synchronization of the neural network; for this reason, the analysis of the network synchronizability is of great importance. In recent years, studies have been made on the change in the seizure probability when network nodes are removed, in order to incorporate the network approach in epilepsy surgery and improve its

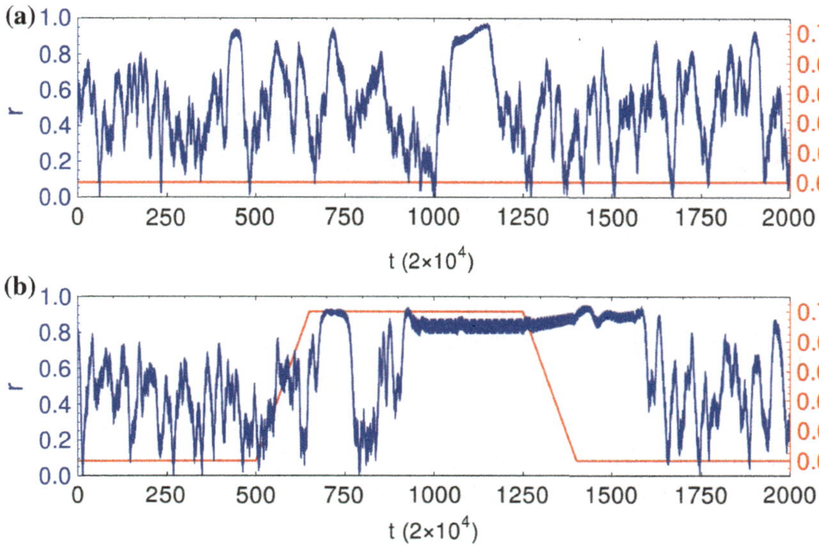

Fig. 4.55 Temporal evolution of the global Kuramoto order parameter r shown in blue for the network with empirical structural connectivity with $a = 0.5$, $N = 90$. The coupling strength σ is shown in red. **a** Chimera state, constant coupling strength $\sigma = 0.6$; **b** controlled dynamics: coupling strength $\sigma = 0.6$ is increased to the value $\sigma = 0.7$ and kept fixed for the time interval $650 < t/2 \times 10^4 \leq 1350$, followed by decrease back to $\sigma = 0.6$, this causes the transitions between the chimera state and solitary or frequency-synchronized states [122]

outcome [390, 391]. We examine the change in synchronizability in the networks with empirical structural and modular fractal connectivity, depending on dynamical and topological properties of the nodes removed.

In our stability analysis of the synchronous solution, we apply the Master Stability approach [395] to the linearized system (4.29):

$$\delta\dot{\xi}(t) = [D\mathbf{F} + \nu D\mathbf{B}]\delta\xi(t) , \tag{4.34}$$

where $\delta\xi = (\delta\mathbf{u}, \delta\mathbf{v})$ is a small perturbation, and $D\mathbf{F}$ and $D\mathbf{B}$ are the Jacobian of the local dynamics and the interaction scheme, respectively. The Master Stability Function is the largest Lyapunov exponent Λ_{max} expressed as a function of the complex parameter ν. The regime of ν where $\Lambda_{max} < 0$ is defined as the synchronous regime S. Due to the diffusive coupling in system (4.29), we consider the Laplacian matrix $\mathbf{L} = \mathbf{D} - \mathbf{G}$, where $\mathbf{D}$ is the degree matrix. Both investigated connectivity matrices are undirected and therefore have real eigenvalues. The longitudinal Laplacian eigenvalue λ_1 is equal to zero, and all transversal Laplacian eigenvalues λ_i, $i = 2, \ldots, N$, are greater than zero: $0 = \lambda_1 < \lambda_2 \leq \cdots \leq \lambda_N$. The synchronous solution is stable if all transversal eigenmodes lie in S, i.e., $-\sigma\lambda_i \in S$, $i = 2, \ldots, N$. For (4.29) $S = (-\infty, \nu_c)$, where $\nu_c < 0$ is the critical value for which $\Lambda_{max}(\nu_c) = 0$. Thus, the smallest transversal eigenvalue λ_2 is sufficient for determining the stability

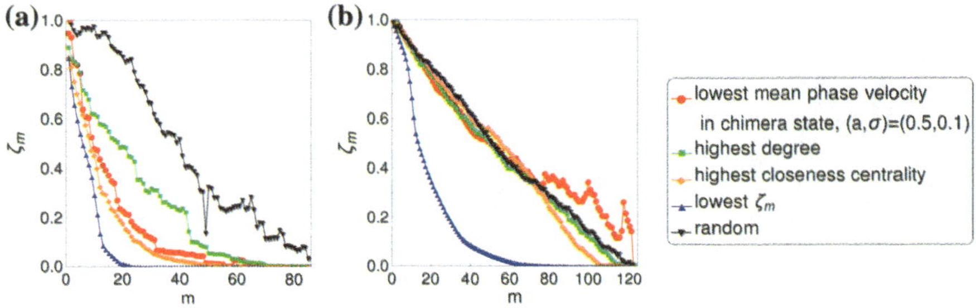

Fig. 4.56 Synchronizability index ζ_m depending on the number m of progressively removed nodes. **a** Empirical structural connectivity; **b** modular fractal connectivity. All measures except the mean phase velocities are recalculated after each removal [122]

of the synchronous solution and the condition $-\sigma\lambda_2 \in S$ has to be fulfilled. The value of λ_2 is therefore used as a synchronizability index indicating the system's ability to synchronize; the closer it is to zero, the less synchronizable is the system.

Removing network nodes perturbs the topological eigenvalue spectrum of the Laplacian matrix and can cause a shift of the smallest transversal eigenvalue altering the system's synchronizability index. The synchronizability index $\zeta_m = \lambda_{2,m}/\lambda_{2,0}$, i.e., the ratio of the smallest transversal eigenvalue when m nodes are removed and in the unperturbed case, measures the change in synchronizability. The incremental synchronizability index $\zeta_m^k = \lambda_{2,m}^k/\lambda_{2,m-1}$, $k \in 1, \ldots, N$ is defined as the change in synchronizability when the removal of $m-1$ nodes is increased to m by removing one further node, $k \in 1, \ldots, N$. If $\zeta_m^k > 1$, the smallest transversal Laplacian eigenvalue shifted to right when the kth node was removed, increasing the systems' ability to synchronize. If $\zeta_m^k < 1$, the opposite is the case. Removing the node k with the lowest value ζ_m^k, thus, most effectively decreases the network's ability to synchronize.

In Fig. 4.56, the change of the synchronizability index ζ_m is shown, when nodes of either highest degree, or highest closeness centrality, or lowest ζ_m^k, or lowest mean phase velocities in the chimera state, or randomly chosen nodes are removed progressively. All measures apart from the mean phase velocities are recalculated at every step. Removing the nodes of lowest ζ_m^k gives the optimal results.

In the case of the network with empirical topology, removing nodes by degree or closeness centrality gives better results than random removal. Evidently, nodes that have a high influence in the network through direct and indirect links enabling quick information flow are promoting synchronization. The same result has been obtained from studies of real-world networks of yeast–protein interaction, e-mail contacts, and Internet autonomous system (AS) relationships [396]. Furthermore, removing nodes of high degree most effectively reduces the mean and standard deviation of the network degrees. This finding is in line with studies reporting an increased mean and standard deviation of degrees in functional networks of epilepsy patients [397, 398] and suggests that considering the Laplacian eigenvalue spectra will provide an explanation for this phenomenon. Counterintuitively, nodes with maximal mean phase velocities in the chimera state promote synchronous behavior.

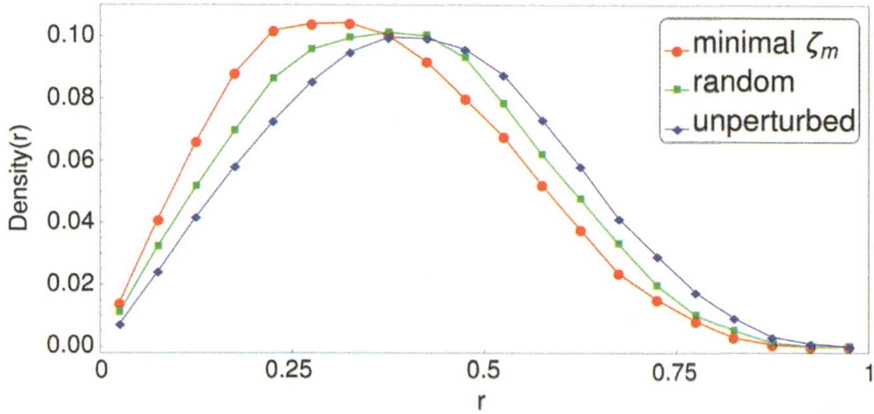

Fig. 4.57 Invariant density of the global Kuramoto order parameter r of the network with empirical structural connectivity and the chimera state with $a = 0.5$, $\sigma = 0.5$, when 5 nodes with the lowest change in the synchronizability index ζ_m^k are removed in comparison to the removal of five random nodes and the unperturbed case [122]

In the case of the modular fractal topology, the applied measures are not sensitive, with the exception of the highest eigenvalue shift. The network degree distribution is narrower and, due to the iterative construction of the topology, the node properties are not strongly distinct (Fig. 4.49).

In the following, we discuss how knowledge on the change of the synchronizability index can be used to control the order in the network. In Fig. 4.57, the asymptotic invariant probability distribution of the global Kuramoto order parameter r of 30 realizations with random initial conditions is depicted for fixed parameters $(a, \sigma) = (0.5, 0.5)$ in the regime of chimera states. The unperturbed case is compared to cases where five nodes with minimum ζ_m^k, $k \in 1, \ldots, N$, $m \in 1, \ldots, 5$, as well as five random nodes are removed. As expected, removing nodes with minimum ζ_m^k improves the results of using random nodes and more successfully lowers the order parameter, i.e., decreases the level of synchronization.

4.5.3.7 Summary

Neural networks are usually characterized by a complex topology. In order to get deeper insight into their structure and functionality, we have compared the dynamics of a network with empirical structural neural connectivity derived from diffusion-weighted magnetic resonance imaging with a mathematically constructed network with modular fractal connectivity generated by a hierarchical iterative algorithm. Although the latter topology is more regular, we observe similar dynamical scenarios of complex partially synchronized states for both networks, when we apply the FitzHugh-Nagumo model to describe the individual node dynamics.

In the plane of system parameters, we uncover the regimes of existence for chimera states which are coexisting domains of coherent and incoherent dynamics. Chimera states can be observed for weak coupling strength, and further increasing the coupling strength leads the system towards the completely coherent synchronized state. In between, we observe solitary states, where only single oscillators have different frequencies, as well as frequency-synchronized, coherent traveling wave-like states, where all oscillators have identical frequencies, but not yet synchronized amplitudes. We demonstrate that both empirical and artificially constructed networks can be characterized by qualitatively similar maps of dynamic regimes, and tuning the strength of the coupling in the network can result in the stabilization or destabilization of the completely synchronized state, as well as in partially synchronized patterns.

The knowledge of the possible dynamical regimes and transitions between them can be applied to the study of brain diseases, where synchronized behavior is pathological, and to find possible ways to avoid it. As a prominent example, we have simulated epileptic seizures where the increased coupling strength leads to pathological synchrony, initiated or terminated via chimera states, and have discussed the suppression of synchrony by the removal of nodes. We have analyzed the synchronizability of the neural networks depending on dynamical and topological properties of the nodes removed, and have shown that a topologically defined synchronizability index can be a good instrument for optimizing which nodes should be removed to decrease the synchronizability. These results from our network approach might be applied for the improvement of epilepsy surgery.

4.6 Conclusions

In the present chapter, we have discussed the role of complex topologies, such as multilayer networks, networks with power-law coupling kernel, and networks with fractal connectivities for the dynamics of coupled nonlinear systems. In particular, we have focused on neural networks where local dynamics is described by the paradigmatic FitzHugh-Nagumo model. This model can operate in two different regimes: oscillatory and excitable. For the excitable regime, we have studied the phenomenon of coherence resonance and developed control strategies based on multiplexing. For the oscillatory regime, we have investigated the impact of multilayer network structure on the dynamics and control of partial synchronization patterns, such as chimera states and solitary states. These states have been also analyzed for the power-law coupling kernel.

Long-range interactions characterized by power-law coupling kernel lead to the formation of various spatiotemporal patterns including fully synchronized regime, traveling waves, complete incoherence, chimera states, tree-like perturbations, and solitary states. The values of power-law exponent and coupling strength determine which of these states is realized by the network. For power-law coupling, chimera states are observed for the values of coupling strength $\sigma < 0.35$ similar to those in the case of a nonlocally coupled ring. However, in contrast to the case of a nonlocally coupled ring, we do not detect chimera states with more than one incoherent domain for a network with power-law coupling kernel. Interestingly, solitary states that we find for networks with power-law coupling kernel and multilayer networks have similar properties. They have a rather flat mean phase velocity profile indicating that the frequency of the solitary nodes is close to that of the synchronized cluster. Further, the solitary nodes and the synchronized elements have different amplitudes. Finally, the solitary nodes and the synchronized cluster are characterized by a phase shift. We believe that these features are characteristic for networks of coupled FitzHugh-Nagumo oscillators.

For multilayer networks, we have developed a control strategy based on multiplexing and successfully applied it to coherence resonance and chimera states. This control scheme allows to achieve the desired states in parameter regimes where they are absent for isolated networks. This control mechanism can be used to induce various network states including coherence resonance, chimera states, and solitary states. The control of the dynamics for one of the layers is realized without manipulating the internal parameters of its elements or the connections between them. The desired state is achieved by adjusting the parameters of the other layer and varying the coupling strength between the layers. A significant advantage of this method is that it works for weak multiplexing, i.e., when the coupling between the layers is much smaller than that within the layers. We demonstrate that weak multiplexing has a dramatic effect on the behavior of the network and represents a powerful tool for controlling dynamic patterns in neural networks. From the point of view of applications, the multiplexing-based control is of particular importance, since in the real-world networks it is not always possible to directly access the desired layer, while the layer it is multiplexed with may be adaptable. We believe that these results can be especially useful for the modeling of brain multiplex networks, where the adjustment of physical connection is feasible due to advances of modern brain surgery, while the manipulation of the functional connectivity appears to be much more complicated. Since multilayer structures are inherent in biological and man-made systems, we expect wide-range applications of our findings to neural and technological networks.

For ring networks of Van der Pol oscillators with hierarchical connectivities, we have identified the clustering coefficient and symmetry properties of the base pattern as crucial factors in classifying different topologies with respect to the occurrence of chimera states. It turns out that symmetric topologies with large clustering coefficients promote the emergence of chimera states, while they are suppressed by slight topological asymmetries or small clustering coefficients. Further, in ring networks of FitzHugh-Nagumo oscillators with hierarchical connectivities, we have explored the role of time delay in the coupling term and its influence on the chimera states. An appropriate choice of time delay allows to stabilize several types of chimera states. The interplay of complex hierarchical network topology and time delay results in a plethora of patterns going beyond regular two-population or nonlocally coupled ring networks: we observe chimera states with coherent and incoherent domains of non-identical sizes and non-equidistantly distributed in space. Moreover, traveling and non-traveling chimera states can be obtained for a proper choice of time delay. We also demonstrate that time delay can induce patterns which are not observed in the undelayed case. In addition, we have shown analytically the influence of τ upon the period, i.e., the phase velocity, a piecewise linear dependence in regimes with coherent states, whereas a nonlinear dependence upon τ is found for incoherent states. Therefore, time delay can play the role of a powerful control mechanism which allows either to promote or to destroy chimera patterns. Finally, for coupled FitzHugh-Nagumo oscillators, we have compared the dynamics of a network with empirical structural neural connectivity derived from diffusion-weighted magnetic resonance imaging with a mathematically constructed network with 2D modular fractal connectivity generated by a hierarchical iterative algorithm. We found regions of existence for chimera states and other complex patterns. In the context of application to the study of epileptic seizures, we qualitatively simulated the dynamics of epileptic seizures and analyzed the removal of nodes and its influence on the general synchronizability of the networks. We have demonstrated that a topologically defined synchronizability index can be exploited for optimizing the selection of nodes to be removed in order to decrease the synchronizability. We believe that these findings might be useful for epilepsy surgery.

References

1. H. Haken, *Synergetics, An Introduction*, 3rd edn. (Springer, Berlin, 1983)
2. A. Pikovsky, M.G. Rosenblum, J. Kurths, *Synchronization: A Universal Concept in Nonlinear Sciences* (Cambridge University Press, Cambridge, 2001)
3. S. Boccaletti, J. Kurths, G. Osipov, D.L. Valladares, C.S. Zhou, The synchronization of chaotic systems. Phys. Rep. **366**, 1–101 (2002)
4. E. Mosekilde, Y. Maistrenko, D. Postnov, *Chaotic Synchronization: Applications to Living Systems* (World Scientific, Singapore, 2002)
5. A.G. Balanov, N.B. Janson, D.E. Postnov, O.V. Sosnovtseva, *Synchronization: From Simple to Complex* (Springer, Berlin, 2009)
6. A.F. Taylor, M.R. Tinsley, F. Wang, Z. Huang, K. Showalter, Dynamical quorum sensing and synchronization in large populations of chemical oscillators. Science **323**(5914), 614–617 (2009)
7. E. Schöll, S.H.L. Klapp, P. Hövel, *Control of Self-organizing Nonlinear Systems* (Springer, Berlin, 2016)
8. S. Boccaletti, A.N. Pisarchik, C.I. del Genio, A. Amann, *Synchronization: From Coupled Systems to Complex Networks* (Cambridge University Press, Cambridge, 2018)
9. W. Singer, Neuronal synchrony: a versatile code review for the definition of relations? Neuron **24**, 49–65 (1999)
10. P. Tass, M.G. Rosenblum, J. Weule, J. Kurths, A. Pikovsky, J. Volkmann, A. Schnitzler, H.J. Freund, Detection of n:m phase locking from noisy data: application to magnetoencephalography. Phys. Rev. Lett. **81**, 3291 (1998)
11. J. Buck, E. Buck, Mechanism of rhythmic synchronous flashing of fireflies: fireflies of southeast asia may use anticipatory time-measuring in synchronizing their flashing. Science **159**, 1319–1327 (1968)
12. A.E. Motter, S.A. Myers, M. Anghel, T. Nishikawa, Spontaneous synchrony in power-grid networks. Nat. Phys. **9**, 191–197 (2013)
13. K.M. Cuomo, A.V. Oppenheim, Circuit implementation of synchronized chaos with applications to communications. Phys. Rev. Lett. **71**(1), 65–68 (1993)
14. I. Kanter, E. Kopelowitz, W. Kinzel, Public channel cryptography: chaos synchronization and Hilbert's tenth problem. Phys. Rev. Lett. **101**(8), 84102 (2008)
15. B. Blasius, A. Huppert, L. Stone, Complex dynamics and phase synchronization in spatially extended ecological systems. Nature **399**, 354 (1999)
16. M.G. Rosenblum, A. Pikovsky, J. Kurths, C. Schäfer, P. Tass, Phase synchronization: from theory to data analysis, volume 4 of *Handbook of Biological Physics*, chapter 9, pp. 279–321. Elsevier Science, Amsterdam, 1. illustrated edition (2001)
17. D.J. DeShazer, R. Breban, E. Ott, R. Roy, Detecting phase synchronization in a chaotic laser array. Phys. Rev. Lett. **87**(4), 044101 (2001)

A. Zakharova, *Chimera Patterns in Networks*, Understanding Complex Systems,
https://doi.org/10.1007/978-3-030-21714-3

18. Y. Aviad, I. Reidler, W. Kinzel, I. Kanter, M. Rosenbluh, Phase synchronization in mutually coupled chaotic diode lasers. Phys. Rev. E **78**(2), 025204 (2008)
19. C.A.S. Batista, E.L. Lameu, A.M. Batista, S.R. Lopes, T. Pereira, G. Zamora-López, J. Kurths, R.L. Viana, Phase synchronization of bursting neurons in clustered small-world networks. Phys. Rev. E **86**(1), 016211 (2012)
20. R.G. Andrzejak, G. Ruzzene, I. Malvestio, K. Schindler, E. Schöll, A. Zakharova, Mean field phase synchronization between chimera states. Chaos **28**(9), 091101 (2018)
21. N.F. Rulkov, M.M. Sushchik, L.S. Tsimring, H.D.I. Abarbanel, Generalized synchronization of chaos in directionally coupled chaotic systems. Phys. Rev. E **51**(2), 980–994 (1995)
22. A. Uchida, R.E. McAllister, R. Meucci, R. Roy, Generalized synchronization of chaos in identical systems with hidden degrees of freedom. Phys. Rev. Lett. **91**(17), 174101 (2003)
23. R.G. Andrzejak, G. Ruzzene, I. Malvestio, Generalized synchronization between chimera states. Chaos **17**, 053114 (2017)
24. M. Golubitsky, I. Stewart, *The symmetry perspective* (Birkhäuser, Basel, 2002)
25. J. Kestler, W. Kinzel, I. Kanter, Sublattice synchronization of chaotic networks with delayed couplings. Phys. Rev. E **76**(3), 035202 (2007)
26. I. Kanter, E. Kopelowitz, R. Vardi, M. Zigzag, W. Kinzel, M. Abeles, D. Cohen, Nonlocal mechanism for cluster synchronization in neural circuits. Europhys. Lett. **93**(6), 66001 (2011)
27. T. Dahms, J. Lehnert, E. Schöll, Cluster and group synchronization in delay-coupled networks. Phys. Rev. E **86**(1), 016202 (2012)
28. L. Lücken, S. Yanchuk, Two-cluster bifurcations in systems of globally pulse-coupled oscillators. Phys. D **241**, 350–359 (2012)
29. E. Schöll, Synchronization, in *delay-coupled complex networks, in Advances in Analysis and Control of Time-Delayed Dynamical Systems*, ed. by J.-Q. Sun, Q. Ding, chapter 4, (World Scientific, Singapore, 2013), pp. 57–83
30. L.M. Pecora, F. Sorrentino, A.M. Hagerstrom, T.E. Murphy, R. Roy, Symmetries, cluster synchronization, and isolated desynchronization in complex networks. Nat. Commun. **5**, 4079 (2014)
31. J. Lehnert, *Controlling synchronization patterns in complex networks* (Springer Theses, Springer, Heidelberg, 2016)
32. W. Poel, A. Zakharova, E. Schöll, Partial synchronization and partial amplitude death in mesoscale network motifs. Phys. Rev. E **91**, 022915 (2015)
33. S. Krishnagopal, J. Lehnert, W. Poel, A. Zakharova, E. Schöll, Synchronization patterns: From network motifs to hierarchical networks. Phil. Trans. R. Soc. A **375**, 20160216 (2017)
34. L. Illing, C.D. Panda, L. Shareshian, Isochronal chaos synchronization of delay-coupled optoelectronic oscillators. Phys. Rev. E **84**, 016213 (2011)
35. Y. Aviad, I. Reidler, M. Zigzag, M. Rosenbluh, I. Kanter, Synchronization in small networks of time-delay coupled chaotic diode lasers. Opt. Express **20**(4), 4352–4359 (2012)
36. K. Blaha, J. Lehnert, A. Keane, T. Dahms, P. Hövel, E. Schöll, J.L. Hudson, Clustering in delay-coupled smooth and relaxational chemical oscillators. Phys. Rev. E **88**, 062915 (2013)
37. C.R.S. Williams, T.E. Murphy, R. Roy, F. Sorrentino, T. Dahms, E. Schöll, Experimental observations of group synchrony in a system of chaotic optoelectronic oscillators. Phys. Rev. Lett. **110**(6), 064104 (2013)
38. D.P. Rosin, *Dynamics of Complex Autonomous Boolean Networks* (Springer Theses, Springer, Heidelberg, 2015)
39. J. Totz, R. Snari, D. Yengi, M.R. Tinsley, H. Engel, K. Showalter, Phase-lag synchronization in networks of coupled chemical oscillators. Phys. Rev. E **92**, 022819 (2015)
40. A. Ijspeert, Central pattern generators for locomotion control in animals and robots: a review. Neural Netw. **21**(4), 642–653 (2008)
41. F. Sorrentino, E. Ott, Network synchronization of groups. Phys. Rev. E **76**(5), 056114 (2007)
42. G. Saxena, A. Prasad, R. Ramaswamy, Amplitude death: the emergence of stationarity in coupled nonlinear systems. Phys. Rep. **521**, 205 (2012)
43. A. Koseska, E. Volkov, J. Kurths, Oscillation quenching mechanisms: amplitude vs. oscillation death. Phys. Rep. **531**(4), 173–199 (2013)

44. A. Gjurchinovski, A. Zakharova, E. Schöll, Amplitude death in oscillator networks with variable-delay coupling. Phys. Rev. E **89**, 032915 (2014)
45. I. Schneider, M. Kapeller, S.A.M. Loos, A. Zakharova, B. Fiedler, E. Schöll, Stable and transient multi-cluster oscillation death in nonlocally coupled networks. Phys. Rev. E **92**, 052915 (2015)
46. A. Zakharova, I. Schneider, Y.N. Kyrychko, K.B. Blyuss, A. Koseska, B. Fiedler, E. Schöll, Time delay control of symmetry-breaking primary and secondary oscillation death. Europhys. Lett. **104**, 50004 (2013)
47. M. Kim, R. Roy, J.L. Aron, T.W. Carr, I.B. Schwartz, Scaling behavior of laser population dynamics with time-delayed coupling: theory and experiment. Phys. Rev. Lett. **94**, 088101 (2005)
48. M. Heinrich, T. Dahms, V. Flunkert, S.W. Teitsworth, E. Schöll, Symmetry breaking transitions in networks of nonlinear circuit elements. New J. Phys. **12**, 113030 (2010)
49. S.M. Crook, G.B. Ermentrout, J.M. Bower, Spike frequency adaptation affects the synchronization properties of networks of cortical oscillators. Neural Comput. **10**(4), 837–854 (1998)
50. M. Dolnik, M. Marek, Extinction of oscillations in forced and coupled reaction cells. J. Phys. Chem. **92**, 2452–2455 (1988)
51. R. Curtu, Singular Hopf bifurcations and mixed-mode oscillations in a two-cell inhibitory neural network. Phys. D **239**(9), 504 (2010)
52. B. Gallego, P. Cessi, Dynamical variability of two oceans and an atmosphere. J. Clim. **14**, 2815–2832 (2001)
53. P. Kumar, A. Prasad, R. Gosh, Stable phase-locking of an external-cavity diode laser subjected to external optical injection. J. Phys. B **41**, 135402 (2008)
54. G.B. Ermentrout, N. Kopell, Oscillator death in systems of coupled neural oscillators. SIAM J. Appl. Math. **50**, 125–146 (1990)
55. C. Cakan, J. Lehnert, E. Schöll, Heterogeneous delays in neural networks. Eur. Phys. J. B **87**, 54 (2014)
56. N. Suzuki, C. Furusawa, K. Kaneko, Oscillatory protein expression dynamics endows stem cells with robust differentiation potential. PLOS ONE **6**, e27232 (2011)
57. F.M. Atay, Total and partial amplitude death in networks of diffusively coupled oscillators. Phys. D **183**, 1–18 (2002)
58. L.L. Rubchinsky, M.M. Sushchik, G.V. Osipov, Patterns in networks of oscillators formed via synchronization and oscillator death. Math. Comp. Simul **58**, 443–467 (2002)
59. A.L. Do, J.M. Höfener, T. Gross, Engineering mesoscale structures with distinct dynamical implications. New J. Phys. **14**(11), 115022 (2012)
60. F. Sorrentino, L.M. Pecora, A.M. Hagerstrom, T.E. Murphy, R. Roy, Complete characterization of the stability of cluster synchronization in complex dynamical networks. Sci. Adv. **2**, e1501737 (2016)
61. Y. Kuramoto, D. Battogtokh, Coexistence of coherence and incoherence in nonlocally coupled phase oscillators. Nonlin. Phen. Complex Sys. **5**(4), 380–385 (2002)
62. D.M. Abrams, S.H. Strogatz, Chimera states for coupled oscillators. Phys. Rev. Lett. **93**(17), 174102 (2004)
63. M.J. Panaggio, D.M. Abrams, Chimera states: coexistence of coherence and incoherence in networks of coupled oscillators. Nonlinearity **28**, R67 (2015)
64. E. Schöll, Synchronization patterns and chimera states in complex networks: interplay of topology and dynamics. Eur. Phys. J. Spec. Top. **225**, 891–919 (2016)
65. O.E. Omel'chenko, The mathematics behind chimera states. Nonlinearity **31**(5), R121 (2018)
66. S. Majhi, B.K. Bera, D. Ghosh, M. Perc, Chimera states in neuronal networks: a review. Phys. Life Rev. **26**, 100–121 (2019)
67. E. Schöll, A. Zakharova, Chasing chimeras: control of complex networks. Res. Features **123**, 34–37 (2018)
68. E. Schöll, A. Zakharova, R.G. Andrzejak, *Chimera States in Complex Networks*. Research Topics, Front. Appl. Math. Stat. (Frontiers Media SA, Lausanne, 2020). Ebook

69. I. Omelchenko, Y. Maistrenko, P. Hövel, E. Schöll, Loss of coherence in dynamical networks: spatial chaos and chimera states. Phys. Rev. Lett. **106**, 234102 (2011)
70. N. Semenova, A. Zakharova, E. Schöll, V.S. Anishchenko, Does hyperbolicity impede emergence of chimera states in networks of nonlocally coupled chaotic oscillators? Europhys. Lett. **112**, 40002 (2015)
71. N. Semenova, G. Strelkova, V.S. Anishchenko, A. Zakharova, Temporal intermittency and the lifetime of chimera states in ensembles of nonlocally coupled chaotic oscillators. Chaos **27**, 061102 (2017)
72. I.A. Shepelev, A.A. Bukh, T.E. Vadivasova, V.S. Anishchenko, A. Zakharova, Double-well chimeras in 2D lattice of chaotic bistable elements. Commun. Nonlinear Sci. Numer. Simul. **54**, 50–61 (2017)
73. I. Omelchenko, B. Riemenschneider, P. Hövel, Y. Maistrenko, E. Schöll, Transition from spatial coherence to incoherence in coupled chaotic systems. Phys. Rev. E **85**, 026212 (2012)
74. D.P. Rosin, D. Rontani, N. Haynes, E. Schöll, D.J. Gauthier, Transient scaling and resurgence of chimera states in coupled Boolean phase oscillators. Phys. Rev. E **90**, 030902(R) (2014)
75. A. Zakharova, M. Kapeller, E. Schöll, Chimera death: symmetry breaking in dynamical networks. Phys. Rev. Lett. **112**, 154101 (2014)
76. A. Zakharova, M. Kapeller, E. Schöll, Amplitude chimeras and chimera death in dynamical networks. J. Phys. Conf. Ser. **727**, 012018 (2016)
77. S. Olmi, E.A. Martens, S. Thutupalli, A. Torcini, Intermittent chaotic chimeras for coupled rotators. Phys. Rev. E **92**, 030901(R) (2015)
78. I. Omelchenko, A. Zakharova, P. Hövel, J. Siebert, E. Schöll, Nonlinearity of local dynamics promotes multi-chimeras. Chaos **25**, 083104 (2015)
79. I. Omelchenko, O.E. Omel'chenko, P. Hövel, E. Schöll, When nonlocal coupling between oscillators becomes stronger: patched synchrony or multichimera states. Phys. Rev. Lett. **110**, 224101 (2013)
80. I. Omelchenko, A. Provata, J. Hizanidis, E. Schöll, P. Hövel, Robustness of chimera states for coupled FitzHugh-Nagumo oscillators. Phys. Rev. E **91**, 022917 (2015)
81. J. Hizanidis, V. Kanas, A. Bezerianos, T. Bountis, Chimera states in networks of nonlocally coupled Hindmarsh-Rose neuron models. Int. J. Bifurc. Chaos **24**(03), 1450030 (2014)
82. A. Vüllings, J. Hizanidis, I. Omelchenko, P. Hövel, Clustered chimera states in systems of type-I excitability. New J. Phys. **16**, 123039 (2014)
83. J. Hizanidis, E. Panagakou, I. Omelchenko, E. Schöll, P. Hövel, A. Provata, Chimera states in population dynamics: networks with fragmented and hierarchical connectivities. Phys. Rev. E **92**, 012915 (2015)
84. T. Banerjee, P.S. Dutta, A. Zakharova, E. Schöll, Chimera patterns induced by distance-dependent power-law coupling in ecological networks. Phys. Rev. E **94**, 032206 (2016)
85. V.M. Bastidas, I. Omelchenko, A. Zakharova, E. Schöll, T. Brandes, Quantum signatures of chimera states. Phys. Rev. E **92**, 062924 (2015)
86. V.M. Bastidas, I. Omelchenko, A. Zakharova, E. Schöll, T. Brandes, Chimera states in quantum mechanics, in *Control of Self-Organizing Nonlinear Systems*, ed. by E. Schöll, S.H.L. Klapp, P. Hövel (Springer, Berlin, Heidelberg, 2016), pp. 315–336
87. S. Nkomo, M.R. Tinsley, K. Showalter, Chimera states in populations of nonlocally coupled chemical oscillators. Phys. Rev. Lett. **110**, 244102 (2013)
88. M.R. Tinsley, S. Nkomo, K. Showalter, Chimera and phase cluster states in populations of coupled chemical oscillators. Nat. Phys. **8**, 662–665 (2012)
89. M. Wickramasinghe, I.Z. Kiss, Spatially organized dynamical states in chemical oscillator networks: synchronization, dynamical differentiation, and chimera patterns. PLoS ONE **8**(11), e80586 (2013)
90. J.F. Totz, J. Rode, M.R. Tinsley, K. Showalter, H. Engel, Spiral wave chimera states in large populations of coupled chemical oscillators. Nat. Phys. **14**, 282–285 (2017)
91. E.A. Martens, S. Thutupalli, A. Fourriere, O. Hallatschek, Chimera states in mechanical oscillator networks. Proc. Natl. Acad. Sci. USA **110**, 10563 (2013)

92. L. Larger, B. Penkovsky, Y. Maistrenko, Laser chimeras as a paradigm for multistable patterns in complex systems. Nat. Commun. **6**, 7752 (2015)
93. F. Böhm, A. Zakharova, E. Schöll, K. Lüdge, Amplitude-phase coupling drives chimera states in globally coupled laser networks. Phys. Rev. E **91**(4), 040901(R) (2015)
94. D. Brunner, B. Penkovsky, R. Levchenko, E. Schöll, L. Larger, Y. Maistrenko, Two-dimensional spatiotemporal complexity in dual-delayed nonlinear feedback systems: chimeras and dissipative solitons. Chaos **28**, 103106 (2018)
95. L.V. Gambuzza, A. Buscarino, S. Chessari, L. Fortuna, R. Meucci, M. Frasca, Experimental investigation of chimera states with quiescent and synchronous domains in coupled electronic oscillators. Phys. Rev. E **90**, 032905 (2014)
96. J.D. Hart, K. Bansal, T.E. Murphy, R. Roy, Experimental observation of chimera and cluster states in a minimal globally coupled network. Chaos **26**, 094801 (2016)
97. P. Ashwin, O. Burylko, Weak chimeras in minimal networks of coupled phase oscillators. Chaos **25**, 013106 (2015)
98. J.E. Franke, J.F. Selgrade, Abstract x-limit sets, chain recurrent sets, and basic sets for flows. Proc. Am. Math. Soc. **60**, 309–316 (1976)
99. G.C. Sethia, A. Sen, G.L. Johnston, Amplitude-mediated chimera states. Phys. Rev. E **88**(4), 042917 (2013)
100. G.C. Sethia, A. Sen, Chimera states: the existence criteria revisited. Phys. Rev. Lett. **112**, 144101 (2014)
101. A. Zakharova, S.A.M. Loos, J. Siebert, A. Gjurchinovski, J.C. Claussen, E. Schöll, Controlling chimera patterns in networks: interplay of structure, noise, and delay, in *Control Self-organizing Nonlinear Systems*, ed. by E. Schöll, S.H.L. Klapp, P. Hövel (Springer, Berlin, Heidelberg, 2016)
102. S.A.M. Loos, J.C. Claussen, E. Schöll, A. Zakharova, Chimera patterns under the impact of noise. Phys. Rev. E **93**, 012209 (2016)
103. L. Tumash, A. Zakharova, J. Lehnert, W. Just, E. Schöll, Stability of amplitude chimeras in oscillator networks. Europhys. Lett. **117**, 20001 (2017)
104. A. Gjurchinovski, E. Schöll, A. Zakharova, Control of amplitude chimeras by time delay in dynamical networks. Phys. Rev. E **95**(4), 042218 (2017)
105. P.S. Dutta, T. Banerjee, Spatial coexistence of synchronized oscillation and death: a chimera-like state. Phys. Rev. E **92**, 042919 (2015)
106. T. Banerjee, D. Ghosh, D. Biswas, E. Schöll, A. Zakharova, Networks of coupled oscillators: from phase to amplitude chimeras. Chaos **28**, 113124 (2018)
107. T. Banerjee, B. Bandyopadhyay, A. Zakharova, E. Schöll, Filtering suppresses amplitude chimeras. Front. Appl. Math. Stat. **5**, 8 (2019)
108. A. Yeldesbay, A. Pikovsky, M. Rosenblum, Chimeralike states in an ensemble of globally coupled oscillators. Phys. Rev. Lett. **112**, 144103 (2014)
109. L. Schmidt, K. Krischer, Clustering as a prerequisite for chimera states in globally coupled systems. Phys. Rev. Lett. **114**, 034101 (2015)
110. L. Schmidt, K. Schönleber, K. Krischer, V. García-Morales, Coexistence of synchrony and incoherence in oscillatory media under nonlinear global coupling. Chaos **24**(1), 013102 (2014)
111. C.R. Laing, Chimeras in networks with purely local coupling. Phys. Rev. E **92**, 050904(R) (2015)
112. I.A. Shepelev, A. Zakharova, T. Vadivasova, Chimera regimes in a ring of oscillators with local nonlinear interaction. Commun. Nonlinear Sci. Numer. Simul. **44**, 277 (2016)
113. T.W. Ko, G.B. Ermentrout, Partially locked states in coupled oscillators due to inhomogeneous coupling. Phys. Rev. E **78**, 016203 (2008)
114. M. Shanahan, Metastable chimera states in community-structured oscillator networks. Chaos **20**(1), 013108 (2010)
115. C.R. Laing, K. Rajendran, Y.G. Kevrekidis, Chimeras in random non-complete networks of phase oscillators. Chaos **22**(1), 043104 (2012)
116. N. Yao, Z.G. Huang, Y.C. Lai, Z. Zheng, Robustness of chimera states in complex dynamical systems. Sci. Rep. **3**, 3522 (2013)

117. Y. Zhu, Z. Zheng, J. Yang, Chimera states on complex networks. Phys. Rev. E **89**, 022914 (2014)
118. S. Ulonska, I. Omelchenko, A. Zakharova, E. Schöll, Chimera states in networks of Van der Pol oscillators with hierarchical connectivities. Chaos **26**, 094825 (2016)
119. N.D. Tsigkri-DeSmedt, J. Hizanidis, P. Hövel, A. Provata, Multi-chimera states and transitions in the leaky integrate-and-fire model with excitatory coupling and hierarchical connectivity. Eur. Phys. J. ST **225**(6), 1149 (2016)
120. J. Sawicki, I. Omelchenko, A. Zakharova, E. Schöll, Chimera states in complex networks: interplay of fractal topology and delay. Eur. Phys. J. Spec. Top. **226**(9), 1883–1892 (2017)
121. A. zur Bonsen, I. Omelchenko, A. Zakharova, E. Schöll, Chimera states in networks of logistic maps with hierarchical connectivities. Eur. Phys. J. B **91**, 65 (2018)
122. T. Chouzouris, I. Omelchenko, A. Zakharova, J. Hlinka, P. Jiruska, E. Schöll, Chimera states in brain networks: empirical neural vs. modular fractal connectivity. Chaos **28**(4), 045112 (2018)
123. J. Sawicki, I. Omelchenko, A. Zakharova, E. Schöll, Delay-induced chimeras in neural networks with fractal topology. Eur. Phys. J. B **92**, 54 (2019)
124. S. Ghosh, A. Kumar, A. Zakharova, S. Jalan, Birth and death of chimera: interplay of delay and multiplexing. Europhys. Lett. **115**, 60005 (2016)
125. S. Ghosh, A. Zakharova, S. Jalan, Non-identical multiplexing promotes chimera states. Chaos Solitons Fractals **106**, 56–60 (2018)
126. M. Mikhaylenko, L. Ramlow, S. Jalan, A. Zakharova, Weak multiplexing in neural networks: switching between chimera and solitary states. Chaos **29**, 023122 (2019)
127. Y. Maistrenko, B. Penkovsky, M. Rosenblum, Solitary state at the edge of synchrony in ensembles with attractive and repulsive interactions. Phys. Rev. E **89**, 060901 (2014)
128. P. Jaros, Y. Maistrenko, T. Kapitaniak, Chimera states on the route from coherence to rotating waves. Phys. Rev. E **91**, 022907 (2015)
129. M. Wolfrum, O.E. Omel'chenko, Chimera states are chaotic transients. Phys. Rev. E **84**(1), 015201 (2011)
130. P. Kalle, J. Sawicki, A. Zakharova, E. Schöll, Chimera states and the interplay between initial conditions and non-local coupling. Chaos **27**, 033110 (2017)
131. N. Semenova, A. Zakharova, V.S. Anishchenko, E. Schöll, Coherence-resonance chimeras in a network of excitable elements. Phys. Rev. Lett. **117**, 014102 (2016)
132. A. Zakharova, N. Semenova, V. S. Anishchenko, and E. Schöll. Noise-induced chimera states in a neural network, in *Patterns of Dynamics*, ed. by P. Gurevich, J. Hell, B. Sandstede, volume 205 of Springer Proceedings in Mathematics & Statistics (Springer, 2018), p. 44
133. A. Zakharova, N. Semenova, V.S. Anishchenko, E. Schöll, Time-delayed feedback control of coherence resonance chimeras. Chaos **27**, 114320 (2017)
134. G. Hu, T. Ditzinger, C.Z. Ning, H. Haken, Stochastic resonance without external periodic force. Phys. Rev. Lett. **71**, 807 (1993)
135. A. Pikovsky, J. Kurths, Coherence resonance in a noise-driven excitable system. Phys. Rev. Lett. **78**, 775 (1997)
136. A. Zakharova, T. Vadivasova, V.S. Anishchenko, A. Koseska, J. Kurths, Stochastic bifurcations and coherencelike resonance in a self-sustained bistable noisy oscillator. Phys. Rev. E **81**, 011106 (2010)
137. A. Zakharova, A. Feoktistov, T. Vadivasova, E. Schöll, Coherence resonance and stochastic synchronization in a nonlinear circuit near a subcritical Hopf bifurcation. Eur. Phys. J. ST **222**(10), 2481–2495 (2013)
138. V. Semenov, A. Feoktistov, T. Vadivasova, E. Schöll, A. Zakharova, Time-delayed feedback control of coherence resonance near subcritical Hopf bifurcation: theory versus experiment. Chaos **25**, 033111 (2015)
139. P.M. Geffert, A. Zakharova, A. Vüllings, W. Just, E. Schöll, Modulating coherence resonance in non-excitable systems by time-delayed feedback. Eur. Phys. J. B **87**, 291 (2014)

140. W. Just, P.M. Geffert, A. Zakharova, E. Schöll, Noisy dynamical systems with time delay: some basic analytical perturbation schemes with applications, in *Control of Self-Organizing Nonlinear Systems*, ed. by E. Schöll, S.H.L. Klapp, P. Hövel (Springer, Berlin, Heidelberg, 2016)
141. M. Masoliver, N. Malik, E. Schöll, A. Zakharova, Coherence resonance in a network of FitzHugh-Nagumo systems: interplay of noise, time-delay and topology. Chaos **27**, 101102 (2017)
142. N. Semenova, A. Zakharova, Weak multiplexing induces coherence resonance. Chaos **28**(5), 051104 (2018)
143. V. Semenov, A. Zakharova, Y. Maistrenko, E. Schöll, Delayed-feedback chimera states: Forced multiclusters and stochastic resonance. Europhys. Lett. **115**, 10005 (2016)
144. J. Sieber, O.E. Omel'chenko, M. Wolfrum, Controlling unstable chaos: stabilizing chimera states by feedback. Phys. Rev. Lett. **112**, 054102 (2014)
145. C. Bick, E.A. Martens, Controlling chimeras. New J. Phys. **17**(3), 033030 (2015)
146. I. Omelchenko, O.E. Omel'chenko, A. Zakharova, M. Wolfrum, E. Schöll, Tweezers for chimeras in small networks. Phys. Rev. Lett. **116**, 114101 (2016)
147. I. Omelchenko, O.E. Omel'chenko, A. Zakharova, E. Schöll, Optimal design of tweezer control for chimera states. Phys. Rev. E **97**, 012216 (2018)
148. G. Ruzzene, I. Omelchenko, E. Schöll, A. Zakharova, R.G. Andrzejak, Controlling chimera states via minimal coupling modification. Chaos **29**, 051103 (2019)
149. J. Sawicki, I. Omelchenko, A. Zakharova, E. Schöll, Delay controls chimera relay synchronization in multiplex networks. Phys. Rev. E **98**, 062224 (2018)
150. J. Sawicki, S. Ghosh, S. Jalan, A. Zakharova, Chimeras in multiplex networks: interplay of inter- and intra-layer delays. Front. Appl. Math. Stat. **5**, 19 (2019)
151. A.M. Hagerstrom, T.E. Murphy, R. Roy, P. Hövel, I. Omelchenko, E. Schöll, Experimental observation of chimeras in coupled-map lattices. Nat. Phys. **8**, 658–661 (2012)
152. T. Kapitaniak, P. Kuzma, J. Wojewoda, K. Czolczynski, Y. Maistrenko, Imperfect chimera states for coupled pendula. Sci. Rep. **4**, 6379 (2014)
153. L. Larger, B. Penkovsky, Y. Maistrenko, Virtual chimera states for delayed-feedback systems. Phys. Rev. Lett. **111**, 054103 (2013)
154. M. Wickramasinghe, I.Z. Kiss, Spatially organized partial synchronization through the chimera mechanism in a network of electrochemical reactions. Phys. Chem. Chem. Phys. **16**, 18360–18369 (2014)
155. E.A. Viktorov, T. Habruseva, S.P. Hegarty, G. Huyet, B. Kelleher, Coherence and incoherence in an optical comb. Phys. Rev. Lett. **112**(22), 224101 (2014)
156. N.C. Rattenborg, C.J. Amlaner, S.L. Lima, Behavioral, neurophysiological and evolutionary perspectives on unihemispheric sleep. Neurosci. Biobehav. Rev. **24**, 817–842 (2000)
157. M. Tamaki, J.W. Bang, T. Watanabe, Y. Sasaki, Night watch in one brain hemisphere during sleep associated with the first-night effect in humans. Curr. Biol. **26**(9), 1190–1194 (2016)
158. P. Jiruska, M. de Curtis, J.G.R. Jefferys, C.A. Schevon, S.J. Schiff, K. Schindler, Synchronization and desynchronization in epilepsy: controversies and hypotheses. J. Physiol. **591**(4), 787–797 (2013)
159. V.K. Jirsa, W.C. Stacey, P.P. Quilichini, A.I. Ivanov, C. Bernard, On the nature of seizure dynamics. Brain **137**, 2210 (2014)
160. A. Rothkegel, K. Lehnertz, Irregular macroscopic dynamics due to chimera states in small-world networks of pulse-coupled oscillators. New J. Phys. **16**, 055006 (2014)
161. R.G. Andrzejak, C. Rummel, F. Mormann, K. Schindler, All together now: analogies between chimera state collapses and epileptic seizures. Sci. Rep. **6**, 23000 (2016)
162. A.R. Nikolaev, S. Gepshtein, P. Gong, C. van Leeuwen, Duration of coherence intervals in electrical brain activity in perceptual organization. Cerebral Cortex **20**, 365 (2010)
163. S. Ahn, L.L. Rubchinsky, Short desynchronization episodes prevail in synchronous dynamics of human brain rhythms. Chaos **23**, 013138 (2013)
164. S. Ahn, L.L. Rubchinsky, C.C. Lapish, Dynamical reorganization of synchronous activity patterns in prefrontal cortex-hippocampus networks during behavioral sensitization. Cerebral Cortex **24**, 2553–2561 (2014)

165. J.C. Gonzalez-Avella, M.G. Cosenza, M.S. Miguel, Localized coherence in two interacting populations of social agents. Physica A **399**, 24–30 (2014)
166. O.E. Omel'chenko, M. Wolfrum, Y. Maistrenko, Chimera states as chaotic spatiotemporal patterns. Phys. Rev. E **81**(6), 065201(R) (2010)
167. M. Wolfrum, O.E. Omel'chenko, S. Yanchuk, Y. Maistrenko, Spectral properties of chimera states. Chaos **21**, 013112 (2011)
168. F.P. Kemeth, S.W. Haugland, L. Schmidt, Y.G. Kevrekidis, K. Krischer, A classification scheme for chimera states. Chaos **26**, 094815 (2016)
169. I.A. Shepelev, A.V. Bukh, G.I. Strelkova, T.E. Vadivasova, V.S. Anishchenko, Chimera states in ensembles of bistable elements with regular and chaotic dynamics. Nonlinear Dyn. **90**, 2317 (2017)
170. D.M. Abrams, R.E. Mirollo, S.H. Strogatz, D.A. Wiley, Solvable model for chimera states of coupled oscillators. Phys. Rev. Lett. **101**(8), 084103 (2008)
171. C.R. Laing, The dynamics of chimera states in heterogeneous Kuramoto networks. Physica D **238**(16), 1569–1588 (2009)
172. C.R. Laing, Chimera states in heterogeneous networks. Chaos **19**(1), 013113 (2009)
173. S. Bogomolov, A. Slepnev, G. Strelkova, E. Schöll, V.S. Anishchenko, Mechanisms of appearance of amplitude and phase chimera states in a ring of nonlocally coupled chaotic systems. Commun. Nonlinear Sci. Numer. Simul. **43**, 25 (2017)
174. A.K. Malchow, I. Omelchenko, E. Schöll, P. Hövel, Robustness of chimera states in nonlocally coupled networks of nonidentical logistic maps. Phys. Rev. E **98**, 012217 (2018)
175. O.E. Omel'chenko, Y. Maistrenko, P. Tass, Chimera states: the natural link between coherence and incoherence. Phys. Rev. Lett. **100**(4), 044105 (2008)
176. O.E. Omel'chenko, Y. Maistrenko, P. Tass, Chimera states induced by spatially modulated delayed feedback. Phys. Rev. E **82**, 066201 (2010)
177. G. Bordyugov, A. Pikovsky, M. Rosenblum, Self-emerging and turbulent chimeras in oscillator chains. Phys. Rev. E **82**(3), 035205 (2010)
178. T. Banerjee, Mean-field-diffusion-induced chimera death state. Europhys. Lett. **110**, 60003 (2015)
179. C.R. Laing, Chimeras in networks of planar oscillators. Phys. Rev. E **81**(6), 066221 (2010)
180. S. Olmi, Chimera states in coupled kuramoto oscillators with inertia. Chaos **25**, 123125 (2015)
181. T. Bountis, V. Kanas, J. Hizanidis, A. Bezerianos, Chimera states in a two-population network of coupled pendulum-like elements. Eur. Phys. J. Spec. Top. **223**(4), 721–728 (2014)
182. M.G. Clerc, S. Coulibaly, A.M. Ferreira, R. Rojas, Chimera states in a duffing oscillators chain coupled to nearest neighbors. Chaos **28**, 083126 (2018)
183. G. Argyropoulos, T. Kasimatis, A. Provata, Chimera patterns and subthreshold oscillations in two dimensional networks of fractally coupled Leaky Integrate-and-Fire neurons (Preprint, 2018)
184. A. Calim, P. Hövel, M. Ozer, M. Uzuntarla, Chimera states in networks of type-I Morris-Lecar neurons. Phys. Rev. E **98**(062217) (2018)
185. T.A. Glaze, S. Lewis, S. Bahar, Chimera states in a hodgkin-huxley model of thermally sensitive neurons. Chaos **26**, 083119 (2016)
186. J. Hizanidis, N.E. Kouvaris, C. Antonopoulos, Metastable and chimera-like states in the c. elegans brain network. Cybern. Phys. **4**(1), 17–20 (2015)
187. J. Hizanidis, N.E. Kouvaris, G. Zamora-López, A. Díaz-Guilera, C. Antonopoulos, Chimera-like states in modular neural networks. Sci. Rep. **6**, 19845 (2016)
188. S. Olmi, A. Politi, A. Torcini, Chimera states and collective chaos in pulse-coupled neural networks. BMC Neurosci. **12**(Suppl 1), P336 (2011)
189. S. Olmi, A. Torcini, Chimera states in pulse coupled neural networks: the influence of dilution and noise. arXiv:1606.0861 (2016. submitted)
190. J. Xie, E. Knobloch, H.C. Kao, Multicluster and traveling chimera states in nonlocal phase-coupled oscillators. Phys. Rev. E **90**, 022919 (2014)
191. J. Xie, H.C. Kao, E. Knobloch, Chimera states in systems of nonlocal nonidentical phase-coupled oscillators. Phys. Rev. E **91**, 032918 (2015)

192. L. Bauer, J. Bassett, P. Hövel, Y.N. Kyrychko, K.B. Blyuss, Chimera states in multi-strain epidemic models with temporary immunity. Chaos **27**(11), 114317 (2017)
193. J. Hizanidis, N. Lazarides, G.P. Tsironis, Robust chimera states in SQUID metamaterials with local interactions. Phys. Rev. E **94**, 032219 (2016)
194. Z.G. Nicolaou, H. Riecke, A.E. Motter, Chimera states in continuous media: existence and distinctness. Phys. Rev. Lett. **119**, 244101 (2017)
195. O.E. Omel'chenko, Coherence-incoherence patterns in a ring of non-locally coupled phase oscillators. Nonlinearity **26**(9), 2469 (2013)
196. O.E. Omel'chenko, M. Wolfrum, C.R. Laing, Partially coherent twisted states in arrays of coupled phase oscillators. Chaos **24**(2), 023102 (2014)
197. O.E. Omel'chenko, M. Wolfrum, Is there an impact of small phase lags in the kuramoto model? Chaos **26**, 094806 (2016)
198. O.E. Omel'chenko, M. Sebek, I.Z. Kiss, Universal relations of local order parameters for partially synchronized oscillators. Phys. Rev. E **97**, 062207 (2018)
199. M. Wolfrum, O.E. Omel'chenko, J. Sieber, Regular and irregular patterns of self-localized excitation in arrays of coupled phase oscillators. Chaos **25**, 053113 (2015)
200. D.M. Abrams, Two coupled oscillator models: the Millennium bridge and the Chimera state. Ph.D. thesis, Cornell University, 2006
201. A. Blechschmidt, Chimera states in a network of FitzHugh-Nagumo oscillators with power-law coupling. Master's thesis, Technische Universität Berlin, 2018
202. Y. Suda, K. Okuda, Persistent chimera states in nonlocally coupled phase oscillators. Phys. Rev. E **92**, 060901 (2015)
203. M.J. Panaggio, D.M. Abrams, P. Ashwin, C.R. Laing, Chimera states in networks of phase oscillators: the case of two small populations. Phys. Rev. E **93**, 012218 (2016)
204. K. Premalatha, V.K. Chandrasekar, M. Senthilvelan, M. Lakshmanan, Imperfectly synchronized states and chimera states in two interacting populations of nonlocally coupled Stuart-Landau oscillators. Phys. Rev. E **94**(1), 012311 (2016)
205. E.A. Martens, Bistable chimera attractors on a triangular network of oscillator populations. Phys. Rev. E **82**(1), 016216 (2010)
206. E.A. Martens, Chimeras in a network of three oscillator populations with varying network topology. Chaos **20**(4), 043122 (2010)
207. R. Ma, J. Wang, Z. Liu, Robust features of chimera states and the implementation of alternating chimera states. Europhys. Lett. **91**(4), 40006 (2010)
208. G.C. Sethia, A. Sen, F.M. Atay, Clustered chimera states in delay-coupled oscillator systems. Phys. Rev. Lett. **100**(14), 144102 (2008)
209. L. Schmidt, K. Krischer, Chimeras in globally coupled oscillatory systems: from ensembles of oscillators to spatially continuous media. Chaos **25**, 064401 (2015)
210. B.B. Mandelbrot, *The Fractal Geometry of Nature*, 3rd edn. (W. H. Freeman and Comp, New York, 1983)
211. J. Feder, *Fractals* (Plenum Press, New York, 1988)
212. Y. Cho, T. Nishikawa, A.E. Motter, Stable chimeras and independently synchronizable clusters. Phys. Rev. Lett. **119**, 084101 (2017)
213. S. Boccaletti, G. Bianconi, R. Criado, C. I. del Genio, J. Gómez-Gardeñes, M. Romance, I. Sendiña Nadal, Z. Wang, M. Zanin, The structure and dynamics of multilayer networks. Phys. Rep. **544**(1), 1–122 (2014)
214. M. Kivelä, A. Arenas, M. Barthélemy, J.P. Gleeson, Y. Moreno, M.A. Porter, Multilayer networks. J. Complex Netw. **2**(3), 203–271 (2014)
215. G. Bianconi, Multilayer networks: dangerous liaisons? Nat. Phys. **10**(10), 712–714 (2014)
216. M.V. Goremyko, V.A. Maksimenko, V.V. Makarov, D. Ghosh, B. Bera, S.K. Dana, A.E. Hramov, Interaction of chimera states in a multilayered network of nonlocally coupled oscillators. Tech. Phys. Lett. **43**(8), 712–715 (2017)
217. S. Majhi, M. Perc, D. Ghosh, Chimera states in a multilayer network of coupled and uncoupled neurons. Chaos **27**, 073109 (2017)

218. A. Bukh, E. Rybalova, N. Semenova, G. Strelkova, V. Anishchenko, New type of chimera and mutual synchronization of spatiotemporal structures in two coupled ensembles of nonlocally interacting chaotic maps. Chaos **27**, 111102 (2017)
219. E. Rybalova, T. Vadivasova, G. Strelkova, V. Anishchenko, A. Zakharova, Forced synchronization of a multilayer heterogeneous network of chaotic maps in the chimera state mode. Chaos **29**, 033134 (2019)
220. J. Sawicki, I. Omelchenko, A. Zakharova, E. Schöll, Synchronization scenarios of chimeras in multiplex networks. Eur. Phys. J. Spec. Top. **227**, 1161 (2018)
221. M. Winkler, J. Sawicki, I. Omelchenko, A. Zakharova, V. Anishchenko, E. Schöll, Relay synchronization in multiplex networks of discrete maps. EPL **126**, 50004 (2019)
222. I. Omelchenko, T. Hülser, A. Zakharova, E. Schöll, Control of chimera states in multilayer networks. Front. Appl. Math. Stat. **4**, 67 (2019)
223. B.K. Bera, S. Majhi, D. Ghosh, M. Perc, Chimera states: effects of different coupling topologies. Europhys. Lett. **118**(1), 10001 (2017)
224. A. Buscarino, M. Frasca, L.V. Gambuzza, P. Hövel, Chimera states in time-varying complex networks. Phys. Rev. E **91**(2), 022817 (2015)
225. D.V. Kasatkin, S. Yanchuk, E. Schöll, V.I. Nekorkin, Self-organized emergence of multilayer structure and chimera states in dynamical networks with adaptive couplings. Phys. Rev. E **96**(6), 062211 (2017)
226. D.V. Kasatkin, V.I. Nekorkin, Synchronization of chimera states in a multiplex system of phase oscillators with adaptive couplings. Chaos **28**, 093115 (2018)
227. D.V. Kasatkin, V. Klinshov, V.I. Nekorkin, Itinerant chimeras in an adaptive network of pulse-coupled oscillators. Phys. Rev. E **99**, 022203 (2019)
228. S. Jalan, S. Ghosh, B. Patra, Is repulsion good for the health of chimeras? Chaos **27**, 101104 (2017)
229. M.J. Panaggio, D.M. Abrams, Chimera states on a flat torus. Phys. Rev. Lett. **110**, 094102 (2013)
230. S. Kundu, S. Majhi, B.K. Bera, D. Ghosh, M. Lakshmanan, Chimera states in two-dimensional networks of locally coupled oscillators. Phys. Rev. E **97**, 022201 (2018)
231. A. Schmidt, T. Kasimatis, J. Hizanidis, A. Provata, P. Hövel, Chimera patterns in two-dimensional networks of coupled neurons. Phys. Rev. E **95**(3), 032224 (2017)
232. M.J. Panaggio, D.M. Abrams, Chimera states on the surface of a sphere. Phys. Rev. E **91**, 022909 (2015)
233. Y. Maistrenko, O. Sudakov, O. Osiv, V.L. Maistrenko, Chimera states in three dimensions. New J. Phys. **17**, 073037 (2015)
234. O.E. Omel'chenko, M. Wolfrum, S. Yanchuk, Y. Maistrenko, O. Sudakov, Stationary patterns of coherence and incoherence in two-dimensional arrays of non-locally-coupled phase oscillators. Phys. Rev. E **85**, 036210 (2012)
235. E. Rybalova, G.I. Strelkova, V.S. Anishchenko, Mechanism of realizing a solitary state chimera in a ring of nonlocally coupled chaotic maps. Chaos Solitons Fractals **115**, 300–305 (2018)
236. E. Rybalova, V.S. Anishchenko, G.I. Strelkova, A. Zakharova, Solitary states and solitary state chimera in neural networks. Chaos **29**, 071106 (2019)
237. S. Shima, Y. Kuramoto, Rotating spiral waves with phase-randomized core in nonlocally coupled oscillators. Phys. Rev. E **69**(3), 036213 (2004)
238. G. Tanaka, K. Morino, H. Daido, K. Aihara, Dynamical robustness of coupled heterogeneous oscillators. Phys. Rev. E **89**, 052906 (2014)
239. J. Xie, E. Knobloch, H.C. Kao, Twisted chimera states and multicore spiral chimera states on a two-dimensional torus. Phys. Rev. E **92**, 042921 (2015)
240. E.A. Martens, C.R. Laing, S.H. Strogatz, Solvable model of spiral wave chimeras. Phys. Rev. Lett. **104**(4), 044101 (2010)
241. C. Gu, G. St-Yves, J. Davidsen, Spiral wave chimeras in complex oscillatory and chaotic systems. Phys. Rev. Lett. **111**, 134101 (2013)

242. O.E. Omel'chenko, M. Wolfrum, E. Knobloch, Stability of spiral chimera states on a torus. SIAM J. Appl. Dyn. Syst. **17**(1), 97–127 (2018)
243. S.W. Haugland, L. Schmidt, K. Krischer, Self-organized alternating chimera states in oscillatory media. Sci. Rep. **5**, 9883 (2015)
244. C. Bick, Isotropy of angular frequencies and weak chimeras with broken symmetry. J. Nonlinear Sci. **27**(2), 605–626 (2016)
245. E.A. Martens, M.J. Panaggio, D.M. Abrams, Basins of attraction for chimera states. New J. Phys. **18**, 022002 (2016)
246. C. Bick, Heteroclinic switching between chimeras. Phys. Rev. E **97**(5), 050201 (2018)
247. T. Banerjee, Chimera death induced by the mean-field diffusive coupling. EPL **110**(6), 60003 (2015)
248. T.M. Isele, J. Hizanidis, A. Provata, P. Hövel, Controlling chimera states: the influence of excitable units. Phys. Rev. E **93**(2), 022217 (2016)
249. L.V. Gambuzza, M. Frasca, Pinning control of chimera states. Phys. Rev. E **94**, 022306 (2016)
250. V.A. Maksimenko, V.V. Makarov, B.K. Bera, D. Ghosh, S.K. Dana, M.V. Goremyko, N.S. Frolov, A.A. Koronovskii, A.E. Hramov, Excitation and suppression of chimera states by multiplexing. Phys. Rev. E **94**, 052205 (2016)
251. S. Ghosh, L. Schülen, A.D. Kachhvah, A. Zakharova, S. Jalan, Taming chimeras in networks through multiplexing delays. EPL **127**, 30002 (2019)
252. N.C. Rattenborg, B. Voirin, S.M. Cruz, R. Tisdale, G. Dell'Omo, H.P. Lipp, M. Wikelski, A.L. Vyssotski, Evidence that birds sleep in mid-flight. Nat. Commun. **7**, 12468 (2016)
253. L. Ramlow, J. Sawicki, A. Zakharova, J. Hlinka, J. C. Claussen, E. Schöll, Partial synchronization in empirical brain networks as a model for unihemispheric sleep. EPL **126**, 50007 (2019). Highlighted in https://phys.org/news/2019-07-unihemispheric-humans.html and in Europhys. News 50, no. 5–6 (2019)
254. C.C. Chow, S. Coombes, Existence and wandering of bumps in a spiking neural network model. SIAM J. Appl. Dyn. Syst. **5**, 552–574 (2006)
255. M.S. Santos, J.D. Szezech, F.S. Borges, K.C. Iarosz, I.L. Caldas, A.M. Batista, R.L. Viana, J. Kurths, Chimera-like states in a neuronal network model of the cat brain. Chaos Solitons Fractals **101**, 86 (2017)
256. A. Röhm, F. Böhm, K. Lüdge, Small chimera states without multistability in a globally delay-coupled network of four lasers. Phys. Rev. E **94**, 042204 (2016)
257. F.M. Atay, Distributed delays facilitate amplitude death of coupled oscillators. Phys. Rev. Lett. **91**, 094101 (2003)
258. B. Fiedler, V. Flunkert, P. Hövel, E. Schöll, Delay stabilization of periodic orbits in coupled oscillator systems. Philos. Trans. R. Soc. London, Ser. A **368**(1911), 319–341 (2010)
259. C.U. Choe, T. Dahms, P. Hövel, E. Schöll, Controlling synchrony by delay coupling in networks: from in-phase to splay and cluster states. Phys. Rev. E **81**(2), 025205(R) (2010)
260. Y.N. Kyrychko, K.B. Blyuss, E. Schöll, Amplitude and phase dynamics in oscillators with distributed-delay coupling. Phil. Trans. R. Soc. A **371**, 20120466 (2013)
261. I. Schneider, Delayed feedback control of three diffusively coupled Stuart-Landau oscillators: a case study in equivariant Hopf bifurcation. Phil. Trans. R. Soc. A **371**, 20120472 (2013)
262. C.M. Postlethwaite, G. Brown, M. Silber, Feedback control of unstable periodic orbits in equivariant Hopf bifurcation problems. Philos. Trans. R. Soc. London, Ser. A **371**, 20120467 (2013)
263. L. Illing, Amplitude death of identical oscillators in networks with direct coupling. Phys. Rev. E **94**, 022215 (2016)
264. E. Panteley, A. Loria, Effects of network topology on the synchronized behaviour of coupled nonlinear oscillators: a case study, in *Proceedings 6th IFAC Workshop on Periodic Control Systems (PSYCHO 2016), IFAC-PapersOnLine*, vol. 49, Issue 14, pp. 90–92 (2016). https://doi.org/10.1016/j.ifacol.2016.07.990
265. H. Risken, *The Fokker-Planck Equation*, 2nd edn. (Springer, Berlin, 1996)
266. K. Bar-Eli, On the stability of coupled chemical oscillators. Physica D **14**, 242 (1985)

267. K. Tsaneva-Atanasova, C.L. Zimliki, R. Bertram, A. Sherman, Diffusion of calcium and metabolites in pancreatic islets: killing oscillations with a Pitchfork. Biophys. J. **90**, 3434 (2006)
268. E. Ullner, A. Zaikin, E. Volkov, J. García-Ojalvo, Multistability and clustering in a population of synthetic genetic oscillators via phase-repulsive cell-to-cell communication. Phys. Rev. Lett. **99**, 148103 (2007)
269. A. Kuznetsov, M. Kaern, N. Kopell, Synchrony in a population of hysteresis-based genetic oscillators. SIAM J. Appl. Math. **65**(2), 392 (2004)
270. M.F. Crowley, I.R. Epstein, Experimental and theoretical studies of a coupled chemical oscillator: phase death, multistability and in-phase and out-of-phase entrainment. J. Phys. Chem. **93**, 2496 (1989)
271. M. Toiya, V.K. Vanag, I.R. Epstein, Diffusively coupled chemical oscillators in a microfluidic assembly. Angew. Chem. Int. Ed. **47**, 7753 (2008)
272. K.P. Zeyer, M. Mangold, E.D. Gilles, Experimentally coupled thermokinetic oscillators: phase death and rhythmogenesis. J. Phys. Chem. A **105**, 7216 (2001)
273. A. Koseska, E. Volkov, J. Kurths, Parameter mismatches and oscillation death in coupled oscillators. Chaos **20**, 023132 (2010)
274. C.R. Hens, O.I. Olusola, P. Pal, S.K. Dana, Oscillation death in diffusively coupled oscillators by local repulsive link. Phys. Rev. E **88**, 034902 (2013)
275. W. Zou, D.V. Senthilkumar, M. Zhan, J. Kurths, Reviving oscillations in coupled nonlinear oscillators. Phys. Rev. Lett. **111**, 014101 (2013)
276. R. Karnatak, R. Ramaswamy, U. Feudel, Conjugate coupling in ecosystems: cross-predation stabilizes food webs. Chaos Solitons Fractals **68**, 48–57 (2014)
277. W. Liu, E. Volkov, J. Xiao, W. Zou, M. Zhan, Inhomogeneous stationary and oscillatory regimes in coupled chaotic oscillators. Chaos **22**, 033144 (2012)
278. A. Koseska, J. Kurths, Topological structures enhance the presence of dynamical regimes in synthetic networks. Chaos **20**, 045111 (2010)
279. K. Premalatha, V.K. Chandrasekar, M. Senthilvelan, M. Lakshmanan, Impact of symmetry breaking in networks of globally coupled oscillators. Phys. Rev. E **91**, 052915 (2015)
280. J.A. Acebrón, L.L. Bonilla, C.J. Pérez Vicente, F. Ritort, R. Spigler, The Kuramoto model: a simple paradigm for synchronization phenomena. Rev. Mod. Phys. **77**, 137–185 (2005)
281. Y.C. Lai, T. Tél, *Transient Chaos* (Springer, Berlin, 2011)
282. T. Tél, The joy of transient chaos. Chaos **25**, 097619 (2015)
283. A. Wacker, S. Bose, E. Schöll, Transient spatio-temporal chaos in a reaction-diffusion model. Europhys. Lett. **31**, 257 (1995)
284. E. Schöll, H.G. Schuster (Eds.), *Handbook of Chaos Control* (Wiley-VCH, Weinheim, 2008). Second completely revised and enlarged edition
285. A. Koseska, E. Volkov, J. Kurths, Transition from amplitude to oscillation death via turing bifurcation. Phys. Rev. Lett. **111**, 024103 (2013)
286. Q. Ouyang, R. Li, G. Li, H.L. Swinney, Dependence of turing pattern wavelength on diffusion rate. J. Chem. Phys. **102**, 2551 (1995)
287. T.M. Isele, E. Schöll, Effect of small-world topology on wave propagation on networks of excitable elements. New J. Phys. **17**, 023058 (2015)
288. A. Gjurchinovski, T. Jüngling, V. Urumov, E. Schöll, Delayed feedback control of unstable steady states with high-frequency modulation of the delay. Phys. Rev. E **88**, 032912 (2013)
289. W. Michiels, V. van Assche, S.I. Niculescu, Stabilization of time-delay systems with a controlled time-varying delay and applications. IEEE Trans. Autom. Control **50**(4), 493–504 (2005)
290. P.E. Kloeden, E. Platen, *Numerical Solution of Stochastic Differential Equations* (Springer, Berlin, 1992)
291. A.B. Neiman, P.I. Saparin, L. Stone, Coherence resonance at noisy precursors of bifurcations in nonlinear dynamical systems. Phys. Rev. E **56**, 270 (1997)
292. C. Masoller, Noise-induced resonance in delayed feedback systems. Phys. Rev. Lett. **88**, 034102 (2002)

293. A.A. Zaikin, J. García-Ojalvo, R. Báscones, E. Ullner, J. Kurths, Doubly stochastic coherence via noise-induced symmetry in bistable neural models. Phys. Rev. Lett. **90**(3), 030601 (2003)
294. I. Franović, V. Klinshov, Slow rate fluctuations in a network of noisy neurons with coupling delay. Europhys. Lett. **116**, 48002 (2016)
295. B. Lindner, J. García-Ojalvo, A.B. Neiman, L. Schimansky-Geier, Effects of noise in excitable systems. Phys. Rep. **392**, 321–424 (2004)
296. O.V. Ushakov, H.J. Wünsche, F. Henneberger, I.A. Khovanov, L. Schimansky-Geier, M.A. Zaks, Coherence resonance near a Hopf bifurcation. Phys. Rev. Lett. **95**, 123903 (2005)
297. A. Zakharova, J. Kurths, T. Vadivasova, A. Koseska, Analysing dynamical behavior of cellular networks via stochastic bifurcations. PLoS ONE **6**(5), e19696 (2011)
298. B. Hu, C. Zhou, Phase synchronization in coupled nonidentical excitable systems and array-enhanced coherence resonance. Phys. Rev. E **61**(2), 1001(R)–1004(R) (2000)
299. N.B. Janson, A.G. Balanov, E. Schöll, Delayed feedback as a means of control of noise-induced motion. Phys. Rev. Lett. **93**, 010601 (2004)
300. A.G. Balanov, N.B. Janson, E. Schöll, Control of noise-induced oscillations by delayed feedback. Physica D **199**, 1–12 (2004)
301. E. Schöll, A.G. Balanov, N.B. Janson, A.B. Neiman, Controlling stochastic oscillations close to a Hopf bifurcation by time-delayed feedback. Stoch. Dyn. **5**, 281 (2005)
302. B. Hauschildt, N.B. Janson, A.G. Balanov, E. Schöll, Noise-induced cooperative dynamics and its control in coupled neuron models. Phys. Rev. E **74**, 051906 (2006)
303. R. Aust, P. Hövel, J. Hizanidis, E. Schöll, Delay control of coherence resonance in type-I excitable dynamics. Eur. Phys. J. Spec. Top. **187**, 77–85 (2010)
304. D. Ziemann, R. Aust, B. Lingnau, E. Schöll, K. Lüdge, Optical injection enables coherence resonance in quantum-dot lasers. Europhys. Lett. **103**, 14002 (2013)
305. C. Otto, *Dynamics of Quantum Dot Lasers-Effects of Optical Feedback and External Optical Injection* (Springer Theses, Springer, Heidelberg, 2014)
306. P. Balenzuela, P. Rué, S. Boccaletti, J. García-Ojalvo, Collective stochastic coherence and synchronizability in weighted scale-free networks. New J. Phys. **16**(1), 013036 (2014)
307. P.M. Geffert, *Stochastic Non-Excitable Systems with Time Delay*. BestMasters 2015 (Springer-Spektrum, Wiesbaden, 2015)
308. J. Hizanidis, A.G. Balanov, A. Amann, E. Schöll, Noise-induced front motion: signature of a global bifurcation. Phys. Rev. Lett. **96**, 244104 (2006)
309. J. Hizanidis, E. Schöll, Control of coherence resonance in semiconductor superlattices. Phys. Rev. E **78**, 066205 (2008)
310. R. FitzHugh, Impulses and physiological states in theoretical models of nerve membrane. Biophys. J. **1**, 445–466 (1961)
311. J. Nagumo, S. Arimoto, S. Yoshizawa, An active pulse transmission line simulating nerve axon. Proc. IRE **50**, 2061–2070 (1962)
312. A. Scott, The electrophysics of a nerve fiber. Rev. Mod. Phys. **47**, 487 (1975)
313. V. Klinshov, L. Lücken, D. Shchapin, V.I. Nekorkin, S. Yanchuk, Multistable jittering in oscillators with pulsatile delayed feedback. Phys. Rev. Lett. **114**, 178103 (2015)
314. M. Ciszak, O. Calvo, C. Masoller, C.R. Mirasso, R. Toral, Anticipating the response of excitable systems driven by random forcing. Phys. Rev. Lett. **90**(20), 204102 (2003)
315. D.P. Rosin, K.E. Callan, D.J. Gauthier, E. Schöll, Pulse-train solutions and excitability in an optoelectronic oscillator. Europhys. Lett. **96**(3), 34001 (2011)
316. R. Kozma, Intermediate-range coupling generates low-dimensional attractors deeply in the chaotic region of one-dimensional lattices. Phys. Lett. A **244**(1), 85–91 (1998)
317. E. Hulata, I. Baruchi, R. Segev, Y. Shapira, E. Ben-Jacob, Self-regulated complexity in cultured neuronal networks. Phys. Rev. Lett. **92**(19), 198105 (2004)
318. J.A. Henderson, P.A. Robinson, Geometric effects on complex network structure in the cortex. Phys. Rev. Lett. **107**, 018102 (2011)
319. V.S. Anishchenko, V. Astakhov, A.B. Neiman, T. Vadivasova, L. Schimansky-Geier, *Nonlinear Dynamics of Chaotic and Stochastic Systems: Tutorial and Modern Developments* (Springer, Berlin, 2007)

320. I.A. Shepelev, A. Slepnev, T. Vadivasova, Different synchronization characteristics of distinct types of traveling waves in a model of active medium with periodic boundary conditions. Commun. Nonlinear Sci. Numer. Simul. **38**, 206 (2016)
321. O.A. Rosso, C. Masoller, Detecting and quantifying stochastic and coherence resonances via information-theory complexity measurements. Phys. Rev. E **79**, 040106(R) (2009)
322. J.L.A. Dubbeldam, B. Krauskopf, D. Lenstra, Excitability and coherence resonance in lasers with saturable absorber. Phys. Rev. E **60**(6), 6580 (1999)
323. G. Giacomelli, M. Giudici, S. Balle, J.R. Tredicce, Experimental evidence of coherence resonance in an optical system. Phys. Rev. Lett. **84**, 3298 (2000)
324. J.F.M. Avila, H.L.D. de S. Cavalcante, J.R. Rios Leite. Experimental deterministic coherence resonance. Phys. Rev. Lett. **93**(14), 144101 (2004)
325. Y. Huang, H. Qin, W. Li, S. Lu, J. Dong, H.T. Grahn, Y. Zhang, Experimental evidence for coherence resonance in a noise-driven GaAs/AlAs superlattice. Europhys. Lett. **105**(4), 47005 (2014)
326. M. Schanz, A. Pelster, Synergetic system analysis for the delay-induced Hopf bifurcation in the Wright equation. SIAM J. Appl. Dyn. Syst. **2**, 277 (2003)
327. J. Hizanidis, R. Aust, E. Schöll, Delay-induced multistability near a global bifurcation. Int. J. Bifur. Chaos **18**(6), 1759–1765 (2008)
328. K. Pyragas, Continuous control of chaos by self-controlling feedback. Phys. Lett. A **170**, 421 (1992)
329. P. Hövel, E. Schöll, Control of unstable steady states by time-delayed feedback methods. Phys. Rev. E **72**, 046203 (2005)
330. A. Sen, R. Dodla, G. Johnston, G.C. Sethia, Amplitude death, synchrony, and chimera states, in *delay coupled limit cycle oscillators, in Complex Time-Delay Systems, vol. 16, Understanding Complex Systems*, ed. by F.M. Atay (Springer, Berlin, 2010), pp. 1–43
331. J.H. Sheeba, V.K. Chandrasekar, M. Lakshmanan, Globally clustered chimera states in delay-coupled populations. Phys. Rev. E **79**(5), 055203 (2009)
332. J.H. Sheeba, V.K. Chandrasekar, M. Lakshmanan, Chimera and globally clustered chimera: Impact of time delay. Phys. Rev. E **81**, 046203 (2010)
333. D. Nikitin, I. Omelchenko, A. Zakharova, M. Avetyan, A.L. Fradkov, E. Schöll, Complex partial synchronization patterns in networks of delay-coupled neurons. Phil. Trans. R. Soc. A **377**, 20180128 (2019)
334. P. Jaros, S. Brezetsky, R. Levchenko, D. Dudkowski, T. Kapitaniak, Y. Maistrenko, Solitary states for coupled oscillators with inertia. Chaos **28**, 011103 (2018)
335. L. Schülen, S. Ghosh, A.D. Kachhvah, A. Zakharova, S. Jalan, Delay engineered solitary states in complex networks. Chaos Solitons Fractals **128**, 290 (2019)
336. R.P. Feynman, R.B. Leighton, M. Sands, The Feynman lectures on physics, Vols. 1–3 (Narosa, New Delhi, 2008)
337. Z. Fric, M. Konvicka, Dispersal kernels of butterflies: power-law functions are invariant to marking frequency. Basic Appl. Ecol. **8**, 377 (2007)
338. B.G. Szaro, R. Tompkins, Effect of tetraploidy on dendritic branching in neurons and glial cells of the frog, xenopus laevis. J. Comp. Neurol. **258**, 304 (1987)
339. D.E. Bowler, T.G. Benton, Impact of dispersal on population growth: the role of inter-patch distance. Oikos **118**, 403 (2009)
340. J.H. Brown, V.K. Gupta, B. Li, B.T. Milne, C. Restrepo, G.B. West, The fractal nature of nature: power laws, ecological complexity and biodiversity. Philos. Trans. R. Soc. B **357**, 619 (2002)
341. P. Katsaloulis, D.A. Verganelakis, A. Provata, Fractal dimension and lacunarity of tractography images of the human brain. Fractals **17**(02), 181–189 (2009)
342. P. Expert, T.S. Evans, V.D. Blondel, R. Lambiotte, Uncovering space-independent communities in spatial networks. Proc. Natl. Acad. Sci. USA **108**(19), 7663 (2011)
343. P. Katsaloulis, A. Ghosh, A.C. Philippe, A. Provata, R. Deriche, Fractality in the neuron axonal topography of the human brain based on 3-D diffusion MRI. Eur. Phys. J. B **85**(5), 1–7 (2012)

344. P. Katsaloulis, J. Hizanidis, D.A. Verganelakis, A. Provata, Complexity measures and noise effects on Diffusion Magnetic Resonance imaging of the neuron axons network in human brains. Fluct. Noise Lett. **11**, 1250032 (2012)
345. A. Provata, P. Katsaloulis, D.A. Verganelakis, Dynamics of chaotic maps for modelling the multifractal spectrum of human brain Diffusion Tensor Images. Chaos Solitons Fractals **45**(2), 174–180 (2012)
346. I. Leyva, R. Sevilla-Escoboza, I. Sendiña-Nadal, R. Gutiérrez, J.M. Buldú, S. Boccaletti, Inter-layer synchronization in non-identical multi-layer networks. Sci. Rep. **7**, 45475 (2017)
347. L. Zhang, A.E. Motter, T. Nishikawa, Incoherence-mediated remote synchronization. Phys. Rev. Lett. **118**(17), 174102 (2017)
348. A. Singh, S. Ghosh, S. Jalan, J. Kurths, Synchronization in delayed multiplex networks. Europhys. Lett. **111**(3), 30010 (2015)
349. L.V. Gambuzza, M. Frasca, J. Gómez-Gardeñes, Intra-layer synchronization in multiplex networks. Europhys. Lett. **110**(2), 20010 (2015)
350. N.E. Kouvaris, S. Hata, A. Díaz-Guilera, Pattern formation in multiplex networks. Sci. Rep. **5**, 10840 (2015)
351. S. Ghosh, S. Jalan, Emergence of chimera in multiplex network. Int. J. Bifurc. Chaos **26**(07), 1650120 (2016)
352. M. De Domenico, Multilayer modeling and analysis of human brain networks. Gigascience **6**(5), 1–8 (2017)
353. A.N. Pisarchik, U. Feudel, Control of multistability. Phys. Rep. **540**, 167–218 (2014). https://doi.org/10.1016/j.physrep.2014.02.007
354. A.V. Andreev, V.V. Makarov, A.E. Runnova, A.N. Pisarchik, A.E. Hramov, Coherence resonance in stimulated neuronal network. Chaos Solitons Fractals **106**, 80–85 (2018)
355. O. Kwon, H.T. Moon, Coherence resonance in small-world networks of excitable cells. Phys. Lett. A **298**, 319 (2002)
356. E. Rybalova, N. Semenova, G. Strelkova, V. Anishchenko, Transition from complete synchronization to spatio-temporal chaos in coupled chaotic systems with nonhyperbolic and hyperbolic attractors. Eur. Phys. J. Spec. Top. **226**, 1857 (2017)
357. N. Semenova, T. Vadivasova, V. Anishchenko, Mechanism of solitary state appearance in an ensemble of nonlocally coupled Lozi maps. Eur. Phys. J. Spec. Top. **227**, 1173 (2018)
358. A.V. Bukh, G.I. Strelkova, V.S. Anishchenko, Synchronization of chimera states in coupled networks of nonlinear chaotic oscillators. Russ. J. Nonlinear Dyn. **14**(4), 419 (2018)
359. M. Girvan, M.E.J. Newman, Community structure in social and biological networks. Proc. Natl. Acad. Sci. USA **99**, 7821 (2002)
360. S. Jalan, A. Singh, Cluster synchronization in multiplex networks. Europhys. Lett. **113**(3), 30002 (2016)
361. S.A. Cannas, F.A. Tamarit, Long-range interactions and nonextensivity in ferromagnetic spin models. Phys. Rev. B **54**(18), R12661 (1996)
362. R. Golestanian, J.M. Yeomans, N. Uchida, Hydrodynamic synchronization at low Reynolds number. Soft Matter **7**, 3074 (2011)
363. N. Uchida, R. Golestanian, Generic conditions for hydrodynamic synchronization. Phys. Rev. Lett. **106**, 058104 (2011)
364. C. Anteneodo, S.E. de S. Pinto, A.M. Batista, R.L. Viana, Analytical results for coupled-map lattices with long-range interactions. Phys. Rev. E **68**, 045202 (2003)
365. J.L. Rogers, L.T. Wille, Phase transitions in nonlinear oscillator chains. Phys. Rev. E **54**(3), R2193 (1996)
366. M. Maródi, F. d'Ovidio, T. Vicsek, Synchronization of oscillators with long range interaction: phase transition and anomalous finite size effects. Phys. Rev. E **66**, 011109 (2002)
367. H.Y. Kuo, K.A. Wu, Synchronization and plateau splitting of coupled oscillators with long-range power-law interactions. Phys. Rev. E **92**(6), 062918 (2015)
368. P. Barthold, Transport- und Rauschmessung an einzelnen und vertikal gekoppelten selbstorganisierten Quantenpunktsystemen. Master's thesis, Universität Hannover, 2004
369. B. van der Pol, On relaxation oscillations. Philos. Mag. **2**, 978–992 (1926)

370. D.J. Watts, S.H. Strogatz, Collective dynamics of 'small-world' networks. Nature **393**, 440–442 (1998)
371. A. Ahlborn, U. Parlitz, Stabilizing unstable steady states using multiple delay feedback control. Phys. Rev. Lett. **93**, 264101 (2004)
372. V. Flunkert, I. Fischer, E. Schöll, Dynamics, control and information in delay-coupled systems. Theme Issue of Phil. Trans. R. Soc. A **371**, 20120465 (2013)
373. A. Saha, U. Feudel, Extreme events in FitzHugh-Nagumo oscillators coupled with two time delays. Phys. Rev. E **95**(6), 062219 (2017)
374. C.A. Bachmair, E. Schöll, Nonlocal control of pulse propagation in excitable media. Eur. Phys. J. B **87**(11), 276 (2014)
375. S. Yanchuk, M. Wolfrum, P. Hövel, E. Schöll, Control of unstable steady states by long delay feedback. Phys. Rev. E **74**, 026201 (2006)
376. S.A. Brandstetter, M.A. Dahlem, E. Schöll, Interplay of time-delayed feedback control and temporally correlated noise in excitable systems. Philos. Trans. R. Soc. A **368**(1911), 391 (2010)
377. S. Yanchuk, P. Perlikowski, Delay and periodicity. Phys. Rev. E **79**(4), 046221 (2009)
378. W. Just, T. Bernard, M. Ostheimer, E. Reibold, H. Benner, Mechanism of time-delayed feedback control. Phys. Rev. Lett. **78**, 203 (1997)
379. W. Just, E. Reibold, K. Kacperski, P. Fronczak, J.A. Holyst, H. Benner, Influence of stable Floquet exponents on time-delayed feedback control. Phys. Rev. E **61**, 5045 (2000)
380. W. Just, H. Benner, C.v. Loewenich, On global properties of time-delayed feedback control: weakly nonlinear analysis. Phys. D **199**, 33 (2004)
381. V.K. Jirsa, A.R. McIntosh, Handbook of brain connectivity, in *Handbook of Brain Connectivity*, ed. by V.K. Jirsa, A.R. McIntosh, vol. 1 (Berlin, Springer, 2007)
382. C.J. Honey, J.P. Thivierge, O. Sporns, Can structure predict function in the human brain? NeuroImage **52**(3), 766–776 (2010)
383. A. Daffertshofer, B.C.M. van Wijk, On the influence of amplitude on the connectivity between phases. Front. Neuroinform. **5**, 6 (2011)
384. J. Hlinka, S. Coombes, Using computational models to relate structural and functional brain connectivity. Eur. J. Neurosci. **36**(2), 2137–2145 (2012)
385. J. Cabral, M.L. Kringelbach, G. Deco, Exploring the network dynamics underlying brain activity during rest. Prog. Neurobiol. **114**, 102–131 (2013)
386. C.R. Laing, C.C. Chow, Stationary bumps in networks of spiking neurons. Neural Comput. **13**(7), 1473–1494 (2001)
387. C.R. Laing, Travelling waves in arrays of delay-coupled phase oscillators. Chaos **26**, 094802 (2016)
388. H. Sakaguchi, Instability of synchronized motion in nonlocally coupled neural oscillators. Phys. Rev. E **73**(3), 031907 (2006)
389. N.D. Tsigkri-DeSmedt, J. Hizanidis, E. Schöll, P. Hövel, A. Provata, Chimeras in leaky integrate-and-fire neural networks: effects of reflecting connectivities. Eur. Phys. J. B **90**(7), 139 (2017)
390. N. Sinha, J. Dauwels, M. Kaiser, S.S. Cash, M. Brandon Westover, Y. Wang, P.N. Taylor, Predicting neurosurgical outcomes in focal epilepsy patients using computational modelling. Brain **140**(2), 319–332 (2016)
391. M. Goodfellow, C. Rummel, E. Abela, M.P. Richardson, K. Schindler, J.R. Terry, Estimation of brain network ictogenicity predicts outcome from epilepsy surgery. Sci. Rep. **6**, 29215 (2016)
392. T. Melicher, J. Horacek, J. Hlinka, F. Spaniel, J. Tintera, I. Ibrahim, P. Mikolas, T. Novak, P. Mohr, C. Hoschl, White matter changes in first episode psychosis and their relation to the size of sample studied: a DTI study. Schizophr. Res. **162**(1), 22–28 (2015)
393. T.E. Behrens, H.J. Berg, S. Jbabdi, M.F. Rushworth, M.W. Woolrich, Probabilistic diffusion tractography with multiple fibre orientations: what can we gain? Neuroimage **34**(1), 144–155 (2007)

394. N. Tzourio-Mazoyer, B. Landeau, D. Papathanassiou, F. Crivello, O. Etard, N. Delcroix, B. Mazoyer, M. Joliot, Automated anatomical labeling of activations in SPM using a macroscopic anatomical parcellation of the MNI MRI single-subject brain. Neuroimage **15**(1), 273–289 (2002)
395. L.M. Pecora, T.L. Carroll, Master stability functions for synchronized coupled systems. Phys. Rev. Lett. **80**(10), 2109–2112 (1998)
396. J.G. Restrepo, E. Ott, B. Hunt, Characterizing dynamical importance of nodes and links. Phys. Rev. Lett. **97**(094102) (2006)
397. F.A. Chowdhury, W. Woldman, T.H.B. FitzGerald, R.D.C. Elwes, L. Nashef, J.R. Terry, M.P. Richardson, Revealing a brain network endophenotype in families with idiopathic generalised epilepsy. PLOS ONE **9**(10), e110136 (2014)
398. G. Petkov, M. Goodfellow, M.P. Richardson, J.R. Terry, A critical role for network structure in seizure onset: a computational modeling approach. Front. Neurol. **5**, 261 (2014)

Index

A. Zakharova, *Chimera Patterns in Networks*, Understanding Complex Systems,
https://doi.org/10.1007/978-3-030-21714-3

The manufacturer's authorised representative in the EU is Springer Nature Customer Service Centre GmbH, Europaplatz 3, 69115 Heidelberg, Germany. If you have any concerns regarding our products, please contact ProductSafety@springernature.com

Printed and bound by CPI Group (UK) Ltd, Croydon, CR0 4YY

30/09/2026

02225229-0006